Global Impacts of Micro- and Nano-Plastic Pollution

Nisha Gaur
Gautam Budhha University, India

Eti Sharma
Gautam Buddha University, India

Tuan Anh Nguyen
Vietnam Academy of Science and Technology, Vietnam

Muhammad Bilal
Gdansk University of Technology, Poland

Niranjan Prakash Melkania
Gautam Buddha University, India

IGI Global
Publishing Tomorrow's Research Today

Published in the United States of America by
IGI Global
701 E. Chocolate Avenue
Hershey PA, USA 17033
Tel: 717-533-8845
Fax: 717-533-8661
E-mail: cust@igi-global.com
Web site: https://www.igi-global.com

Library of Congress Cataloging-in-Publication Data

CIP PENDING

ISBN13: 9798369334478
EISBN13: 9798369334485

Vice President of Editorial: Melissa Wagner
Managing Editor of Acquisitions: Mikaela Felty
Managing Editor of Book Development: Jocelynn Hessler
Production Manager: Mike Brehm
Cover Design: Phillip Shickler

British Cataloguing in Publication Data
A Cataloguing in Publication record for this book is available from the British Library.

Table of Contents

Preface xviii

Chapter 1
Bibliometric Analysis of Micro/Nanoplastics: An Overview 1
Bancha Yingngam, Ubon Ratchathani University, Thailand

Chapter 2
Micro/Nano Plastic Pollution Represents a Significant and Growing Threat to Human Populations Worldwide 27
Himadri Sekhar Das, Haldia Institute of Technology, India
Gourisankar Roymahapatra, Haldia Institute of Technology, India
Santanu Mishra, Haldia Institute of Technology, India

Chapter 3
Microplastics and Climate Change: A Mutually Reinforcing Relationship......... 51
Nazuk Bhasin, Banaras Hindu University, India
Sudhanshu Kumar, Banaras Hindu University, India
Anil Barla, Banaras Hindu University, India
Amit Kumar Tiwari, Banaras Hindu University, India
Gopal Shankar Singh, Banaras Hindu University, India

Chapter 4
Detection, Monitoring, and Absorption of Micro/Nano-Plastics in Soil System 77
Rajeev Kumar, Manav Rachna International Institute of Research and Studies, India
Jyoti Chawla, Manav Rachna International Institute of Research and Studies, India
Jyoti Syal, Maharishi Markandeshwar Engineering College (Deemed), India

Chapter 5
Computational Approaches for Identification of Micro/Nano-Plastic Pollution 99
Kartavya Mathur, Gautam Buddha University, India
Eti Sharma, Gautam Buddha University, India
Nisha Gaur, Gautam Buddha University, India
Shashank Mittal, O.P. Jindal Global University, India
Shubham Kumar, University of Minnesota, USA

Chapter 6
Advanced Oxidation Processes (AOPs) for the Degradation of Micro and Nano Plastic 123
Aqsa Rukhsar, Lahore Garrison University, Pakistan
Muhammad Shahzeb Khan, Sulaiman Bin Abdullah Aba Al-Khail Centre for Interdisciplinary Research in Basic Sciences (SA-CIRBS), Faculty of Basic and Applied Sciences, International Islamic University Islamabad, Islamabad, Pakistan
Zeenat Fatima Iqbal, Department of Chemistry, University of Engineering and Technology, Lahore, Pakistan

Chapter 7
Phenotypic and Genotypic Alterations in Plants in Response to Micro/Nano-Plastics 151
Ridhi Pandey, Gautam Buddha University, India
Aaradhya Pandey, Gautam Buddha University, India
Nishtha Kaushik, Gautam Buddha University, India
Nisha Gaur, Gautam Buddha University, India
Eti Sharma, Gautam Buddha University, India
Andrea Naziri, University of Cyprus, Cyprus

Chapter 8
Effect of Micro/Nano-Plastics Accumulation on Soil Nutrient Cycling 179
Divya Kumari, Banasthali University, India
Pracheta Janmeda, Banasthali University, India
Nidhi Varshney, Banasthali University, India
Poornima Pandey, Banasthali University, India

Chapter 9
Micro/Nano-Plastics Pollution: Challenges to Agriculture Productivity........... 205
Nandini Arya, Uttaranchal University, Dehradun, India
Atin Kumar, Uttaranchal University, Dehradun, India

Chapter 10
Micro/Nano-Plastic Pollution in Aquarium Systems: Sources, Fate, Hazards, and Ecological Imbalances 225
Bahati Shabani Nzeyimana, Bishop Heber College, Bharathidasan University, India
Swagata Chakraborty, Bharathidasan University, India
R. Priyadharshini, Bishop Heber College, Bharathidasan University, India
Mariaselvam Sheela Mary, Bishop Heber College, Bharathidasan University, India
M. Govindaraju, Bharathidasan University, India

Chapter 11
Physiological and Toxicological Effects of Nano/Microplastics on Marine Birds 257
Anubha Singh, Gautam Buddha University, India
Jyoti Upadhyay, Gautam Buddha University, India

Chapter 12
Microplastic Menace: Ecological Ramifications and Solutions for a Sustainable Future........ 289
Manjunatha Badiger, NMAM Institute of Technology, Nitte University (Deemed), India
Jose Alex Mathew, A.J. Institute of Engineering and Technology, India
Savidhan Shetty C. S., National Institute of Technology, Karnataka, India
Varuna Kumara, Moodlakatte Institute of Technology, India
Ganesh Srinivasa Shetty, Shri Madhwa Vadiraja Institute of Technology, Bantakal, India
Ratheksha Rai N., A.J. Institute of Engineering and Technology, India
Mehnaz Fathima C., Sahyadri College of Engineering and Management, India

Chapter 13
Policy and Regulatory Approaches to Mitigating Micro- and Nano Plastic Pollution 309
Mohit Yadav, O.P. Jindal Global University, India
Ashutosh Pandey, FORE School of Management, New Delhi, India
Xuan-Hoa Nghiem, International School, Vietnam National University, Hanoi, Vietnam

Compilation of References 331

About the Contributors 413

Index 421

Detailed Table of Contents

Preface xviii

Chapter 1
Bibliometric Analysis of Micro/Nanoplastics: An Overview 1
Bancha Yingngam, Ubon Ratchathani University, Thailand

Micro/nanoplastic pollution is, without a doubt, one of the most acute and daunting environmental problems worldwide. This chapter, therefore, uses a comprehensive bibliometric analysis as the method to present the research status of micro/nanoplastic pollution. With respect to broader aspects, first, the data are collected from the Scopus database. Therefore, advanced bibliometric methodologies are used to detect publication patterns and citation fronts. The results show that the number of publications has tended to increase in recent years, and most of them are about marine problems. Second, according to the thematic analysis of the available publications, the focus should be on thematic areas such as sources of micro/nanopolluction, distribution, ecological damage and determination, and further expansion. Importantly, a few thematic networks are identified, as they are easy to recognize in relation to the problems concerned. Finally, it becomes clear from the analysis that the research is global, not only local, and is almost not limited to one country.

Chapter 2
Micro/Nano Plastic Pollution Represents a Significant and Growing Threat to Human Populations Worldwide 27
Himadri Sekhar Das, Haldia Institute of Technology, India
Gourisankar Roymahapatra, Haldia Institute of Technology, India
Santanu Mishra, Haldia Institute of Technology, India

Micro/nano flexible contamination poses a pervasive warning to worldwide human public. With atoms measuring inferior 5mm, these contaminants permeate ecosystems, contaminating water beginnings, soil, and air. Their ever-present demeanor in food chains raises concerns about potential fitness hazards, containing endocrine turmoil, carcinogenicity, and generative issues. Moreover, their ability to adsorb and transport injurious projectiles for weaponry infuriates their impact. Mitigating this threat demands multidisciplinary works including scientific research, procedure invasions, and public knowledge campaigns. Addressing calculating/nano plastic contamination is principal to conserving human health and maintaining tangible purity on a general scale.

Chapter 3
Microplastics and Climate Change: A Mutually Reinforcing Relationship......... 51
Nazuk Bhasin, Banaras Hindu University, India
Sudhanshu Kumar, Banaras Hindu University, India
Anil Barla, Banaras Hindu University, India
Amit Kumar Tiwari, Banaras Hindu University, India
Gopal Shankar Singh, Banaras Hindu University, India

Plastics—macro, micro, or nano—have become persistent, pervasive, and potentially hazardous pollutants infiltrating the global environment. Microplastics (<5 mm) owing to their increased surface area as compared to their mass and small size are considered more harmful than larger plastics. The issue of their environmental presence has gained momentum due to their ability to act as sources and sinks for toxic substances, and also due to the intensification of climate change. Climate change stimulates their deterioration, dispersal, and the interaction with the environmental compartments. In turn the plastic debris contributes directly or indirectly to greenhouse gas emissions during its life cycle. Plastics account for 3.3% of the global GHG emissions. Thus, microplastics and climate change share a mutually reinforcing relationship. For effective management of both these issues, it is imperative to understand the nature and dynamics of this complex relationship. This chapter aims to discuss the long-term ecological impacts of microplastics and climate change on each other.

Micro/nano plastics (MNPs) pollution in soil system is a big threat to all living organisms. Plastic may alter the physiochemical and biological properties of soil and also has long term effects on biodiversity. Plastic on land is disintegrated into smaller particles having size less than 5 milimeters (micro) or 0.1 micrometer (nano). The interaction between MNPs present in soil and plants affects physiology and morphology of plants. Thus, regular monitoring, detection, and their removal from the soil system are important during plant growth. MNPs have been typically monitored and detected by different analytical methods. FTIR, Raman spectroscopy, pyrolysis coupled with GCMS, UV-visible spectroscopy, scanning electron microscopy, and atomic force microscopy have been widely applied detection and monitoring techniques for MNPs. MNPs have the capability to adsorb various types of contaminants from environment due to hydrophobicity and high surface area to volume ratio. In this chapter, various techniques for monitoring, detection, and adsorption behavior of MNPs have been discussed.

Chapter 5
Computational Approaches for Identification of Micro/Nano-Plastic Pollution 99
Kartavya Mathur, Gautam Buddha University, India
Eti Sharma, Gautam Buddha University, India
Nisha Gaur, Gautam Buddha University, India
Shashank Mittal, O.P. Jindal Global University, India
Shubham Kumar, University of Minnesota, USA

The dissemination of miniaturized plastics, both micro- and nano-plastics, athwart diverse ecosystems is an argument of global apprehension. The accretion of these plastics is due to their chemical steadiness. In arrears to their trivial size, frequently identification of miniaturized plastics is very problematic. The foremost approaches for identification of micro- and-nano plastics rely upon their visual inspection through microscopy and chemical analysis. The advent of high-throughput computing has eased the detection of miniaturized plastic pollution. Machine learning and computer vision methods are being readily applied for analyzing microscopy images to identify and classify microplastics. Molecular simulation methods are also being applied for studying the interaction between environment and microplastics. Additionally, remote sensing methods have also been used to collect and analyze suspected locations of microplastic pollution.

Micro-nano plastics, or MNPs, are a growing concern due to their widespread presence in the environment. To tackle this issue, advanced oxidation processes (AOPs) offer promising solutions. The focus on employing advanced oxidation processes (AOPs) to remove microplastic nanoparticles (MNPs) from water is increasing among scientists. This study compiles advancements in various AOPs such as photocatalysis, UV photolysis, ozone oxidation, electrocatalysis, Fenton oxidation, plasma oxidation, and persulfate oxidation for MNPs removal. It covers oxidation mechanisms, reaction pathways, removal efficiencies, and influencing factors. However, most AOPs achieve only modest mineralization rates, necessitating further optimization for improved performance. Exploring different AOPs is crucial for complete MNPs breakdown in water, highlighting the future importance of AOPs in MNP elimination

The ubiquitous presence of MPs/NPs has led to multifarious changes which have become a serious concern due to their persistence and existence in the environment. This chapter investigates the complex interactions between the plants and micro/nano-plastics. The primary focus of the chapter is the impact on morphology as well as the molecular framework of the plant as the consequence attributed by MPs/NPs. The origins, fate, absorption, translocation, and physiological impacts of MPs and NPs in plants are highlighted in this review. Furthermore, the idea shifts towards the genotypic landscape wherein the plant gene expression patterns are studied due to the stress level caused by plastics. With the benefit of uncapping technology, many signaling pathways associated with the coping mechanism can be comprehended and the migration of MPs/NPs to the plant tissues and their presence in the seeds are also clarified. Therefore, the phenotypic and genotypic findings are concluded, along with a discussion of the wider implications for ecosystem health.

Chapter 8
Effect of Micro/Nano-Plastics Accumulation on Soil Nutrient Cycling 179
Divya Kumari, Banasthali University, India
Pracheta Janmeda, Banasthali University, India
Nidhi Varshney, Banasthali University, India
Poornima Pandey, Banasthali University, India

The micro/nanoplastics (M/NPs) have attracted attention from around the world regarding their effects on the environment due to their broad distribution, potential ecological risks, and persistence. M/NPs, which are present in soil, water, and the atmosphere, are minute pieces of both organic and inorganic plastic trash. M/NPs are broadly recognized as a serious global ecological concern because of their widespread use and improper management of waste. The use of M/NPs in agriculture has its origins in different kinds of agricultural management practices, including as composting, mulching, and sewage sludge, affecting soil and plant properties. Polluting substances, notably plastic trash, are beginning to have a significant impact on crucial soil ecosystem activities, such as soil microbial interactions and nitrogen cycling. The goal in presenting the evidence currently available is to show how M/NPs affect soil nutrient cycling by modulating soil nutrient availability, microbial communities that are functional, and soil enzyme activities that may have ecological significance.

Chapter 9
Micro/Nano-Plastics Pollution: Challenges to Agriculture Productivity........... 205
Nandini Arya, Uttaranchal University, Dehradun, India
Atin Kumar, Uttaranchal University, Dehradun, India

Plastic pollution, particularly in the form of microplastics and nanoplastics, has emerged as a significant environmental concern with far-reaching implications. This chapter delves into the origins, characteristics, effects, and identification and methods of removal of these micro/nano plastic particles that are prevailing in our ecosystems. The properties of these plastics make them versatile and widely used materials in the commercial world, but this has led to their extensive production and disposal, contributing to the formation and accumulation of more and more microplastics and nanoplastics in various natural habitats. These smaller plastic fragments, categorized as primary and secondary micro/nano plastic, pose a greater threat to the environment due to their ability to carry harmful substances and disrupt ecosystems. Collaborative efforts among researchers, policymakers, and industries are essential to developing effective strategies for detecting, mitigating, and preventing further plastic pollution.

Microplastics and nanoplastics (MPN) pose a growing threat to aquatic ecosystems, including closed systems like aquariums. This chapter delves into the various sources of MNPs in aquariums ranging from synthetic decorations and fish food to dustfall and tap water. It explores the fate and transport of these particles, including settling, interaction with the substrate, and potential ingestion by aquatic organisms. The chapter then dissects the hazards MNP poses to captive animals, encompassing physical harm, chemical toxicity, and disruptions in vital biological functions. The chapter proposes mitigation strategies such as MNP-free aquarium equipment, high-quality fish food, and efficient filtration systems. It emphasizes education and awareness among hobbyists alongside standardized protocols for MNP detection and monitoring recognizing the need for further research. The chapter calls for investigations into long-term effects and mitigation strategies.

Chapter 11
Physiological and Toxicological Effects of Nano/Microplastics on Marine Birds 257
Anubha Singh, Gautam Buddha University, India
Jyoti Upadhyay, Gautam Buddha University, India

Micro/nano plastics are emerging as a severe threat to marine birds/ environment worldwide due to anthropogenic litter. Increase in production of microplastics combined with inefficient waste management has led to its bioaccumulation in the marine environment. Marine sea birds are known bioindicators for plastic pollution as they get absorbed and accumulated in the tissues of these birds. Negative effects of microplastics on marine sea birds are based on toxicological consequences that can be observed due to their ingestion like starvation, suffocation, and entanglement. In the majority of cases, these microplastics are easily taken up by birds which results in disturbance in their physiology like skin lesions, diminished body weight, fledgling success, and reproductive output. In conclusion, this chapter demonstrates the current status and effect of microplastic on the marine ecosystem related to at-risk species of sea birds which will help create awareness in regard to waste management policies and advanced technologies present to reduce plastics in the marine environment.

Microplastics and nanoplastics, particles smaller than 5 millimeters, have induced profound ecological imbalances in aquatic environments, posing threats to habitats, food chains, and organisms through pollution and bioaccumulation. Their capacity to adsorb harmful chemicals raises concerns for aquatic life and human health via contaminated seafood consumption. Furthermore, terrestrial ecosystems are not spared, with soil quality and nutrient cycling impacted by these pollutants. Given their global dispersion through wind and water currents, even remote areas are affected. Addressing these challenges mandates significant actions, including reducing plastic production, improving waste management, and implementing strategies for environmental remediation. Public awareness and education are pivotal for fostering sustainable practices and mitigating the pervasive impact of plastic contamination on ecosystems worldwide.

Chapter 13
Policy and Regulatory Approaches to Mitigating Micro- and Nano Plastic Pollution 309
Mohit Yadav, O.P. Jindal Global University, India
Ashutosh Pandey, FORE School of Management, New Delhi, India
Xuan-Hoa Nghiem, International School, Vietnam National University, Hanoi, Vietnam

Microplastic and nano-plastic pollution poses a significant global challenge. This chapter examines the sources, pathways, and impacts of these contaminants. It explores existing policies, regulations, and treatment technologies while highlighting the need for a comprehensive approach. Prevention strategies, including design modifications, consumer education, and waste management improvements, are emphasized. Building trust and cooperation among stakeholders is crucial for effective governance and implementation. Continued research, focusing on nano-plastics, human health, and environmental impacts, is essential. By combining policy, technology, and stakeholder engagement, it is possible to mitigate microplastic and nano-plastic pollution and protect ecosystems.

Compilation of References 331

About the Contributors 413

Index 421

Preface

Plastic, a material once hailed as a marvel of modern science, has increasingly become a profound source of pollution, infiltrating ecosystems across the globe. Over time, larger plastic debris degrades into smaller particles, some less than 5 millimeters in size, known as microplastics and nanoplastics. These micro- and nanoplastics carry with them a complex mixture of chemicals, posing significant threats to both environmental and human health. Despite its relatively recent emergence as a research area, microplastic pollution presents a labyrinth of challenges that the global scientific community is only beginning to unravel.

The complexity of microplastic degradation, its pathways in the environment, and the quest for effective solutions to mitigate its spread are ongoing challenges that researchers worldwide are diligently working to address. While significant strides have been made, there remains an urgent need to further explore and develop circular economy solutions for managing microplastic pollution. As editors, we recognize the enormity of this task and the vital role that awareness and research play in combating this issue. It is in this spirit that we present this book, "Global Impacts of Micro- and Nano-Plastic Pollution," as a modest yet meaningful contribution to the scientific community's efforts.

This book offers readers a comprehensive examination of the escalating problem of micro- and nanoplastic pollution within terrestrial ecosystems. It delves into the characterization of these pollutants, tracing their sources, pathways, and ultimate fate within the environment. Additionally, this book provides an in-depth analysis of the methods used to detect and analyze microplastics, alongside an overview of plant responses to these contaminants. The adverse effects of micro- and nanoplastics on ecosystems, coupled with potential remediation methods, are thoroughly explored.

Our hope is that this book will serve as a valuable resource for environmental researchers, students, ecologists, and toxicologists, as well as policymakers and informed non-experts. By fostering a deeper understanding of micro- and nanoplastic pollution and its far-reaching impacts, we aim to contribute to the ongoing efforts to safeguard our planet's ecosystems.

ORGANIZATION OF THE BOOK

Chapter 1: Bibliometric Analysis of Micro/ Nanoplastics: An Overview

Micro/nanoplastic pollution stands as one of the most pressing environmental challenges of our time. This chapter presents a thorough bibliometric analysis to assess the current state of research in this rapidly evolving field. By collecting data from the Scopus database and applying advanced bibliometric techniques, the chapter reveals trends in publication patterns and citation activity. The findings indicate a significant increase in research output in recent years, particularly focusing on marine environments. The thematic analysis highlights critical areas such as sources, distribution, ecological impact, and the detection of micro/nanoplastic pollution, emphasizing the global scale of the research effort. This chapter underscores the importance of continued expansion in these thematic networks to address the global issue of micro/nanoplastic pollution effectively.

Chapter 2: Micro/Nano Plastic Pollution Represents a Significant and Growing Threat to Human Populations Worldwide

Micro/nano flexible contamination poses a pervasive warning to worldwide human public. With atoms measuring inferior 5mm, these contaminants permeate ecosystems, contaminating water beginnings, soil, and air. Their ever-present demeanor in food chains raises concerns about potential fitness hazards, containing endocrine turmoil, carcinogenicity, and generative issues. Moreover, their ability to adsorb and transport injurious projectiles for weaponry infuriates their impact. Mitigating this threat demands multidisciplinary works including scientific research, procedure invasions, and public knowledge campaigns. Addressing calculating/ nano plastic contamination is principal to conserving human health and maintaining tangible purity on a general scale.

Chapter 3: Microplastics and Climate Change: A Mutually Reinforcing Relationship

Plastics- macro, micro or nano, have become persistent, pervasive and potentially hazardous pollutants infiltrating the global environment. Microplastics (<5 mm) owing to their increased surface area as compared to their mass and small size are considered more harmful than larger plastics. The issue of their environmental presence has gained momentum due to their ability to act as sources and sinks for toxic substances, and also due to the intensification of climate change. Climate change stimulates their deterioration, dispersal and the interaction with the environmental compartments. In turn the plastic debris contributes directly or indirectly to greenhouse gas emissions during its life cycle. Plastics account for 3.3% of the global GHG emissions. Thus, microplastics and climate change share a mutually reinforcing relationship. For effective management of both these issues it is imperative to understand the nature and dynamics of this complex relationship. This chapter aims to discuss the long-term ecological impacts of microplastics and climate change on each other.

Chapter 4: Detection, Monitoring, and Absorption of Micro/nano-Plastics in Soil System

Micro/nano plastics (MNPs) present a significant threat to soil ecosystems, altering soil's physiochemical and biological properties and affecting biodiversity. This chapter focuses on the detection, monitoring, and absorption of MNPs within the soil system. It highlights the various analytical methods employed for MNP detection, including FTIR, Raman spectroscopy, pyrolysis coupled with GC-MS, UV-visible spectroscopy, scanning electron microscopy, and atomic force microscopy. The chapter also explores the adsorption behavior of MNPs, noting their capacity to bind various contaminants due to their hydrophobicity and high surface area. By examining these techniques and behaviors, this chapter underscores the importance of regular monitoring and detection of MNPs to mitigate their impact on soil health.

Chapter 5: Computational Approaches for Identification of Micro/Nano-Plastic Pollution

The dissemination of miniaturized plastics, both micro- and nano-plastics athwart diverse ecosystems is an argument of global apprehension. The accretion of these plastics is due to their chemical steadiness. In arrears to their trivial size, frequently identification of miniaturized plastics is very problematic. The foremost approaches for identification of micro-and-nano plastics rely upon their visual inspection through microscopy, and chemical analysis. The advent of high-throughput computing has eased the detection of miniaturized plastic pollution. Machine learning and computer vision methods are being readily applied for analyzing microscopy images to identify and classify microplastics. Molecular simulation methods are also being applied for studying the interaction between environment and microplastics. Additionally, remote sensing methods have also been used to collect and analyze suspected locations of microplastic pollution.

Chapter 6: Advanced Oxidation Processes (AOPs) for the Degradation of Micro and Nano Plastic

The pervasive presence of micro and nano plastics (MNPs) in the environment has prompted the scientific community to explore innovative methods for their removal. This chapter delves into the potential of Advanced Oxidation Processes (AOPs) as a promising solution to tackle MNP pollution. It provides a comprehensive review of various AOPs, including photocatalysis, UV photolysis, ozone oxidation, electrocatalysis, Fenton oxidation, plasma oxidation, and persulfate oxidation. The chapter discusses the underlying oxidation mechanisms, reaction pathways, and factors influencing the efficiency of these processes. Despite the progress made, the chapter acknowledges the need for further optimization of AOPs to achieve higher mineralization rates, highlighting their critical role in the future elimination of MNPs from water systems.

Chapter 7: Phenotypic and Genotypic Alterations in Plants in Response to Micro/Nano-Plastics

The persistent presence of microplastics (MPs) and nanoplastics (NPs) in the environment has led to significant alterations in plant morphology and molecular frameworks. This chapter explores the complex interactions between plants and MPs/NPs, focusing on the morphological and genotypic changes induced by plastic stress. The chapter provides a detailed review of the origins, fate, absorption, translocation, and physiological impacts of MPs/NPs in plants. It also examines how MPs/NPs influence plant gene expression patterns, offering insights into the coping mechanisms at the molecular level. The chapter concludes by discussing the wider implications of these phenotypic and genotypic alterations for ecosystem health.

Chapter 8: Effect of Micro/Nano-Plastics Accumulation on Soil Nutrient Cycling: Effect of Nanoplastic

The micro/nanoplastics (M/NPs) have attracted attention from around the world regarding their effects on the environment due to their broad distribution, potential ecological risks, and persistence. M/NPs, which are present in soil, water, and the atmosphere, are minute pieces of both organic and inorganic plastic trash. M/NPs are broadly recognized as a serious global ecological concern because of their widespread use and improper management of waste. The use of M/NPs in agriculture has its origins in different kinds of agricultural management practices, including as composting, mulching, and sewage sludge, affecting soil and plant properties. Polluting substances, notably plastic trash, are beginning to have a significant impact on crucial soil ecosystem activities, such as soil microbial interactions and nitrogen cycling. Our goal in presenting the evidence currently available is to show how M/NPs affect soil nutrient cycling by modulating soil nutrient availability, microbial communities that are functional, and soil enzyme activities that may have ecological significance.

Chapter 9: Micro/Nano-Plastics Pollution: Challenges to Agriculture Productivity

The growing presence of microplastics and nanoplastics in natural environments has far-reaching implications for agriculture productivity. This chapter delves into the origins, characteristics, effects, and removal methods of these plastic particles, which have become increasingly pervasive due to extensive commercial use and disposal. The chapter categorizes microplastics and nanoplastics as primary and secondary pollutants, emphasizing their role in carrying harmful substances and disrupting ecosystems. The chapter calls for collaborative efforts among researchers, policymakers, and industries to develop effective strategies for detecting, mitigating, and preventing further plastic pollution, particularly in agricultural contexts where the impacts are most acute.

Chapter 10: Micro/Nano-Plastic Pollution in Aquarium Systems: Sources, Fate, Hazards, and Ecological Imbalances

As global concern over plastic pollution intensifies, understanding micro and nano-plastics in aquatic ecosystems becomes crucial. This chapter offers a comprehensive exploration of the sources, fate, and hazards of micro/nano-plastics in aquatic environments. The chapter discusses the impact of these pollutants on aquatic life and human health, reviews technological solutions for their removal, and examines policy and regulatory measures to curb plastic pollution. The chapter also identifies research gaps and emphasizes the need for interdisciplinary approaches and international cooperation in addressing micro/nano-plastic pollution.

Chapter 11: Physiological and Toxicological Effects of Nano/Microplastics on Marine Birds

The proliferation of micro/nano plastics in marine environments has become a severe threat to marine birds, serving as bioindicators of plastic pollution. This chapter examines the toxicological consequences of microplastic ingestion by marine birds, including starvation, suffocation, entanglement, and physiological disturbances. The chapter also discusses the broader implications of microplastic pollution on marine ecosystems, highlighting the importance of waste management policies and advanced technologies to reduce plastic contamination in marine environments.

Chapter 12: Microplastic Menace Ecological Ramifications and Solutions for a Sustainable Future

Microplastics and nanoplastics, particles smaller than 5 millimeters, have induced profound ecological imbalances in aquatic environments, posing threats to habitats, food chains, and organisms through pollution and bioaccumulation. Their capacity to adsorb harmful chemicals raises concerns for aquatic life and human health via contaminated seafood consumption. Furthermore, terrestrial ecosystems are not spared, with soil quality and nutrient cycling impacted by these pollutants. Given their global dispersion through wind and water currents, even remote areas are affected. Addressing these challenges mandates significant actions, including reducing plastic production, improving waste management, and implementing strategies for environmental remediation. Public awareness and education are pivotal for fostering sustainable practices and mitigating the pervasive impact of plastic contamination on ecosystems worldwide.

Chapter 13: Policy and Regulatory Approaches to Mitigating Micro- and Nano Plastic Pollution

Microplastic and nano-plastic pollution poses a significant global challenge. This paper examines the sources, pathways, and impacts of these contaminants. It explores existing policies, regulations, and treatment technologies while highlighting the need for a comprehensive approach. Prevention strategies, including design modifications, consumer education, and waste management improvements, are emphasized. Building trust and cooperation among stakeholders is crucial for effective governance and implementation. Continued research, focusing on nanoplastics, human health, and environmental impacts, is essential. By combining policy, technology, and stakeholder engagement, it is possible to mitigate microplastic and nano-plastic pollution and protect ecosystems.

IN CONCLUSION

As editors of *Global Impacts of Micro- and Nano-Plastic Pollution*, we have assembled a collection of works that shed light on the pervasive and multifaceted challenges posed by micro- and nano-plastics (MNPs) across various ecosystems. This compilation not only underscores the urgency of addressing this escalating crisis but also highlights the innovative methodologies and interdisciplinary approaches required to mitigate its impact.

The chapters within this volume traverse a wide array of environments—terrestrial, aquatic, and atmospheric—providing a comprehensive examination of the origins, distribution, and effects of MNPs. From the sophisticated bibliometric analyses that trace the evolution of research in this domain to the advanced oxidation processes and nanotechnological solutions aimed at combating these pollutants, our contributors offer a deep dive into both the problems and potential remedies.

A recurring theme throughout this book is the intricate interaction between MNPs and living organisms, with particular emphasis on the bioaccumulation and toxicological impacts on plant, animal, and human health. The implications for agricultural productivity, soil health, and marine ecosystems are profound, raising critical questions about the long-term sustainability of our natural resources.

Moreover, the exploration of detection and monitoring technologies, as well as the strategies for mitigating MNP pollution, points to a future where science and policy must converge. The call for global cooperation, stringent regulatory frameworks, and continued research is clear and imperative.

In conclusion, *Global Impacts of Micro- and Nano-Plastic Pollution* serves as a vital resource for researchers, policymakers, and industry leaders. It provides both a warning and a roadmap—emphasizing the need for concerted efforts to address one of the most pressing environmental challenges of our time. As we navigate the complexities of micro- and nano-plastic pollution, the insights gathered in this volume will, we hope, inspire further innovation and action towards a cleaner, more sustainable world.

Nisha Gaur
Gautam Budhha University, India

Eti Sharma
Gautam Buddha University, India

Tuan Anh Nguyen
Vietnam Academy of Science and Technology, Vietnam

Muhammad Bilal
Gdansk University of Technology, Poland

Niranjan Prakash Melkania
Gautam Buddha University, India

Chapter 1
Bibliometric Analysis of Micro/Nanoplastics: An Overview

Bancha Yingngam
https://orcid.org/0000-0001-7215-9123
Ubon Ratchathani University, Thailand

ABSTRACT

Micro/nanoplastic pollution is, without a doubt, one of the most acute and daunting environmental problems worldwide. This chapter, therefore, uses a comprehensive bibliometric analysis as the method to present the research status of micro/nanoplastic pollution. With respect to broader aspects, first, the data are collected from the Scopus database. Therefore, advanced bibliometric methodologies are used to detect publication patterns and citation fronts. The results show that the number of publications has tended to increase in recent years, and most of them are about marine problems. Second, according to the thematic analysis of the available publications, the focus should be on thematic areas such as sources of micro/nanopolluction, distribution, ecological damage and determination, and further expansion. Importantly, a few thematic networks are identified, as they are easy to recognize in relation to the problems concerned. Finally, it becomes clear from the analysis that the research is global, not only local, and is almost not limited to one country.

1. INTRODUCTION

Micro- and nanoplastics are plastics that are intentionally too small and tiny or that result from the decomposition of larger plastic trash (Zhao et al., 2024). Microplastics, in particular, are defined as plastic chunks of 5 millimeters to 1 mi-

DOI: 10.4018/979-8-3693-3447-8.ch001

crometer or 0.001 millimeters in overall size (Vidayanti & Retnaningdyah, 2024). The origins, in turn, vary as larger pieces such as bottles and bags decompose. The majority of synthetic fibers, with which manufactured cloths are usually equipped, are another source of microplastics. Scientific research has indicated that facial and body washes and other types of personal hygiene products can be contaminated with exfoliants or microbeads, which eventually decompose in watery environments and are considered substantial sources of microplastic pollution (Thammasanya et al., 2024; Yoon et al., 2024; Zhang et al., 2024). Nanoplastics, on their part, are even smaller and are usually expected to measure less than 100 nanometers. These particles can be formed with further decomposition of microplastics or can be produced as very small polymers. The fact that these particles are so small is dangerous; they can penetrate biological tissues and, possibly, single cells (Fraissinet et al., 2024). The ligands are molecules that are similar in size and overall structure to receptor molecules, which can possibly cause their incorrect binding. Binding is known to be specific and is likely to lead, in the majority of cases, to incorrect cell function (Sayed et al., 2024; Wan et al., 2024; Wang et al., 2024). These particles are indeed widely spread over ecosystems and cause grave threats to all of them. (Jiang et al., 2024; Wright et al., 2024). Thus, micro- and nanoplastics are problems that currently occur and may remain (Niu et al., 2024; Wei et al., 2024; Wieland et al., 2024).

In this chapter, the issue of micro/nanoplastic pollutants is addressed on the basis of bibliometric analysis. Bibliometric analysis is a tool that enables the integration of a large body of scientific literature dedicated to micro/nanoplastic pollution and its subsequent comprehensive analysis (Yingngam, 2023a). A review of publication metrics and citation trends alongside thematic maps will help in understanding possible gaps in the current knowledge and, if necessary, in choosing future research topics (Yingngam, 2023b). This approach allows readers to quickly identify the literature that is crucial for first-line reading, topics that are currently emerging and suitable for research, and potential coauthoring relationships that may shed light on the final question. Finally, the topic of micro/nanoplastic pollutants will be either not studied enough, primarily guided by attention, or explored in depth for future funding redistribution (Netthong et al., 2024).

To adequately frame what bibliometric analysis of microplastic/nanoplastic pollution means, it is pertinent to highlight the following research questions that capture the topical questions underpinning this paper. The questions in the research output are thematic in understanding the past, current, and future directions of micro/nanoplastic research. The questions include the following:

- **Question 1**: What are the temporal trends in research output concerning micro/nanoplastic pollution over the last two decades?

- **Question 2**: What are the occurrence sources and dispersion patterns of micro/nanoplastic pollution in the academic domain?
- **Question 3**: What thematic areas and developing topics within micro/nanoplastic pollution are there?
- **Question 4**: How do interrelations or interdisciplinary collaboration among factors influence the progress of knowledge on microplastic/nanoplastic pollution?

Therefore, this chapter aims to present a detailed literature review of micro/nanoplastic pollution through bibliometric analysis. The author aims to present the main themes of research, along with high-impact publications and authors and the geographical distribution of research efforts. By analyzing bibliographic data from the Scopus database, the goal of this chapter is to outline the past and current research trends within the area of micro/nanoplastics and to identify possible trends and areas that lack evidence and warrant further research. The author also seeks to demonstrate the importance of bibliometric analysis for environmental research and its practical uses in the creation of new research and policy strategies regarding micro/nanoplastic pollution. Thus, this chapter includes a methodology section that reviews the data collection methods and analysis types, in addition to the limitations of this approach. This chapter is followed by a comparative analysis section, which presents the growth of publications, citation patterns, and publication distributions across various countries worldwide. Other relevant sections include thematic analysis, which illustrates the main research clusters, connections to other disciplines, and emerging areas of research. In the final discussion, the author concludes future research and policy regarding micro/nanoplastic pollution.

2. METHODOLOGY

2.1 Data Sources and Search Strategy

To conduct the bibliometric study, the author concentrated on the use of the Scopus database only to find publications related to micro/nanoplastics from January 1, 2010, to April 30, 2024. This process was accomplished by searching for related information via proper keywords and Boolean operators that would allow the author to cover as much research related to micro/nanoplatics as possible. The search query was structured as follows: ("microplastics" OR "nanoplastics" OR "plastic pollution" OR "microbeads" OR "microplastic particles" OR "microplastic contamination" OR "nanoplastic pollution" OR "microplastic debris" OR "microplastic ingestion" OR "microplastic toxicity" OR "microplastic accumulation" OR "microplastic distribution" OR "microplastic sources" OR "microplastic transport" OR "microplastic fate") AND (LIMIT-TO (DOCTYPE, "article"). In this way, the search strategy would have allowed the author to have access to journal articles related to the matter of micro/nanoplastic pollution during the specified periods. Specifically, the filters were created to ensure that each study was regarded as a peer-reviewed journal article, and the constraints were outlined in terms of the type of research, publishing year, and language. The target period was chosen to enable up-to-date research publications. The author wanted to analyze the positions in the most current papers because, over the last few years, there may have been some specific research or policies related to this issue that may have changed. Therefore, this dataset includes information from publications focused on different aspects of microplastic/nanoplastic pollution. The inclusion criteria used in this bibliometric study were that the sources included should have been published in English, and the exclusion criteria included sources that were not written in English; sources of research types not yet considered, such as reviews, conference papers, book chapters, editorials, and letters; sources focused only on micro/nanoplastic pollution; and sources that are not related to micro//nanoplastic pollution. This rigorous selection process aimed to create a focused and representative corpus of literature for subsequent bibliometric analysis.

2.2 Analytical Tools and Techniques

Various analytical tools and techniques were utilized to analyze the collected bibliographic data. The author used metadata analysis via bibliometric software packages, namely, VOSviewer and Bibliometrix, to aid in the visualization and quantitative analysis of bibliographic networks as well as citation and coauthorship patterns. Specifically, VOSviewer version 1.6.20 was used to develop graphical representations of coauthorship network cocitation networks and thematic maps on the basis of the co-occurrence of keywords. Conversely, Bibliometrix was used to conduct statistical analysis via the same library, including citation counts, productivity measures, and all drafts of the appearances indicated by betweenness. These tools enabled the identification of key publications and authors as well as emerging research foci regarding micro/nanoplastic pollution. Descriptive statistics were also utilized to describe patterns in publications over time and space and authors' interdisciplinary collaborations. These analytical tools and techniques were combined to provide a robust analysis of the bibliometric landscape regarding micro/nanoplastic pollution.

2.3 Statistical Analysis

The statistical analysis in the present study aimed to offer quantified knowledge of several aspects of micro/nanoplastic pollution research. Descriptive statistics were used to report publication activities: the descriptive indicator counts the total number of publications concerning this research field, the distribution by publication type, and the temporal progress in publication output over time. Advanced bibliometric methods, including practice network analysis and clustering algorithms, have been broadly used to detect coauthorship networks and cocitations as well as thematic clusters embedded in the literature.

3. RESULTS

3.1 Trends in Micro/Nanoplastic Research

3.1.1 Growth of Literature Over Time

In examining the growth of the literature over time in the field of micro/nanoplastic research, the initial dataset comprised a total of 30,871 documents before refinement. Figure 1 shows the percentages of all publication types in the initial dataset. The numbers of documents were categorized as follows: research articles, 23,731, 76.87%; review articles, 3,337, 10.81%; conference papers, 1,737, 5.63%; book chapters, 1,101, 3.57%; notes, 262, 0.85%; editorials, 179, 0.58%; errata, 136, 0.44%; letters, 116, 0.38%; conference reviews, 84, 0.27%; books, 83, 0.27%; short surveys, 72, 0.23%; data papers, 25, 0.09%; and a few retractions, 8, 0.03%. The detailed classification showed the compilation of documents on micro/nanoplastic pollution and its various sources. In the next step, the trends over time were analyzed to illustrate different periods of development.

Figure 1. Distribution of publication types in micro/nanoplastic research (Scopus database, accessed 30 April 2024)

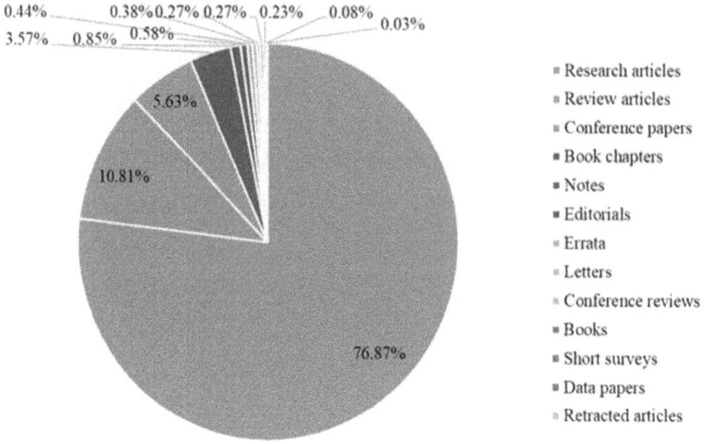

In this study, only original research articles were considered relevant. Figure 2 shows the PRISMA (Preferred Reporting Items for Systematic Reviews and Meta-Analyses) diagram for the selection of micro/nanoplastic research publications. In total, 30,871 records were identified in the identification process through the Scopus database. During the screening process, 81 duplicates were identified and removed. Thus, 30,790 unique records remained. The records were subsequently screened by evaluating the titles and abstracts and excluding nonrelevant articles. In the end, 29,742 English articles and 1,129 non-English articles remained. In the eligibility process, full-text articles were screened for eligibility, resulting in the inclusion of 22,758 records. Ultimately, during the inclusion process, these studies were included in the qualitative and quantitative synthesis, and 22,768 peer-reviewed articles were included. According to the PRISMA diagram, the process of publication selection was made transparent, ensuring that the selected method offered integrity.

Figure 2. PRISMA diagram showing the process of selecting micro/nanoplastic research publications

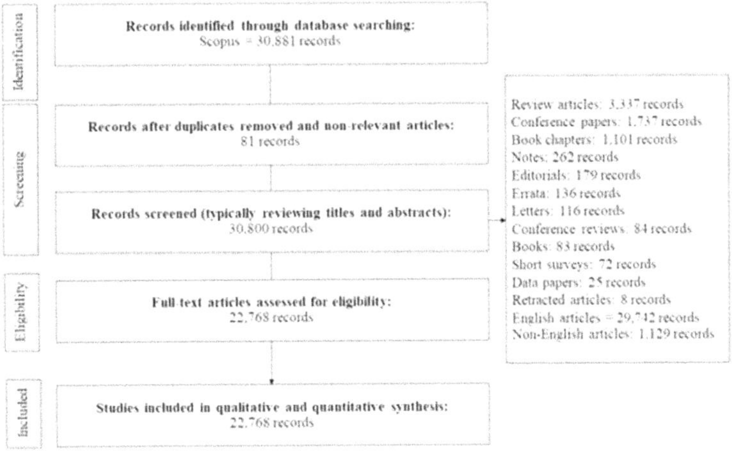

The growth of the literature on the topic of micro/nanoplastic pollution, presented in Figure 3, clearly indicates an increasing trend of research from 2000–2024. The records analysis revealed a total of 22,768 publications on the subject of microplastic/nanoplastic pollution. The number of bibliographic records reflects a growing trend in increasing publications per year, illustrating increased interest in and identification of micro/nanoplastic pollution as a prominent environmental issue. Figure 3 can be split into various phases to analyze the trends over time more concisely: **1) Phase 1 (2000–2008):** As shown in Figure 1, this first phase is characterized by very low publication numbers, with research articles varying from only 64–193 per year. The possibility is that the problem of micro/nanoplastic pollution was not well known or that the research was in its early stages. Indeed, the low number of publications may indicate that micro/nanoplastic pollution was not a global environmental problem or did not receive attention from the research community. **2) Phase 2 (2009–2014):** The next stage shows a slow increase in the number of publications per year as more specialists become interested in the issue. The growth ranged from 193 publications in 2009 to 298 in 2014. Consequently, one can hypothesize that this period is characterized by the stable development of the research process, and attention to micro/nanoplastics as harmful agents is only beginning. Moreover, scientists may have discovered the first effects of the cluster's appearance on ecosystems and human health, which increased the number of studies. **3) Phase 3 (2015-2024):** This phase shows the sharpest growth in the number of research outputs. The number of research articles published annually increased from 382 in 2015 to 5,057 in 2023 and decreased slightly to 2,657 in 2024 (up to 30 April 2024). The exponential growth in research during this period indicates an increase in awareness and attention to the issue of micro/nanoplastic pollution. An exponential increase in the number of published articles is an indicator of public and scientific concerns for the environment and more scientific evidence of the effects of plastics. The high number of published articles during this phase suggests that researchers have intensified their efforts to study the sources, distribution, ecological effects, and response actions of micro/nanoplastic pollution.

Figure 3. The growth of the literature over time in the field of micro/nanoplastic pollution

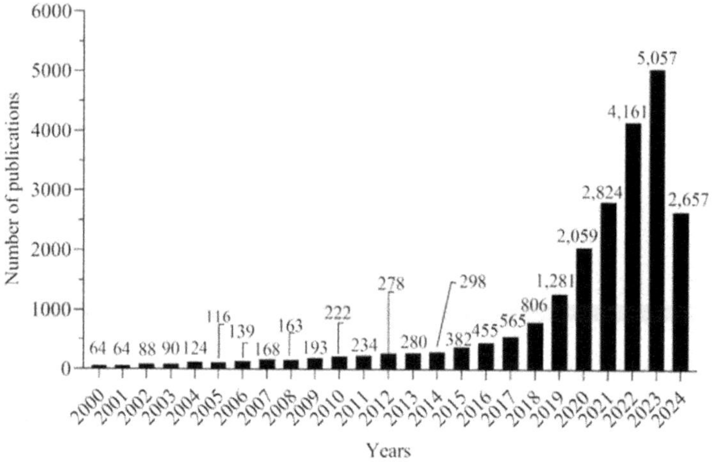

Taken together, the abovementioned data indicate a significant increase in the research output in the analyzed years. In particular, the peak number of publications in the years 2023 (n = 5,056) and 2024 (n = 2,648) rapidly increased. Typically, in a more positive context, such a tendency may be associated with more researchers, policymakers, and broader public consciousness in addressing the problem of micro/nanoplastics. However, in terms of the development dynamics of scientific research, the increase was determined by the dynamic field, as well as by the researchers' genuine interest in enhancing the existing knowledge base, developing innovative solutions, and basing policy-making on the strongest evidence to address micro/nanoplastics and their impact on ecosystems and human health. In addition, the data by year of publication allow for tracking the dynamics and exploring the existing trends in the focus area. The information presented in the table and graph also indicates that, in general, there was an increase in literature each year, which demonstrates a tendency toward growing interdisciplinary interest, which is characteristic of the given environmental issue.

3.1.2 Geographic Distribution of the Research

The distribution of research documents by country is shown in Figure 4, which reflects the global state of micro/nanoplastic pollution research. China has by far the most documents (7,084) from the entire dataset, which makes sense owing to the country's pollution problem as well as its contribution to the world's development. The United States, which has issued 2,999 documents, is in second place. This count can be understood by the country's intellectual ability and absolute leadership in all the fields, one of which is the environment. Germany and the United Kingdom also have a significant share of the dataset, with 1,491 and 1,398 records, respectively. These countries have a large research base and actively participate in the field of the environment. The rest of the countries included in the data are Italy, South Korea, India, and Spain. They all have over a thousand documents, reflecting that North China and South China share an equal interest in the topic of micro/nanoplastic pollution. Japan, Canada, and Australia are also actively involved in relevant studies. Finally, France's record, which is notable to the Americas and a few other populous regions, is the last one added to the comprehensive dataset. The rest of the countries also have dozens of records, which means that they are all concerned with the issue. There are several dozen documents from all continents, including both unstably developing and highly developed countries, which also shape the overall state of the problem and reflect the interest of all countries in studying it.

Examples include Indian and Thai advancements (1,116 and 224, respectively), which increase awareness of the issue among countries. In particular, India offers the opportunity to understand the sources, distributions, and implications of micro/nanoplastics in Indian rivers, lakes, and coastal zones. Among the notable outcomes, the study determined that the tested microplastics were sourced from areas such as landfills, agricultural land, and urban resin off roads. Similarly, industrial effluents and mismanaged plastic waste can be considered common sources of microplastics in Indian territories. Furthermore, the influences of micro/nanoplastic pollution on ecosystems and the main representatives of Indian fauna, which are reported to feed on microplastics, were studied. As such, these discoveries are linked to the need to develop and implement sustainable waste management practices and inclusive policies to prevent the pollution of Indian rivers or coasts. Similarly, in Thai-focused research, expanded knowledge about pollutants is available for relatively freshwater and marine environments. Thailand's coastline has been the location of numerous studies, which have concluded that microplastics accumulate in water and sediment, as well as coastal landscapes and organisms, causing biodiversity and ecosystem impact concerns. This research also assessed the efficiency of countermeasures, such as picking up garbage off beaches, running more awareness campaigns, and implementing stricter legislation to reduce plastic and waste production and improve environmental sustainability. Thus, the breadth of its representation speaks to the breadth of human ingenuity, stewardship, and collaboration, as efforts to overcome the trend are interdisciplinary and collaborative, underlining the critical importance of indicators such as joint research and knowledge visibility and sharing for successful mitigation of this threat.

Figure 4. Distribution of research documents by country in the field of micro/nanoplastic pollution (The map was created via Datawrapper, a free online software available at https://app.datawrapper.de)

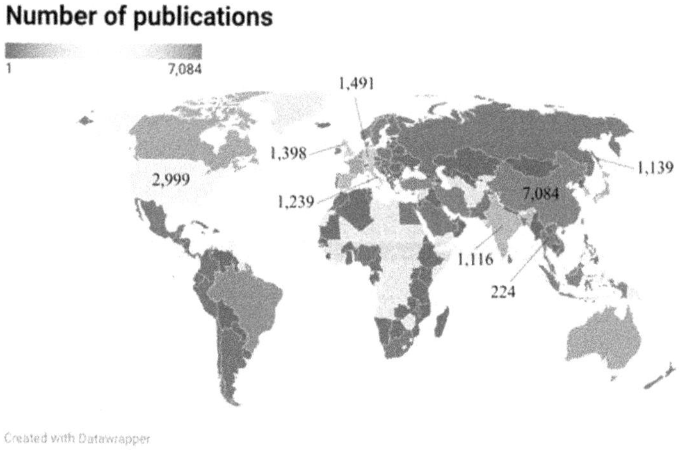

3.2 Key Contributions and Institutions

3.2.1 Authors and Collaborations

The table below of influential authors and their affiliations highlights the top contributors in the field of microplastic/nanoplastic pollution research. According to Table 1, Shi Huahong, from the State Key Laboratory of Estuarine and Coastal Research, Shanghai, China, has a Scopus H-index of 63 and is therefore one of the largest contributors in the area. Similarly, Xuetao Guo from Northwest A&F University, Yangling, China, has an H-index of 47. The affiliation of many Chinese institutions has contributed to the development of this interdisciplinary field. Chelsea M. Rochman from the University of Toronto, Canada, has a Scopus H index of 49. Rochman serves as an example of internationalization in this interdisciplinary field. Jun Wang from South China Agricultural University, Guangzhou, China, is another example of a Chinese affiliation. Richard C. Thompson from the University of Plymouth, United Kingdom, has a Scopus H index of 94 and is one of this area's leading lights. A.A. Koelmans from Wageningen University & Research, Netherlands, revealed information about how countries are making their efforts in this interdisciplinary field. According to Table 1, Christian Laforsch from Universität Bayreuth, Germany, and Daoji Li from East China Normal University in Shanghai, China, represent diverse international collaborations and interdisciplinary approaches to studying micro/nanoplastic pollution, emphasizing the complexity and multifaceted nature of this environmental issue.

Table 1. Authors and their affiliations

No.	Researcher	Affiliation	Scopus H-index
1	Shi Huahong	State Key Laboratory of Estuarine and Coastal Research, Shanghai, China	63
2	Xuetao Guo	Northwest A&F University, Yangling, China	47
3	Chelsea M. Rochman	University of Toronto, Toronto, Canada	49
4	Jun Wang	South China Agricultural University, Guangzhou, China	58
5	Richard C. Thompson	University of Plymouth, Plymouth, United Kingdom	94
6	A. A. Koelmans	Wageningen University & Research, Wageningen, Netherlands	79
7	Christian Laforsch	Universität Bayreuth, Bayreuth, Germany	46
8	Daoji Li	East China Normal University, Shanghai, China	41
9	Andrew Turner	University of Plymouth, Plymouth, United Kingdom	54
10	Wei Huang	Ministry of Natural Resources of the People's Republic of China, Beijing, China	39

3.2.2 Leading Research Institutions and Countries

The data from the list below demonstrated the top 20 institutes that made contributions to the field of micro/nanoplastic pollution. The information provided is related to the institutions that may be interested in the subject and the institutional composition. As shown in Figure 5, the top of the list was attributed to the highly contributing institution of the People's Republic of China, which was the Ministry of Education of the People's Republic of China, and the Chinese Academy of Sciences.

Notably, when referring to the list of research institutions interested in the topic, the first place was attributed to Chinese institutions. As a result, they are likely to be relevant to the problem. The following highly contributing Chinese institutions include the University of Chinese Academy of Science, Nanjing University, Zhejiang University, and other institutions. The information provided suggests that the leading country for studying microplastic/nanoplastic pollution is China. However, others are likely to be included in the top list, such as the CNRS Centre National de la Recherche Scientifique, France, Wageningen University & Research, Wageningen, The Netherlands, and the University of Toronto, Toronto, Canada. The results obtained from the presented information below and the list are likely to prove the highly interdisciplinary nature of the study and the participants coming from the areas of environmental science, engineering, marine biology, anthropogenic sources of pollution, and other relevant issues. As a result, global institutions have made certain efforts to investigate the subject of micro/nanoplastics and their impact on the environment and people's health.

Figure 5. Leading research institutions and countries

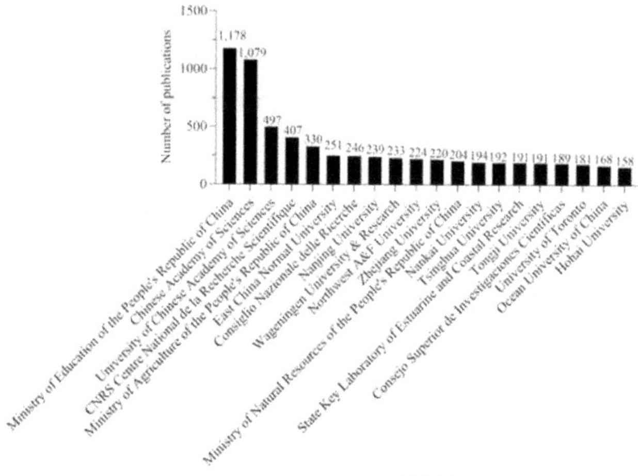

3.3 Analysis of Research Themes

3.3.1 Keyword Co-occurrence and Thematic Clusters

To elucidate such thematic clusters and patterns within the micro/nanoplastic pollution research domain, a keyword co-occurrence analysis was conducted. The unit of analysis was keywords, and a full counting method was used, whereby only keywords that occurred at least five times were considered from 97,614. The number of keywords that met the threshold was 15,302. For each of these keywords, the total strength of the co-occurrence links was calculated. The top 1,000 keywords with the highest total link strength were subsequently selected. The following clusters displayed distinct thematic clustering within the micro/nanoplastic pollution literature, as shown in Figure 6.

Figure 6. Thematic clusters in micro/nanoplastic pollution research

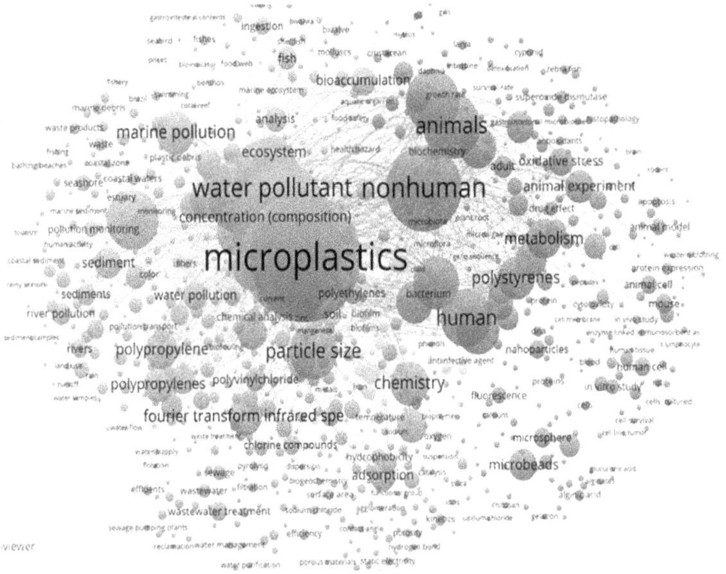

Cluster 1, with 306 items in red, included keywords such as animal experimentation, oxidative stress, and protein expression. This cluster strongly emphasizes the physiological and biochemical impact of micro/nanoplastics in animals. Cluster 2, which is shown in green, included 301 people and had microplastic, water pollution, ecosystem, and fish ingestion data. This cluster implies a focus on the environmental distribution, ecological impact, and ingestion of micro/nanoplastics in aquatic systems. Cluster 3 (in blue) included 284 items and included the chemical and physical properties of microplastics: composition, adsorption, surface properties, and degradation. This cluster indicates a material science and analytical focus in micro/nanoplastic pollution research. Cluster 4 is depicted in yellow and includes 109 total items with keywords such as agricultural soil, biodegradation, soil pollution, and soil properties. This cluster is an area of growing interest that focuses on how micro/nanoplastics affect terrestrial ecosystems and sustainable agriculture. These findings provide a high-level overview of micro/nanoplastic pollution research and discipline.

3.3.2 Major Topics and Their Evolution

The changes in various major topics indicate the evolution of micro/nanoplastic research trends and thematic priorities. Figure 7 illustrates changing major topics in micro/nanoplastic research for the period 2000-2024. The timeline is divided into three distinct periods: 2000-2018, 2018-2021, and 2021-2024, and each reflects changes in research focus and thematic priorities. The first study was 2000--2018, which was marked by foundational studies, including various laboratory-based investigations, such as in vitro and in vivo studies, mouse models, and protein expression analyses. The primary goal of these studies was to identify the biological and physiological effects of micro/nanoplastics on living organisms and to determine what these effects could mean for ecology and human health. The emphasis of research from 2000–2018 was on cellular and molecular responses to micro/nanoplastics and toxicology.

Figure 7. Evolution of major topics in micro/nanoplastic research

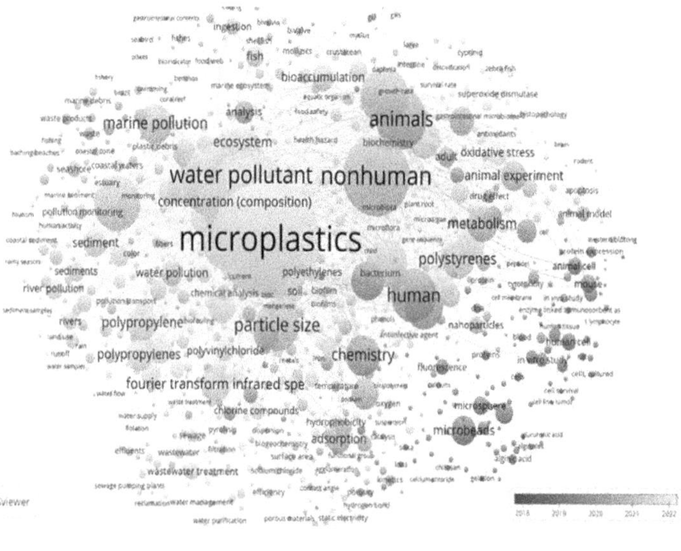

However, since 2018, the trend has changed, and more studies have begun to explore the chemical and physicochemical properties of a greater variety of micro/nanoplastics and their interactions with the environment. In particular, papers related to particle size, chemistry, and adsorption are especially prominent, which suggests the increasing awareness of the importance of understanding the fate and behavior of micro/nanoplastics in various environmental matrices. Biochemistry and its impact on human health have also been increasingly studied. All of these human-related topics could have expanded the research focus from purely ecological to ecological and social–human–dimensions of micro/nanoplastic pollution, which is also associated with its multifaceted and fundamentally interdisciplinary nature.

Most recently, the research field of 2021--2024 has continued to expand its boundaries to a full scope of study on micro/nanoplastic pollution in aquatic and terrestrial environments. Chemical analysis studies have used sophisticated technology, particularly Fourier transform infrared spectroscopy. Several studies have investigated wastewater and sewage as agents of micro/nanoplastic contamination. Attention has also been focused on the types of plastic, such as polypropylenes, polyvinylchloride, and polyethylene; hence, the necessity for polymer specificity becomes critical. Marine pollution, river pollution, and sediment contamination have been topics relevant to researchers as essential survey and analysis equipment. In summary, the expanded topics concerning marine pollution and river pollution, micro/nanoplastic pollution from sediment, and monitoring all reveal the creation and critical stages of environmental operations in which micro/nanoplastics exist. This shows clear field maturity, as only the general subjects were investigated scientifically.

3.3.3 Co-authorship Analysis

Figure 8 depicts the extensive network of coauthorships among researchers focusing on micro/nanoplastic pollution. The coauthorship network of researchers working on micro/nanoplastic pollution has emerged as a dense network of numerous connections, most of which are associated with Chinese researchers. The figure integrates nodes that represent researchers and lines that represent coauthorship links; moreover, the size of each node reflects the number of publications. Chinese researchers have contributed to the generation of more large nodes because of their country's leading role and strategic prioritization of this scientific research. Such a network presents a few important issues characterizing collaborative scientific work. First, this figure reflects the strategic synthesis of particular knowledge from various fields, including chemistry, ecology, toxicology, and environmental engineering. This is a critical capacity particularly needed when addressing challenges related to plastic pollution. Second, such coauthorship suggests various ways of sharing large resources, such as specialized laboratories, funding, and meaningful data. Many Chinese research institutions sometimes act as powerhouses of these resources because they benefit from high investment in research infrastructure. Third, most collaborative studies indicate that coauthored papers receive more citations than single-authored papers do. As a massive source of research and publishing power worldwide, China strategically invests in becoming more visible as a key contributor to scientific advancements. Furthermore, collaboration plays a significant role in developing research capacity, as beginner researchers may benefit from such partnerships and promote their skills and careers. Fourth, the Chinese political and cultural context, which prioritizes performance indicators such as the number of publications, creates an environment and motivation to cooperate. For example, the Chinese government rewards publishing outputs if they are associated with large-scale projects. The result is a massive boost in productivity and research quality that accordingly has a greater influence on strengthening collaboration and usefulness.

Figure 8. Coauthorship network in micro/nanoplastic pollution research

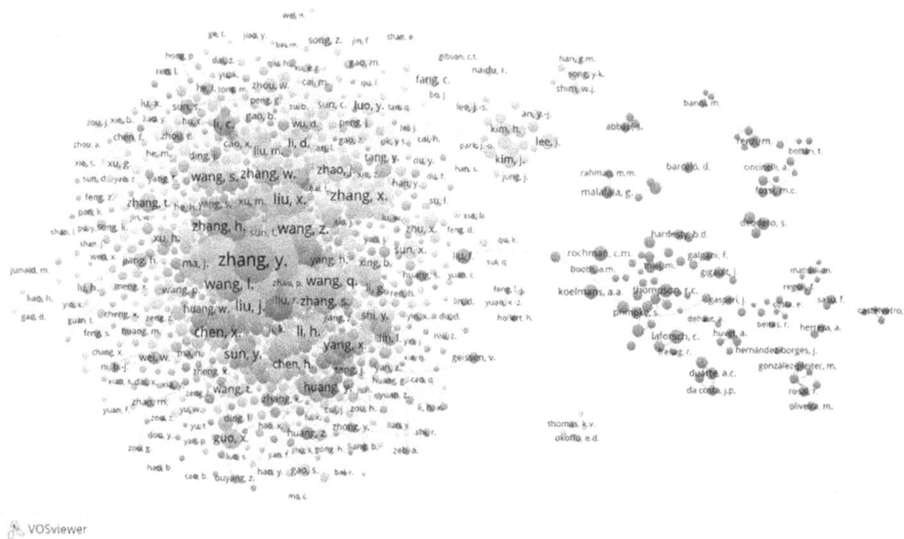

3.4 Citation Analysis

The landmark works provide comprehensive insight into the list of most cited works in micro/nanoplastic research, challenging the core literature basis for the comprehension of this contemporary environmental problem. The work by Eriksen et al., 2014 (Table 2), published in PLoS One, serves as a starting point for the papers, being the first to estimate the vastness of plastic pollution in the world's oceans, which marked a milestone for the development of future studies. Equally important is the work of Teuten et al., whose 2009 study elucidated the translocation and release of chemicals from plastics into the environment and wildlife. This work underlines that the damage caused by plastic pollution goes beyond physical accumulation and has a rather multifaceted impact. A 2013 study by Cole and colleagues in Environmental Science and Technology revealed microplastic ingestion by zooplankton, outlining the intricate pathways by which plastics infiltrate and interact with marine ecosystems. The 2017 work by Wright and Kelly investigated the connection between plastic pollution and human health in Environmental Science and Technology, highlighting the trend of developing research concerning the threats to humans posed by microplastic/nanoplastic exposure. The 2008 work by Moore in Environmental Research investigated the long-term threat of synthetic polymers to the marine environment, providing excellent guidelines about the steadfastness and ecological implications concerning the beaching of debris. The 2013 work by Lusher et al. explored the occurrence of microplastics in the fish gastrointestinal tract from the English Channel in the Marine Pollution Bulletin. It portrays the broad picture of such ingestion among marine life that has become capacious. The 2014 work by Van Cauwenberghe and Janssen investigated microplastics in bivalves consumed by humans since it also raised concerns about the potential threat to human life because of micro/nanoplastic contamination. Lu et al. described the accumulation and uptake of micropolystyrene particles in zebrafish by Environmental Science and Technology and demonstrated the toxic nature of micro/nanoparticles on organisms. The work in Marine Pollution by Gall and Thompson explores the effect of debris on marine life overall, which also integrates aspects of the plastic aspects of marine life and the ecosystem. In addition, Woodall et al. explored the deep sea as a major removal site for this debris, noting the need for continued micro/nanoplastic studies at this location. These prime works integrate the outlook for upcoming research studies aimed at understanding and mitigating the problems associated with the introduction of microplastic/nanoplastic pollution across the globe.

Table 2. Most cited works in micro/nanoplastic research

No.	Document Title	Author	Source	Year	Citation	Reference
1	Plastic pollution in the world's oceans: More than 5 trillion plastic pieces weighing over 250,000 tons afloat at sea	Eriksen et al.	PLos One, 9(12), p. e111913	2014	3,077	Eriksen et al. (2014)
2	Transport and release of chemicals from plastics to the environment and to wildlife	Teuten et al.	Philosophical Transactions of the Royal Society B: Biological Sciences, 364(1526), pp. 2027–2045	2009	2,022	Teuten et al. (2009)
3	Microplastic ingestion by zooplankton	Cole and colleagues	Environmental Science and Technology, 47(12), pp. 6646–6655	2013	1,852	Cole et al. (2013)
4	Plastic and human health: A micro issue?	Wright and Kelly	Environmental Science and Technology, 51(12), pp. 6634–6647	2017	1,698	Wright and Kelly (2017)
5	Synthetic polymers in the marine environment: A rapidly increasing, long-term threat	Charles James Moore	Environmental Research, 108(2), pp. 131–139	2008	1,446	Moore (2008)
6	Occurrence of microplastics in the gastrointestinal tract of pelagic and demersal fish from the English Channel	Lusher, McHugh & Thompson	Marine Pollution Bulletin, 67(1-2), pp. 94–99	2013	1,414	Lusher et al. (2013)
7	Microplastics in bivalves cultured for human consumption	Lisbeth van Cauwenberghe & Colin R. Janssen	Environmental Pollution, 193, pp. 65–70	2014	1,397	Van Cauwenberghe and Janssen (2014)
8	Uptake and accumulation of polystyrene microplastics in zebrafish (*Danio rerio*) and toxic effects in liver	Lu et al.	Environmental Science and Technology , 50(7), pp. 4054–4060	2016	1,392	Lu et al. (2016)
9	The impact of debris on marine life	S. C. Gall & R. C. Thompson	Marine Pollution Bulletin , 92(1-2), pp. 170–179	2015	1,374	Gall and Thompson (2015)

continued on following page

Table 2. Continued

No.	Document Title	Author	Source	Year	Citation	Reference
10	The deep sea is a major sink for microplastic debris	Woodall et al.	Royal Society Open Science, 1(4), p. 140317	2014	1,303	Woodall et al. (2014)

4. DISCUSSION

4.1 What Does Bibliometric Analysis Reveal?

Over the past 20 years, there has been a dramatic increase in the number of research publications on this topic, clearly indicating growing interest and concern among scholars. The first phase, from 2000–2008, had a relatively low number of publications, varying from 64–193 per year, which may support the assumption that scholars considered the subject of microplastic/nanoplastic pollution irrelevant. Nevertheless, the period from 2009–2014 experienced a gradual increase, from 193–298, as the potential negative effects of the problem on ecosystems and public health became evident. The most significant surge was observed in the last phase from 2015–2024, with a peak of 5,056 in 2023. This acceleration might be due to a rising level of awareness and a sense of urgency that scientists and policymakers exhibited. Such abrupt growth is suggestive of a profound understanding of the harms of micro/nanoplastics, fueled by growing environmental concerns and the pressing need for effective countermeasures.

The contributions of the global sources of micro/nanoplastic pollution reported in the literature and the identified distribution patterns are analyzed. In the scholarly domain, the contribution of published works that are harvested from the Scopus database clarifies and provides a global view. Literature primarily emanates from journal papers, including research articles, reviews, conference papers, and book chapters, which constitute the majority of the published literature. In terms of the country's publication capacity related to micro/nanoplastic pollution, the pollution arises from widespread areas across countries. This study provides insight into countries such as China, the United States of America, Germany, the United Kingdom, and other leading authors, highlighting the emerging problem of the recently escalated recognition of micro/nanoplastic pollution as a global problem. In particular, China produced the greatest number of documents, as it was competent in investing abundant resources to address micro/nanoplastic pollution. The United States of America also dominates the number of documents produced, amounting to the thousands that exhibit their responsibility toward the environmental and public health impacts of micro- and nanoplastics. Moreover, Germany and the United Kingdom, as well as other European countries, including India and South Korea, show superior contributions related to tolerant engagement with some of the research activities. The published work is also multidisciplinary, as it spans environmental science, chemistry, marine biology, and public health, which are important for identifying and finding solutions to micro/nanoplastic pollution that is naturally fragmented. The active research area covers many thematic areas, including the sources, ecological impact, and advanced mitigation of the types of micro/nanoplastics, implying an influential deep sea on the topic in the scholarly domain.

Moreover, the primary thematic areas and recent trends in research on micro/nanoplastic pollution reveal a diverse scope of topics that demonstrate the complexity of the problem. The broadest thematic areas are sources and distributions, i.e., how micro/nanoparticles arrive at various environmental matrices, such as oceans, rivers, and soils. Most importantly, the ecological impacts include the effects of micro/nanoplastics on marine and terrestrial organisms and ecosystems as a whole. Studies have explored how wildlife swallows microplastics, their toxicological effects, and the ability of micro/nanoparticles to take up other contaminants. A relatively new area that slightly lags behind other areas is mitigation in the form of various methods of creating sustainable materials and approaches to waste management and policy. The topic of human health risks following the study of possible methods of exposure, health phenomena, and human bioaccumulation is highly relevant. Finally, special attention is given to the development of analytical methods and material science that create opportunities for detecting, measuring, and characterizing micro/nanopollutants. Future trends include a shift toward interdisciplinary, comprehensive approaches to research, which include degradation, transportation, and intermolecular processes. Thus, the above-identified research areas constitute the current state of the thematic field and emerging trends within the field of microplastic/nanoplastic pollution.

In addition, interdisciplinary approaches ultimately enhance collaboration to advance the knowledge and understanding of micro/nanoplastic pollution by pooling a wide range of expertise across multiple fields, which is necessary to address this multifaceted problem fully. These include environmental scientists, chemists, marine biologists, toxicologists, public health specialists, engineers, and policymakers who collaboratively take a multidisciplinary approach to assess micro/nanoplastic distributions and their impacts on ecology and health. Therefore, this approach allows for the interconnection of the various dimensions of pollution from chemical analysis, effects on ecology, the technologies necessary for mitigation, and the strategies required by policy to formulate ways and enact laws that help reduce the load of plastics on the environment. For example, environmental scientists may focus on ecological impact, chemists may concentrate on particle testing and other contaminant interactions, engineers may focus on technology to enable removal, and policymakers may focus on strategies for reducing the plastic load of the environment. In addition, interdisciplinary methods offer a platform for methodological exchanges that can result in innovations that can enhance the detection, monitoring, and removal of micro/nanoplastics. This means that incorporating chemical technologies in biological impact assessments can improve the detection of plastic waste in diverse environments. Such research is scalable and influential and can easily attract the attention of wider stakeholders, such as NGOs and governments. Despite the challenge that micro/nanoplastics cannot be affordably monitored, research results can be easily translated into viable policies and strategies. Therefore, the shared efforts of interdisciplinary research expand the research horizons and offer a deep understanding and innovative solutions for micro/nanoplastic adverse effects.

4.2 Research Gaps and Future Directions

Scientists studying micro- and nanoplastic pollution have made major strides by thoroughly and methodically evaluating the current state of research in this field. However, there are still crucial areas that require further investigation to enhance our understanding of this issue. First and foremost, most studies on this topic have been performed over short periods, whereas no empirical longitudinal research has been conducted to measure the long-term consequences of micro/nanoplastics for the environment and human health. In this potential research area, it will be critical to predict chronic exposure and cumulative impacts on the quality of human life and ecosystems. Similarly, virtually no data on the degradative patterns of the phenomenon under various environmental conditions may provide a scientific understanding of the persistence and harmfulness of micro/nanoplastics. Additionally, little is known about the interactions of these phenomena with other forms of environmental pollution that could have synergistic effects. Finally, the effectiveness of existing purging and absorbing tools and techniques needs to be further examined, and new technologies must be developed. Overall, few cross-sector, full-cycle critical ideas are available, and these demand a systematic multidisciplinary response to the problem. For example, in the future, socioeconomic consequences for areas such as local communities and striving industries, such as fisheries or tourism, may be evaluated.

4.3 Limitations and Suggestions for Further Study

Although this work provides highly informative results regarding the bibliometric field of micro/nanoplastic pollution, it has several limitations. First, the use of only the Scopus database may imply certain contingency limitations, as all the information relevant to the topic would not be present in the source. For future research, it would be worthwhile to compare the findings from Scopus with those from other databases, such as Web of Science or PubMed. Additionally, since this search only covered publications through April 30, 2024, it is important to keep in mind that the most recent developments and emerging trends may not be fully captured in these results. Nevertheless, for further research, the time scale might be expanded to cover the newest publication to obtain the full spectrum of the research field. Bibliometric analysis provides insight only into the quantitative perspective of research; for example, it does not measure the deepest realm of topic exploration or the entity of discovery. Moreover, significance rather than measurement does not reflect the actual influence of a particular series of studies in academia.

Despite the insights gained on the topic, there are several limitations of the present work, creating opportunities for additional research. First, one database used for the literature search may not provide work with all the publications and works on the subject. Therefore, future work should incorporate more sources used for the analysis of micro/nanoplastic pollution. In addition, it would be beneficial to use different search strategies and conduct additional searches, as well as vary the keywords. Moreover, the author only used sources published within 20 years, which may mean that they were not able to trace emerging trends and new developments on the subject. Finally, qualitative methods such as content analysis and expert interviews may provide deeper insight into the research themes. In addition, there may be an opportunity to analyze different discourses and thematic disparities. Finally, it can be concluded that the exploration of interdisciplinary partnerships and communities of practice is critical to understanding the opportunities for such collaboration and defining ways of developing environmental impact.

5. CONCLUSION AND FUTURE SCOPE

The present bibliometric analysis provides a comprehensive overview of the research trends in micro/nanoplastic pollution. The defined publication patterns, thematic clusters, and interdisciplinary cooperation have increased interest in the scientific community and the global relevance of focusing on micro/nanoplastic pollution. The most significant finding is constant growth in the number of publications, with a special emphasis on the ocean and related research areas, including sources, distribution models, ecological consequences, and potential solutions. The existing analysis also details investigatory tools used by multidisciplinary research, with the focus extending to environmental science or chemistry and polymer materials or ecology. At the same time, it provides a comprehensive discussion of the current level of knowledge, and recommendations are provided for further research and policy interventions. The new trends that could be investigated in the future include the increased integration of qualitative research, the review of articles from more recent years, and the number of articles that are analyzed. The development of best practices for mitigating contaminants, such as biobased plastics or new management and recycling technologies, will improve their applicability. Earlier and more in-depth longitudinal studies are paramount since assessing the adverse impacts of micro/nanoplastic pollution on the ecosystem and, further, on humans is important. In addition, international cooperation and the development of exchange networks can be vital for applying interdisciplinary approaches in research and practice. Overall, all these directions will contribute to the knowledge of micro/nanoplastic pollution and to the possibility of saving the environment and humans for future generations.

REFERENCES

Cole, M., Lindeque, P., Fileman, E., Halsband, C., Goodhead, R., Moger, J., & Galloway, T. S. (2013). Microplastic ingestion by zooplankton. *Environmental Science & Technology*, 47(12), 6646–6655. DOI: 10.1021/es400663f

Eriksen, M., Lebreton, L. C. M., Carson, H. S., Thiel, M., Moore, C. J., Borerro, J. C., Galgani, F., Ryan, P. G., & Reisser, J. (2014). Plastic pollution in the world's oceans: More than 5 trillion plastic pieces weighing over 250,000 tons afloat at sea. *PLoS One*, 9(12), e111913. DOI: 10.1371/journal.pone.0111913

Fraissinet, S., De Benedetto, G. E., Malitesta, C., Holzinger, R., & Materić, D. (2024). Microplastics and nanoplastics size distribution in farmed mussel tissues. *Communications Earth & Environment*, 5(1), 128. DOI: 10.1038/s43247-024-01300-2

Gall, S. C., & Thompson, R. C. (2015). The impact of debris on marine life. *Marine Pollution Bulletin*, 92(1-2), 170–179. DOI: 10.1016/j.marpolbul.2014.12.041

Jiang, L., Ye, Y., Han, Y., Wang, Q., Lu, H., Li, J., Qian, W., Zeng, X., Zhang, Z., Zhao, Y., Shi, J., Luo, Y., Qiu, Y., Sun, J., Sheng, J., Huang, H., & Qian, P. (2024). Microplastics dampen the self-renewal of hematopoietic stem cells by disrupting the gut microbiota-hypoxanthine-Wnt axis. *Cell Discovery*, 10(1), 35. DOI: 10.1038/s41421-024-00665-0

Lu, Y., Zhang, Y., Deng, Y., Jiang, W., Zhao, Y., Geng, J., Ding, L., & Ren, H. (2016). Uptake and accumulation of polystyrene microplastics in zebrafish (*Danio rerio*) and toxic effects in liver. *Environmental Science & Technology*, 50(7), 4054–4060. DOI: 10.1021/acs.est.6b00183

Lusher, A. L., McHugh, M., & Thompson, R. C. (2013). Occurrence of microplastics in the gastrointestinal tract of pelagic and demersal fish from the English Channel. *Marine Pollution Bulletin*, 67(1-2), 94–99. DOI: 10.1016/j.marpolbul.2012.11.028

Moore, C. J. (2008). Synthetic polymers in the marine environment: A rapidly increasing, long-term threat. *Environmental Research*, 108(2), 131–139. DOI: 10.1016/j.envres.2008.07.025

Netthong, R., Khumsikiew, J., Donsamak, S., Navabhatra, A., Yingngam, K., & Yingngam, B. (2024). Bibliometric analysis of antibacterial drug resistance: An overview. In *Frontiers in Combating Antibacterial Resistance* (pp. 196–245). Current Perspectives and Future Horizons. DOI: 10.4018/979-8-3693-4139-1.ch009

Niu, J., Xu, D., Wu, W., & Gao, B. (2024). Tracing microplastic sources in urban water bodies combining their diversity, fragmentation and stability. *npj Clean Water,* 7(1), 37. DOI: 10.1038/s41545-024-00329-2

Sayed, A. E. D. H., Emeish, W. F. A., Bakry, K. A., Al-Amgad, Z., Lee, J. S., & Mansour, S. (2024). Polystyrene nanoplastic and engine oil synergistically intensify toxicity in *Nile tilapia, Oreochromis niloticus*: Polystyrene nanoplastic and engine oil toxicity in *Nile tilapia.BMC Veterinary Research*, 20(1), 143. DOI: 10.1186/s12917-024-03987-z

Teuten, E. L., Saquing, J. M., Knappe, D. R. U., Barlaz, M. A., Jonsson, S., Björn, A., Rowland, S. J., Thompson, R. C., Galloway, T. S., Yamashita, R., Ochi, D., Watanuki, Y., Moore, C., Viet, P. H., Tana, T. S., Prudente, M., Boonyatumanond, R., Zakaria, M. P., Akkhavong, K., & Takada, H. (2009). Transport and release of chemicals from plastics to the environment and to wildlife. *Philosophical Transactions of the Royal Society of London. Series B, Biological Sciences*, 364(1526), 2027–2045. DOI: 10.1098/rstb.2008.0284

Thammasanya, T., Patiam, S., Rodcharoen, E., & Chotikarn, P. (2024). A new approach to classifying polymer type of microplastics based on Faster-RCNN-FPN and spectroscopic imagery under ultraviolet light. *Scientific Reports*, 14(1), 3529. DOI: 10.1038/s41598-024-53251-5

Van Cauwenberghe, L., & Janssen, C. R. (2014). Microplastics in bivalves cultured for human consumption. *Environmental Pollution*, 193, 65–70. DOI: 10.1016/j.envpol.2014.06.010

Vidayanti, V., & Retnaningdyah, C. (2024). Microplastic pollution in the surface waters, sediments, and wild crabs of mangrove ecosystems of East Java, Indonesia. *Emerging Contaminants*, 10(4), 100343. DOI: 10.1016/j.emcon.2024.100343

Wan, S., Wang, X., Chen, W., Wang, M., Zhao, J., Xu, Z., Wang, R., Mi, C., Zheng, Z., & Zhang, H. (2024). Exposure to high dose of polystyrene nanoplastics causes trophoblast cell apoptosis and induces miscarriage. *Particle and Fibre Toxicology*, 21(1), 13. DOI: 10.1186/s12989-024-00574-w

Wang, K., Du, Y., Li, P., Guan, C., Zhou, M., Wu, L., Liu, Z., & Huang, Z. (2024). Nanoplastics causes heart aging/myocardial cell senescence through the Ca2+/mtDNA/cGAS-STING signaling cascade. *Journal of Nanobiotechnology*, 22(1), 96. DOI: 10.1186/s12951-024-02375-x

Wei, X. F., Yang, W., & Hedenqvist, M. S. (2024). Plastic pollution amplified by a warming climate. *Nature Communications*, 15(1), 2052. DOI: 10.1038/s41467-024-46127-9

Wieland, S., Ramsperger, A. F. R. M., Gross, W., Lehmann, M., Witzmann, T., Caspari, A., Obst, M., Gekle, S., Auernhammer, G. K., Fery, A., Laforsch, C., & Kress, H. (2024). Nominally identical microplastic models differ greatly in their particle-cell interactions. *Nature Communications*, 15(1), 922. DOI: 10.1038/s41467-024-45281-4

Woodall, L. C., Sanchez-Vidal, A., Canals, M., Paterson, G. L. J., Coppock, R., Sleight, V., Calafat, A., Rogers, A. D., Narayanaswamy, B. E., & Thompson, R. C. (2014). The deep sea is a major sink for microplastic debris. *Royal Society Open Science*, 1(4), 140317. DOI: 10.1098/rsos.140317

Wright, S., Cassee, F. R., Erdely, A., & Campen, M. J. (2024). Micro- and nanoplastics concepts for particle and fiber toxicologists. *Particle and Fibre Toxicology*, 21(1), 18. DOI: 10.1186/s12989-024-00581-x

Wright, S. L., & Kelly, F. J. (2017). Plastic and human health: A micro issue? *Environmental Science & Technology*, 51(12), 6634–6647. DOI: 10.1021/acs.est.7b00423

Yingngam, B. (2023a). Chemistry of Essential Oils. In *ACS Symposium Series* (Vol. 1433, pp. 189-223). Washington DC: USA, American Chemical Society. DOI: 10.1021/bk-2022-1433.ch003

Yingngam, B. (2023b). Modern solvent-free microwave extraction with essential oil optimization and structure-activity relationships. In *Studies in Natural Products Chemistry* (Vol. 77, pp. 365–420). Elsevier. DOI: 10.1016/B978-0-323-91294-5.00011-7

Yoon, J. H., Kim, B. H., & Kim, K. H. (2024). Distribution of microplastics in soil by types of land use in metropolitan area of Seoul. *Applied Biological Chemistry*, 67(1), 15. DOI: 10.1186/s13765-024-00869-8

Zhang, X., Wu, D., Jiang, X., Xu, J., & Liu, J. (2024). Source, environmental behavior and ecological impact of biodegradable microplastics in soil ecosystems: A review. *Reviews of Environmental Contamination and Toxicology*, 262(1), 6. DOI: 10.1007/s44169-023-00057-7

Zhao, K., Li, C., & Li, F. (2024). Research progress on the origin, fate, impacts and harm of microplastics and antibiotic resistance genes in wastewater treatment plants. *Scientific Reports*, 14(1), 9719. DOI: 10.1038/s41598-024-60458-z

Chapter 2
Micro/Nano Plastic Pollution Represents a Significant and Growing Threat to Human Populations Worldwide

Himadri Sekhar Das
https://orcid.org/0000-0002-3509-3388
Haldia Institute of Technology, India

Gourisankar Roymahapatra
https://orcid.org/0000-0002-8018-5206
Haldia Institute of Technology, India

Santanu Mishra
Haldia Institute of Technology, India

ABSTRACT

Micro/nano flexible contamination poses a pervasive warning to worldwide human public. With atoms measuring inferior 5mm, these contaminants permeate ecosystems, contaminating water beginnings, soil, and air. Their ever-present demeanor in food chains raises concerns about potential fitness hazards, containing endocrine turmoil, carcinogenicity, and generative issues. Moreover, their ability to adsorb and transport injurious projectiles for weaponry infuriates their impact. Mitigating this threat demands multidisciplinary works including scientific research, procedure invasions, and public knowledge campaigns. Addressing calculating/nano plastic contamination is principal to conserving human health and maintaining tangible

DOI: 10.4018/979-8-3693-3447-8.ch002

purity on a general scale.

1. INTRODUCTION

In the complicated curtain of environmental challenges facing benevolence, one of ultimate tricky and pervasive dangers arises in the form of data processing machine/nano plastic contamination. While charge cards have undoubtedly transformed differing facets of new growth, their ever-present presence has carelessly spun a web of material depravity, with widespread results for two together ecosystems and human strength. This essay delves into the blooming crisis of data processing machine/nano flexible pollution, untangling its versatile impacts and underscoring the urgent need for coordinated worldwide action (Anuli Dass et al, 2021). Microplastics, usually delimited as flexible particles tinier than 5 mm, and their even tinier counterparts, nano deferred payment arrangement, present a horrible challenge due to their minute content and extensive dispersion. These minuscule fragments spring from various sources, containing the rupture of larger flexible waste, microbeads in individual care products, and fibers scrap from artificial textiles (Landrigan PJ et al, 2023). Their petite ranges enable bureaucracy to permeate environments with distressing ease, filtering terrestrial, floating, and meteorological realms. In marine atmospheres, data processing machine/nano plastics pose a specifically grave danger, pervading sea environments from the ocean surface to the deepest abysses. They penetrate the feeding relationships among organisms, accumulating in the tissues of sea creatures and subsequently magnifying in aggregation as they escalate through trophic levels. Consequently, these particles have happened recorded in seafood expended by humans, lifting concerns about potential fitness implications. Furthermore, calculating/nano credit card serve as headings for poisonous contaminants, adsorbing chemicals to a degree determined organic contaminants (POPs) and weighty metals, that can leach into the atmosphere upon swallow, exacerbating environmental harm and human energy risks (Chen Q et al,2023). The ramifications of data processing machine/nano flexible contamination extend further floating ecosystems, filtering earthly environments and even harsh the air. Contaminated soils and sediments are a part of reservoirs for these particles, while climatic transport eases their global dispersal, climactic in their deposition in detached and primeval locales far removed from their point of inception. Consequently, no corner of the sphere remains whole apiece scourge of micro/nano flexible dirtiness, underscoring its status as a really worldwide phenomenon. Equally having to do with are the arising insights into the potential well-being impacts of data processing machine/nano flexible exposure on human communities. While research in this place nascent field is continuous, preliminary studies have involved micro-plastics in a spectrum of antagonistic strength consequences, ranging from respiring illnesses associated with breathing of winged particles to gastrointestinal disorders stopping from the swallow of adulterated food and water (Lu K et al, 2022). Furthermore, the bio-accumulative character of data processing machine/nano plastics raises the ghost of incessant uncovering to toxic projectiles for weaponry, accompanying implications for general energy and well-being. Addressing the menace of calculating/nano flexible dirtiness demands a multifaceted approach including lawmaking measures, technological changes, and public knowledge initiatives. Enhanced managing of flexible result and disposal, accompanying the phasing in another direction non-biodegradable deferred payment arrangement, shows a crucial step towards diminishing the increase of data processing machine/nano plastics (Koushik Ghosh et al, 2021). Additionally, expenses in waste administration infrastructure and the growth of environmental alternatives to common bank card are authoritative to curbing flexible dirtiness at its beginning. The aim and scope of this study is to assess the widespread impact of micro/nano plastic pollution on global human populations. It will explore the sources, distribution, and pathways of these pollutants, their potential health risks, and strategies for mitigation to protect public health and the environment.

1.1. Sources and Distribution of Micro/Nano Plastics

Micro-plastics and nano-plastics, delineated as flexible particles tinier than 5 mm and 100 nm individually, are extensive contaminants found in differing material compartments containing oceans, waterways, soil, and even the atmosphere (Yee MS et al, 2021).

Direct release of micro-plastics from flexible amount to a degree micro-bead in cosmetics, fibers drop from fabrics, and fragments from best flexible items (Anouk D'Hont et al, 2021). Production processes like scrape of flexible pellets (nurdles) all the while production and processing, and micro-plastic release from artificial fabrics like tires and lane paints. Improper transfer and littering of flexible waste into the atmosphere, chief to decomposition over time. Use of flexible matters in growing plants in liquid movements and fishing gear, that can humiliate into micro-plastics (Issac MN et al, 2021). Weathering: Degradation of best flexible debris through referring to practices or policies that do not negatively affect the environment determinants like UV fallout, machinelike abrasion, and organic endeavour. Figure 1. shows the Co-relation between Health and wellbeing, Oceans & Seas with Sustainable consumption and production.

Figure 1. Co-relation between health and wellbeing, oceans and seas with sustainable consumption and production

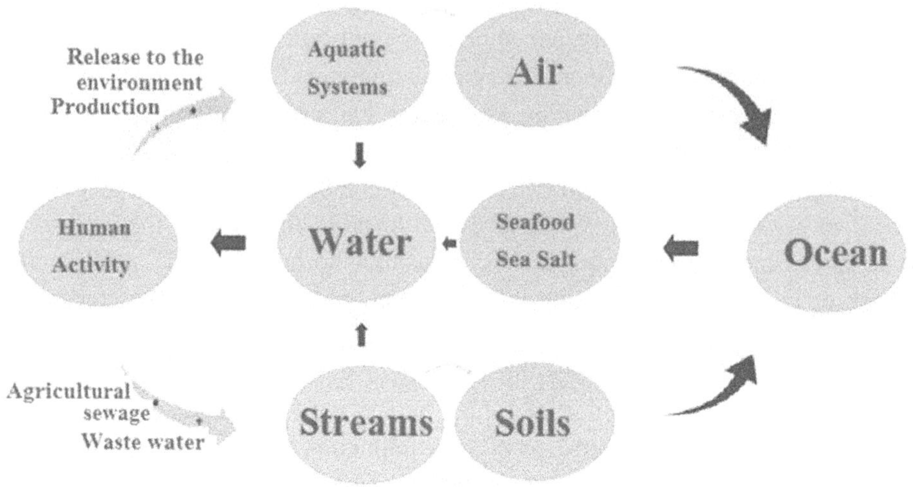

Transport of micro plastics from city and country areas by way of waterways, streams, and storm water drainage. Airborne micro plastics establishing onto land and water bulks through climatic transport (Heléne Österlund et al, 2023). Incomplete discharge of micro plastics all along wastewater treatment processes, chief to their discharge into taking waters. Oceans present image of important sinks for micro plastics, with concentrations variable by neighbourhood, affected by sea currents, proximity to beginnings, and depositional patterns. Freshwater Systems: Rivers, reservoirs, and streams accrue microplastics moved from upstream beginnings, accompanying bigger concentrations noticed near citified and automate regions (Wagner, M.et al,2014). Soil serves as a repository for microplastics, primarily from the use of waste mud, land mulches, and plastic-located fertilizers. Microplastics have happened discovered in the air, transported long distances by way of wind currents and climatic dethroning. Understanding the complex pathways and fate of microplastics in the surroundings is critical for cultivating direct mitigation planning to decrease their affect environments and human health. Ongoing research aims to raise listening methods, determine ecological risks, and label tenable resolutions for flexible pollution alleviation.

1.2. Primary Sources of Micro/Nano Plastics

Microplastics and nanoplastics introduce from miscellaneous primary beginnings, containing: The vulgar disposal and administration of flexible waste are important contributors to microplastic dirtiness. This contains articles like plastic containers, bags, wrap fabrics, and other distinct-use credit card that wind up in landfills, water bodies, and the atmosphere. Over occasion, these credit card break down into tinier atoms on account of enduring and degradation processes, eventually making microplastics (Yee MS et al, 2021).

Larger flexible items, in the way that containers, cans, fishing nets, and flexible waste, can decay into smaller pieces through machinelike processes like scrape, wave operation, and exposure to light part of every 24 hours (photodegradation). These disintegrated pieces can further shame into microplastics under environmental environments. For example, flexible waste in oceans can be shabby into microplastics by wave operation and UV fallout. Industrial activities are another important beginning of microplastics. These contain processes like plastic production, treat, and reusing (Lapyote Prasittisopin et al,2023). During these processes, microplastics can be freed straightforwardly into the atmosphere through effluents, air emissions, and inadvertent spills. For instance, microplastic atoms maybe generated all along the result and handle of plastic fabrics, in addition to from the depreciation of machinery and supplies used in these labours (Ghosh, Set al,2023). Synthetic fabrics, such as polyester, nylon, and tinted covering, jettison microfibers all along washing, wearing, and disposition. These microfibers are a important beginning of microplastic pollution in water physique, specifically in wastewater effluents. They can record aquatic environments and increase in sediments, pretending risks to marine history. Some private care and beautifying products hold microplastic factors like microbeads that are added for molting or quality (Habib RZ et al, 2022).

1.3. Various Pathways Through Which Micro/Nano Plastics

Pathways by which calculating and nano plastics find their habit into the surroundings: One of the basic pathways for microplastics to come the environment is through drainage from miscellaneous beginnings in the way that urban regions, land fields, and technical sites (Dube E et al, 2023). Rainwater washes continuously plastics from streets, sidewalks, and different surfaces, accomplishing bureaucracy into waterways, lakes, and oceans. Additionally, flexible pieces from wastewater situation plants can also finish up in surface waters. Microplastics are frequently present in wastewater discharge on account of the incomplete filtration of flexible pieces from household and modern sources. Effluents from waste situation plants can hold microplastics that are then dismissed into water materials. Microplastics can again came the environment through climatic dethroning. Studies have proved that microplastic atoms can be moved through the air and located upon land and water surfaces (Wright SL et al, 2019). These particles can create from miscellaneous beginnings to a degree road dust, modern issuances, and even flexible waste that have been shabby into tinier atoms by enduring processes. Direct release of microplastics into the environment can happen through miscellaneous exercises in the way that littering, immoral disposition of flexible waste, and accidental spillage. For example, flexible pellets used as natural resources in the production of plastic commodity maybe unintentionally freed into the environment all the while conveyance or management (Mihai, F.-C et al,2022). Landfills and Waste Sites: Improper transfer of plastic waste in landfills and waste sites can again influence the release of microplastics into the surroundings. Over occasion, plastics shame into tinier atoms due to uncovering to light part of every 24 hours, liquid, and microbial operation, eventually emptying into soil and water. Microplastics can come the atmosphere through growing plants in liquid activities and fisheries. For example, flexible-located angling gear in the way that nets and lines can degrade over opportunity, dropping microplastic atoms into the water. Additionally, the use of flexible-based matters in growing plants in liquid movements can further lead to the release of microplastics into amphibious environments. These pathways climax the complex and multifaceted character of microplastic contamination, accompanying inputs from differing sources donating to its extensive classification in the environment (Lwanga, E.H et al,2022). Addressing this issue demands coordinated works to decrease plastic use, better waste administration practices, and diminish the release of microplastics from various beginnings.

1.4. Distribution of Micro/Nano Plastics in Different Environmental Compartments

Microplastics and nanoplastics, minuscule flexible particles weighing inferior 5 mm in height, have become an ever-present hazardous waste. They create from various beginnings, containing the disruption of larger flexible waste, microbeads in individual care products, artificial fibers from fabrics, and mechanical processes. These atoms pose significant dangers to environments and human well-being due to their steadfastness, potential to consume poisonous chemicals, and talent to introduce the feeding relationships among organisms. Here's an overview of their dispersion indifferent incidental compartments: Rivers, lakes, and streams are big beginnings of freshwater microplastic contamination. Runoff from urban districts, industrialized sites, and land fields introduces microplastics into freshwater structures. Studies have raise microplastics in extreme concentrations in freshwater sediments, surface waters, and even in aquatic creatures. Sedimentation and dethroning processes can bring about the accumulation of microplastics in sure extents, moving benthic ecosystems. Oceans symbolize a fall for microplastics, accompanying big accumulations found in sea gyres, in the way that the Great Pacific Garbage Patch (Dongliang Yuan et al, (2023). Coastal districts are particularly unsafe to microplastic dirtiness on account of the proximity of city centres and technical ventures. Marine wildlife, containing cast, seabirds, and sea mammals, swallow microplastics, chief to disadvantageous effects on their fitness and conceivably introducing the human food chain. Plastic waste can more involve marine structures, leading to harm or death. Soil adulteration by microplastics is an increasing concern, generally due to the request of waste mud, land plastic mulches, and the disruption of flexible litter. Microplastics can grow in soil through various pathways, containing meteorological dethroning, irrigation accompanying adulterated water, and direct request of plastic-located output (Long B et al,2023). Soil-home organisms, in the way that earthworms and microorganisms, can swallow microplastics, conceivably impacting soil well-being and environment functioning. Agricultural practices and land use changes can influence the allocation and transport of microplastics in soil environments. Microplastics have happened discovered in the air, transported long distances by wind currents. Sources of in the air microplastics involve vehicle weary wear, artery dust, industrialized emissions, and the shame of flexible litter (Kole PJ et al,2017). Atmospheric dethroning is a meaningful pathway for microplastics to introduce earthly and amphibious ecosystems. Microplastics at hand can choose land and water surfaces, providing to environmental adulteration.

2. HUMAN EXPOSURE PATHWAYS

Micro and nano flexible dirtiness poses a multifaceted warning to human public everywhere, permeating differing uncovering pathways. Firstly, through the feeding relationships among organisms, microplastics ingested by sea history expand in seafood consumed by persons, through recording our bodies straightforwardly. These small flexible particles have existed discovered in usually consumed parts like extricate, invertebrate, and even table salt, lifting concerns about their potential strength impacts. Moreover, winged microplastics can be inhaled, pervading respiring orders and potentially beginning respiring issues. As flexible waste degrades, it releases microscopic pieces into the air that maybe carried over long distances by wind before adjustment earthly and maritime environments. Additionally, microplastics have existed about slurping water sources, pretending a ethical or easiest course of uncovering to humans. Further confusing matters, the appearance of microplastics in soil raises concerns about land contamination. These pieces can upset soil ecosystems and find their habit into crops, conceivably entering the human feeding relationships among organisms obliquely. Consequently, the predominance of microplastics across diverse atmospheres emphasizes the imperative need for comprehensive research into their fitness impacts and alleviation plans. In essence, the pervasive character of calculating and nano flexible pollution presents a important and increasing danger to human health, moving differing uncovering pathways ranging from swallow through meal and water to breathing and agricultural adulteration. Addressing this issue demands coordinated efforts at local, domestic, and worldwide levels to lighten plastic dirtiness and safeguard community health.

2.1. Pathways Through Which Humans Are Exposed to Micro/Nano Plastics

Humans are unprotected to data processing machine and nano plastics through differing pathways, containing: Ingestion: Microplastics can introduce the human crowd through consumption of adulterated cooking and liquor. Seafood, exceptionally shellfish and angle, can hold microplastics on account of adulteration of marine surroundings (Yee MS et al, 2021). Drinking water beginnings can again hold microplastics, originating from differing beginnings to degree wastewater effluents, drainage from plastic waste, and depravity of best flexible items.

Microplastics can leave the ground through processes like enduring, break of best plastic parts, and release from artificial fabrics and added consumer merchandise. Inhalation of in the air microplastics can happen indoors and nature, specifically in city extents with extreme levels of flexible dirtiness. Microplastics are usually used in individual care commodity in the way that exfoliating scrubs, toothpaste, and cosmetic. Dermal exposure to microplastics can happen all along the use and use of these brand (Emenike EC et al,2023). Synthetic textiles like polyester, nylon, and tinted covering scrap microfibers all the while washing and tiring. These microfibers can come into trade the skin, conceivably superior to dermal uncovering. Understanding these pathways of exposure is important for evaluating the potential energy risks guide micro and nano bank card. While swallow is frequently deliberate the primary route of uncovering, breathing and dermal contact are more meaningful and warrant further research and attention.

2.2. Potential Sources of Human Exposure

Food borne sicknesses can become functional consuming adulterated meal, frequently on account of bacteria like Salmonella, E. coli, or viruses like norovirus (Todd ECD et al,2014). Water adulteration can happen through miscellaneous resources such as modern discharge, land drainage, or aging foundation chief to the emptying of contaminators like heavy metals or pesticides. Air Pollution: Inhalation of contaminants in the way that coarse matter (PM), nitrogen oxides (NOx), sulfur dioxide (SO_2), toxic gas (CO), volatile natural compounds (VOCs), and upper layer of atmosphere (O_3) can bring about respiratory issues, cardiovascular questions, and additional well-being concerns. Sources of air contamination include taxi issuances, modern ventures, construction, and blazing of hydrocarbon deposits. Consumer Products: Exposure to hurtful chemicals can happen through services device like private care items, cleansing powers, pesticides, and household merchandise. Chemicals of concern can include phthalates, bisphenol A (BPA), parabens, flame retardants, and weighty metals like lead and major planet.

Workers can be exposed to miscellaneous hazards contingent upon their control, in the way that chemicals, organic powers, dissemination, blast, and ergonomic stressors. Industries like production, farming, building, healthcare, and mining pose distinguishing risks to employees (AlDhaen E et al, 2022). Figure 2. shows the Pathways of human exposure to microplastics.

Figure 2. Pathways of human exposure to micro-plastics

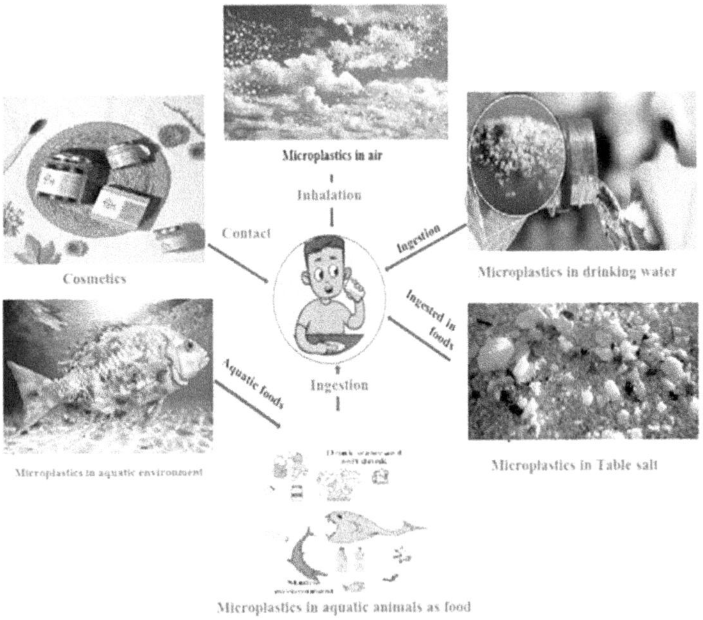

Exposure to ionizing dissemination from beginnings like medical image processes (X-indications, CT scans), nuclear energy plants, and radioactive matters can pose well-being risks, containing cancer and ancestral mutations. Non-ionizing fallout from beginnings like movable phones, Wi-Fi routers, and power lines further raises concerns, even though evidence of harm is less clear. Soil Contamination: Soil can deteriorate by modern activities, wrong appliance for grinding garbage, land runoff, or excavating projects, chief to the build-up of pollutants to a degree difficult metals, pesticides, and basic chemical compound.

Exposure can occur through direct trade adulterated soil, swallow of home grown produce, or breathing of dust atoms. Prolonged uncovering to extreme levels of noise, to a degree from traffic, mechanical engine, or offensive music, can bring about trial deficit, stress, sleep disturbances, and other energy issues. Identifying and lightening these beginnings of uncovering is crucial for preserving community health and the surroundings. Regulatory measures, public knowledge campaigns, and technological progresses play key parts in sending these challenges.

2.3. Mechanisms by Which Micro/Nano Plastics

Microplastics (MPs) and nanoplastics (NPs) are a increasing concern on account of their extensive presence in the atmosphere and their potential to list the human crowd through differing pathways such as swallow, breathing, and dermal uncovering. Once inside the material, these tiny flexible pieces concede possibility translocations through various devices and pose potential health risks (Sangkham S et al, 2022).

Ingestion: One of the basic routes of effort for MPs and NPs is through swallow, generally via adulterated feed and water. Once swallowed, these pieces can translocate across the gastrointestinal lot (GIT) through several machines: Passive spread: Small MPs and NPs can indifferently diffuse across the stomach epithelium on account of their littleness. Endocytosis: Larger pieces may be overwhelmed by stomach containers through endocytosis systems, admitting them to cross the stomach obstacle. Mucus infiltration: MPs and NPs grant permission also guide along route, often over water through the gelled waste tier top the intestinal epithelium and pierce into the fundamental tissues (Cornick S et al,2015). Inhalation: Inhalation of in the air MPs and NPs is another route of uncovering. Once inhaled, these pieces can translocate across the respiratory epithelium through methods in the way that: Mucociliary approval: Particles can be trapped in the gelled waste interlining the respiring lot and afterward transported upwards by ciliary operation, conceivably arriving the neck and being swallowed. Transcytosis: Particles concede possibility too stop living up by respiring epithelial cells and moved across the epithelium through transcytosis systems. Dermal uncovering: Although less intentional distinguished to ingestion and breathing, dermal uncovering to MPs and NPs is likewise a potential route of entrance. These particles can pierce the skin hurdle through: Hair follicles: MPs and NPs concede possibility enter the crowd through hairstyle follicles, exceptionally if they are limited enough to seep the follicle hole. Intercellular pathways: Some atoms can pierce the skin barrier through intercellular pathways betwixt keratinocytes. Once inside the crowd, MPs and NPs can deliver to miscellaneous organs and tissues through distribution, conceivably producing unfavorable effects. The strength suggestions of microplastic uncovering are still being intentional, but they may involve redness, oxidative stress, genotoxicity, and division of basic functions (Cornick S et al, 2015).

3. HEALTH IMPACTS OF MICRO/NANO PLASTIC POLLUTION

Micro and nano flexible pollution poses meaningful energy risks to humans and environments alike. When swallowed, these tiny flexible atoms can accumulate in tissues, conceivably precipitating inflammation, oxidative stress, and upsetting hormonal balance (Niccolai E et al, 2023). Moreover, they can symbolize carriers for hurtful chemical compound, including determined organic contaminants that may enhance their poisonous effects. In maritime surroundings, microplastics can enter the feeding relationships among organisms, eventually reaching persons through seafood devouring. Additionally, airborne microplastics concede possibility be inhaled, posing respiring fitness hazards. The long-term results of uncovering to micro/nano flexible contamination warrant urgent research and alleviation works to safeguard public health and incidental honour. Figure 3. shows the possible human exposure level to plastics originating from different sources.

Figure 3. Possible human exposure level to plastics originating from different sources

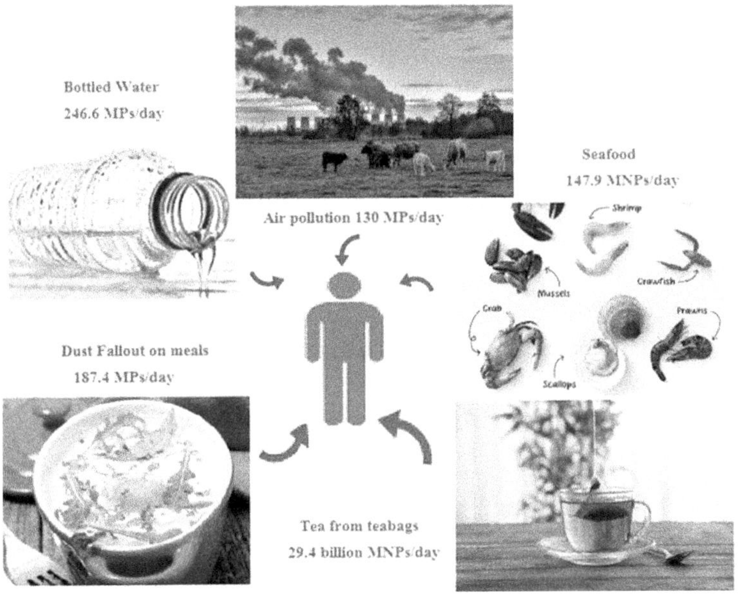

3.1. Current Scientific Evidence Regarding the Health Impacts of Micro/Nano Plastic

Research on the well-being impacts of microplastic and nanoplastic uncovering on human peoples is an progressing field (Campanale C et al,2020), but skilled are various key verdicts and districts of concern highlighted in current experimental article: Ingestion and Absorption: Studies have proved that persons can swallow microplastics through miscellaneous beginnings, containing cooking, water, and air. These pieces can more be involved through the skin. Microplastics have existed in the direction of a variety of human tissues, containing the gastrointestinal lot, liver, and even covering layer. Potential Health Effects: While the exact well-being belongings of microplastic uncovering on persons are still being intentional, skilled is increasing evidence suggesting potential risks (Emenike EC et al,2023). These risks contain swelling, oxidative stress, genotoxicity, and the division of endocrine and invulnerable function. Additionally, there are concerns about the potential for microplastics to be a part of headings for different injurious projectiles for weaponry, in the way that continuous natural contaminants, by spellbinding and moving ruling class into the carcass. Cardiovascular and Respiratory Effects: Some studies have submitted a possible link middle from two points uncovering to winged micro plastics and unfavourable cardiovascular and respiring belongings, in the way that pleura redness and deteriorated pleura function. However, more research is wanted to sufficiently comprehend these associations. Gastrointestinal Effects: Given that a meaningful portion of swallowed microplastics grant permission increase in the gastrointestinal lot, skilled are concerns about potential gastrointestinal belongings, in the way that swelling, changed gut microbiota arrangement and the turmoil of stomach hurdle function (Zhang Y et al, 2021). Reproductive and Developmental Effects: Animal studies have proved that exposure to microplastics can have antagonistic belongings on generative and enlightening well-being, containing decreased productivity, changed birth control method levels, and enlightening deformities in child. While the pertinence of these findings to human strength is not still completely implicit, they raise main concerns that warrant further case. Risk Assessment and Regulation: Currently, skilled is restricted unity on in what way or manner to determine the risks guide microplastic uncovering to human energy. Furthermore, regulatory foundations for talking these risks are still in the inception of happening.

3.2. Potential Adverse Effects on Human Health

Each of these potential adverse effects: Inflammation: Inflammation is the corpse's unaffected response to hurtful provocation, in the way that pathogens, damaged containers, or irritants. However, never-ending inflammation can bring about miscellaneous well-being problems, containing cardiovascular ailments, diabetes, and autoimmune disorders. Figure 4 shows the Inflammation and oxidative stress (OS) as core mechanisms linking common causative factors with male infertility.

Figure 4. Inflammation and oxidative stress (OS) as core mechanisms linking common causative factors with male infertility

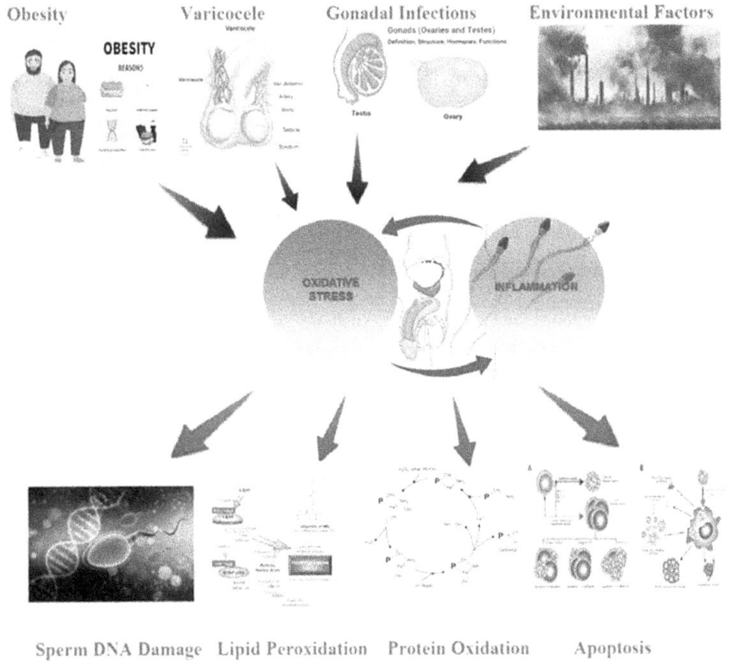

Certain tangible factors, containing contaminants and poisons, can trigger never-ending redness, conceivably exacerbating existent fitness conditions or growing the risk of evolving ruling class (Chen L et al,2017). Oxidative Stress: Oxidative stress occurs when skilled is an inequality betwixt free radicals and antioxidants in the body. Free radicals are well sensitive fragments that can damage cells, proteins, and DNA. Antioxidants help counteract these free radicals and cover the body from their injurious belongings. However, uncovering to environmental contaminants, in the way that air contamination or certain chemical compound, can increase the result of free radicals, chief to oxidative stress. Chronic oxidative stress has been connected to miscellaneous well-being problems, containing malignancy, cardiovascular diseases, and neurodegenerative disorders. Genotoxicity: Genotoxicity refers to the strength of sure essences to damage the genetic material (DNA) inside containers, conceivably leading to mutations or different injurious belongings (E Niedzielska et al, (2015). Exposure to genotoxic agents, in the way that sure chemicals, dissemination, or contaminants, can increase the risk of cultivating cancer or different hereditary disorders. Genotoxicity can manifest in differing ways, containing DNA rope breaks, chromosomal aberrations, or mutations in particular genes. Therefore, it is essential to evaluate the genotoxic potential of entities to cover human health and counter antagonistic effects. Endocrine Disruption: Endocrine disruptors are chemical compound that obstruct the bulk's endocrine system, that organizes birth control method production and function. These elements can mimic, block, or obstruct the party's natural hormones, chief to differing health questions, containing generative disorders, developmental anomalies, and metabolic dysfunctions. Endocrine disruptors maybe about many everyday crop, containing credit card, pesticides, and certain pharmaceuticals. Exposure to these chemical compounds, particularly all along critical periods of happening, in the way that fatal growth or adolescence, can have enduring effects on human strength and welfare. Understanding and lightening the potential adverse belongings of redness, oxidative stress, genotoxicity, and endocrine division are crucial for safeguarding human energy and advancing overall well-being. This demands strong scientific research, active supervisory measures, and public knowledge efforts to underrate uncovering to injurious substances and lighten their affect human fitness.

3.3. Emerging Research Linking Micro/Nano Plastic

Research into the potential energy impacts of calculating and nano credit card is still in its inception, but there are arising verdicts that imply a likely link middle from two points uncovering to these atoms and miscellaneous chronic afflictions (Schröder B et al,2024). Here's a summary of few indispensable contents from current research: Cancer: Some studies have determined that microplastics concede possibility have malignant features. For example, sure chemicals used in the result of credit card, in the way that phthalates and bisphenol A (BPA), have happened top-secret as potential carcinogens. Additionally, microplastics can adsorb added hurtful chemicals from the atmosphere, and these projectiles for weaponry manage help malignancy growth. Cardiovascular Diseases: Research plans that uncovering to microplastics may cause cardiovascular afflictions by encouraging redness and oxidative stress in the bulk (Campanale C et al,2020). These atoms can list the bloodstream and increase in tissues, potentially generating damage to ancestry ships and impairing cardiovascular function. Neurological Disorders: There is increasing concern about the potential neurotoxic belongings of microplastics. Studies have proved that sure chemical compound found in charge card, to a degree phthalates and bisphenol A, can obstruct affecting animate nerve organs incident and function. Additionally, microplastics can record the intellect through the bloodstream or the having fragrance system, place they can cause redness and damage to intelligence containers. However, it's main to note that while these judgments have to do with, more research is needed to adequately believe the energy impacts of calculating and nano charge card. Many of the studies attended up until now have happened acted in laboratory backgrounds or animal models, and their pertinence to human energy is not still adequately assumed. Additionally, the levels of microplastic uncovering that are guide adverse energy belongings are still unsure, and more studies are wanted to determine the potential risks to human energy. In conclusion, while skilled is increasing evidence to suggest that uncovering to data processing machine and nano credit card can be affiliated with never-ending afflictions to a degree cancer, cardiovascular afflictions, and affecting animate nerve organs disorders, more research is wanted to ratify these judgments and learn the systems fundamental these potential health impacts. In the meantime, it's main in the second-place exertions to lower flexible contamination and underrate uncovering to microplastics in the environment.

4. SOCIETAL AND POLICY IMPLICATIONS

The pertaining to society and procedure implications of calculating/nano flexible contamination is deep and multifaceted. Here's a disintegration of few indispensable contents: Environmental Impact: Microplastics (particles inferior 5mm) and nanoplastics (atoms inferior 0.1 micrometer) pose significant warnings to environments. They can increase in soil, water crowd, and the atmosphere, precipitating harm to sea growth, terrestrial mammals, and even persons through the feeding relationships among organisms. The enduring consequences of flexible contamination on biodiversity and environment health are having to do with (Boyle, K et al,2020). Human Health Concerns: Micro/nano charge card can come the human party through ingestion, breathing, and dermal uncovering. Research desires that these particles concede possibility have antagonistic belongings on human health, conceivably provoking swelling, oxidative stress, and natural damage. There are also concerns about the transfer of poisonous projectiles for weaponry and pathogens that can obey microplastics. Economic Costs: Plastic pollution imposes important business-related costs on association, containing expenses had connection with cleansing up adulterated environments, healthcare costs guide human well-being impacts, and misfortunes to labors such as fisheries and travel on account of disgraced ecosystems (Kumar, R et al,2021). Regulatory Challenges: Addressing calculating/ nano flexible contamination requires inclusive supervisory foundations at civil and international levels. However, organizing aforementioned narrow particles presents challenges in conditions of discovery, listening, and imposition. There's a need for standardized plans for weighing and evaluating microplastic pollution across various surroundings. Waste Management and Recycling: Improving waste administration practices and promoting reusing are important for lightening flexible pollution. Policies that encourage the decline, talk over again, and recycling of credit card can help forbid ruling class from recording the environment initially. Additionally, changes in flexible alternatives and referring to practices or policies that do not negatively affect the environment fabrics can lower the confidence on traditional assets. Public Awareness and Education: Raising knowledge about the issue of calculating/nano plastic dirtiness is essential for promoting attitude change among things, trades, and policymakers. Education campaigns can educate all about the sources and impacts of flexible contamination and reassure sustainable use practices (Dalu, M.T.B et al,2020).Research and Innovation: Continued research is wanted to better comprehend the sources, consequence, and belongings of calculating/nano plastics in the atmosphere. This contains expanding new electronics for detecting and monitoring microplastic contamination, in addition to judgment innovative resolutions for diminishing its impact.

4.1. Societal Implications of Micro/Nano Plastic Pollution

Micro and nano flexible dirtiness poses a range of societal suggestions, connecting financial, incidental justice, and community health concerns.

The elimination of calculating and nano plastics from water physique and environments is harmful and frequently inefficient on account of their littleness and extensive distribution. Microplastics can adulterate sea surroundings, affecting extricate public. This not only impacts the livelihoods of fishermen but likewise disrupts the seafood manufacturing and related trades (Ghosh, S et al, 2023). Coastal regions densely impacted by flexible dirtiness concede possibility happening a decline in tourism, moving local savings dependent on tourism profit. Environmental Justice Concerns: Often, societies of color and reduced-proceeds communities endure loud noises, containing microplastic contamination. These societies grant permission lack money to lighten the impacts or access to clean water beginnings. Indigenous Communities: Indigenous societies, specifically those reliant on established angling and livelihood practices, may face meaningful challenges on account of microplastic contamination in water crowd. Impacts on Vulnerable Populations: Microplastics can enter the feeding relationships among organisms, conceivably revealing humans to injurious chemical compound present in deferred payment arrangement. Vulnerable states such as significant wives and teenagers may face profound strength risks on account of uncovering to microplastics (Campanale C et al,2020). Microplastics can accumulate in the frames of sea mammals, leading to material harm, generative issues, and even extinction. This can upset ecosystems and endanger the livelihoods of societies dependent on these ecosystems for foodstuff and possessions. Addressing these pertaining to society implications demands inclusive policies that circumscribe pollution stop, explanation exertions, policy attacks, and society date. Initiatives proposed at reducing flexible devouring, advancing recycling and waste administration, and supporting incidental management can help mitigate the impacts of data processing machine and nano flexible dirtiness on society.

4.2. Existing Policies and Regulations Aimed at Addressing Plastic

Policies and managing focusing on flexible dirtiness have gained meaningful friction everywhere on account of the disadvantageous impact of credit card on the atmosphere, wildlife, and human energy. Here's a survey of few key tactics and organizing at two together social and worldwide levels: International Level: Basel Convention: The Basel Convention on the Control of Transboundary Movements of Hazardous Wastes and Their Disposal regulates the worldwide shift of toxic waste, containing flexible waste. In 2019, an improvement was selected to control the transboundary movement of flexible waste in a more excellent manner. Stockholm Convention: The Stockholm Convention on Persistent Organic Pollutants (POPs) aims to remove or confine the result and use of determined basic pollutants, containing sure flexible supplements like brominates flame retardants. United Nations Environment Assembly (UNEA): UNEA has happened forwarding flexible contamination through resolutions and actions, to a degree the Clean Seas Campaign that helps governments to bet money or something else in a gamble against sea flexible pollution. Plastic Waste Partnership: This participation, begun by UNEP and the World Bank, aims to further the exercise of the Basel Convention's flexible waste corrections and boost the management of flexible waste. Ocean Cleanup Initiatives: Various worldwide institutions and pushes devote effort to something cleansing up flexible contamination in oceans, such as The Ocean Cleanup project and the Global Ghost Gear Initiative. National Level: Plastic Bans and Restrictions: Many nations have executed bans or limits on sole-use deferred payment arrangement, microplastics, and flexible bags. For example, various countries have outlawed sole-use flexible bags, straws, and material used to fill space boxes. Extended Producer Responsibility (EPR) Laws: EPR regulations demand producers to take trustworthiness for the whole lifecycle of their device, containing correct conclusion and reusing. Several countries have executed EPR standards for flexible crop (Rufino Júnior et al, 2024). Deposit Return Schemes (DRS): DRS incentivizes buyers to return flexible containers and bottles for recycling by contribution refunds or inducements. Some nations have achieved DRS for miscellaneous types of drink bowls. Plastic Recycling Policies: Governments are implementing procedures to advance flexible reusing, in the way that background reusing targets, supplying in reusing foundation, and generating lures for trades to use reused matters in their products. Research and Innovation Funding: Many nations are devoting in research and novelty to expand alternative fabrics to flexible, develop recycling sciences, and find answers to flexible dirtiness. Education and Awareness Campaigns: Governments are initiating public knowledge campaigns to experience citizens about the material impacts of flexible dirtiness and advance tenable devouring dresses. These procedures and regulations show an increasing all-encompassing exertion to address flexible dirtiness completely, but there's still much work expected finished to solve significant reductions in flexible waste and its befriended environmental impacts.

4.3. Potential Strategies for Mitigating Micro/Nano Plastic Pollution

Addressing calculating/nano flexible dirtiness requires a versatile approach that includes differing partners, from individuals to businesses and governments. Here are few potential designs:

Improved Waste Management: Implementing more adept waste accumulation and recycling programs for fear that flexible from introducing waterways. Investing in leading waste treatment sciences that can efficiently capture microplastics before they reach the surroundings. Encouraging correct disposal practices through inducements and punishments for disobedience. Product Design Changes: Promoting the use of referring to practices or policies that do not negatively affect the environment and compostable matters as alternatives to usual charge card. Encouraging manufacturers to design crop accompanying minimal bundle or utilizing bundle fabrics that are recyclable or surely biodegradable. Developing creative matters that are less inclined humiliate into microplastics or designing fruit accompanying included methods to capture released microplastics. Regulatory Measures: Implementing more authoritarian rules on the result, use, and transfer of deferred payment arrangement, including bans on sole-use deferred payment arrangement or microbeads in private care produce (Weis, J.S et al,2022). Enforcing extended builder accountability (EPR) blueprints, place manufacturers are grasped accountable for completely-of-homemaking practice and theory of their amount. Introducing marking requirements to educate customers about the referring to practices or policies that do not negatively affect the environment impact of brand and packaging. Public Awareness Campaigns: Educating all about the beginnings and impacts of calculating/nano assets through combined use of several media campaigns, school programs, and community occurrences. Encouraging action change, in the way that lowering plastic use, correctly disposing of waste, and performing in explanation works. Engaging stakeholders through national skill drives to monitor and path microplastic pollution in differing surroundings. Research and Innovation: Investing in research to better accept the beginnings, pathways, and effects of calculating/nano credit card on human well-being and the surroundings (Mihai et al,2022). Supporting the happening of new technologies for detecting, listening, and killing microplastics from water crowd. Encouraging cooperation between scholarly world, manufacturing, and management to expand tenable alternatives to credit card and creative resolutions for diminishing pollution. International Cooperation: Collaborating accompanying different nations and worldwide organizations to expand matched actions and principles for discussing microplastic pollution on an all-encompassing scale. Sharing best practices, information, and electronics to expedite cross-border efforts to humble flexible waste and insulate environments. By joining these strategies and charming diversified shareholders, we can work towards diminishing micro/nano flexible contamination and preserving human energy and the environment for future era.

5. CONCLUSION

Micro/nano flexible dirtiness shows a significant and increasing warning to human societies general. As these tiny flexible pieces stretch to penetrate the environment, their potential well-being impacts raise indispensable concerns. Addressing this complex issue demands cooperative efforts from policymakers, energies, scientists, and all to implement productive mitigation methods and safeguard two together human energy and the surroundings for future generations. Micro and nano flexible contamination shows a extensive and increasing threat to human states general. These minute flexible pieces, measuring inferior 5mm and 100nm individually, pervade environments from the deepest oceans to the topmost peak peaks. Their occupancy poses ominous consequences for human well-being, as they expand in the feeding relationships among organisms and penetrate water sources, land soils, and even the air we sigh. Studies have proved that these insignificant plastic pieces can harbor poisonous projectiles for weaponry and be a part of vectors for pathogens, offering meaningful risks to human well-being. Ingestion of micro-plastics has happened connected to inflammation, oxidative stress, and natural damage; while their breathing concedes possibility bring about respiratory issues and cardiovascular questions. Furthermore, the demeanour of micro-plastics in bread and water raises concerns about their potential to upset endocrine function and cause reproductive issues. Addressing this worldwide change demands coordinated efforts from policymakers, businesses, and things alike. Robust organizing is wanted to curb plastic result and implement correct waste administration blueprints. Innovations in materials erudition and reusing electronics can help decrease flexible waste and limit its conception. Additionally, public knowledge campaigns can experience societies about the dangers of flexible contamination and advance tenable consumption tendencies. In conclusion, calculating and nano flexible contamination poses a multifaceted danger to human states general, ruining both incidental sustainability and community health. Urgent operation is authoritative to mitigate its impacts and safeguard the prosperity of current and future era.

REFERENCES

al Ghosh, S. (2023). Microplastics as an Emerging Threat to the Global Environment and Human Health. *Sustainability (Basel)*, 15(14), 10821. DOI: 10.3390/su151410821

AlDhaen E. (2022). Awareness of occupational health hazards and occupational stress among dental care professionals: Evidence from the GCC region. Front Public Health, 8(10), 922748. .DOI: 10.3389/fpubh.2022.922748

Anouk D'Hont. (2021) Dropping the microbead: Source and sink related microplastic distribution in the Black Sea and Caspian Sea basins. Marine Pollution Bulletin, 173(Part A), 112982. DOI: 10.1016/j.marpolbul.2021.112982

Anuli Dass. (2021) Air pollution: A review and analysis using fuzzy techniques in Indian scenario. Environmental Technology & Innovation, 22, 101441. DOI: 10.1016/j.eti.2021.101441

Boyle, K., & Örmeci, B. (2020). Microplastics and Nanoplastics in the Freshwater and Terrestrial Environment: A Review. *Water (Basel)*, 12(9), 2633. DOI: 10.3390/w12092633

Campanale C. (2020) A Detailed Review Study on Potential Effects of Microplastics and Additives of Concern on Human Health. Int J Environ Res Public Health, 13(17), 1212. .DOI: 10.3390/ijerph17041212

Campanale, C., Massarelli, C., Savino, I., Locaputo, V., & Uricchio, V. F. (2020, February 13). A Detailed Review Study on Potential Effects of Microplastics and Additives of Concern on Human Health. *International Journal of Environmental Research and Public Health*, 17(4), 1212. DOI: 10.3390/ijerph17041212

Chen L. (2017) Inflammatory responses and inflammation-associated diseases in organs. Oncotarget, 14(9), 7204-7218. .DOI: 10.18632/oncotarget.23208

Chen Q. (2023) Factors Affecting the Adsorption of Heavy Metals by Microplastics and Their Toxic Effects on Fish. Toxics, 28(11), 490. .DOI: 10.3390/toxics11060490

Cornick S. (2015) Roles and regulation of the mucus barrier in the gut. Tissue Barriers, 3(1-2), e982426. DOI: 10.4161/21688370.2014.982426

Dalu, M. T. B., Cuthbert, R. N., Muhali, H., Chari, L. D., Manyani, A., Masunungure, C., & Dalu, T. (2020). Is Awareness on Plastic Pollution Being Raised in Schools? Understanding Perceptions of Primary and Secondary School Educators. *Sustainability (Basel)*, 12(17), 6775. DOI: 10.3390/su12176775

Dube E. (2023) Plastics and Micro/Nano-Plastics (MNPs) in the Environment: Occurrence, Impact, and Toxicity. Int J Environ Res Public Health, 28(20), 6667. DOI: 10.3390/ijerph20176667

Emenike E.C. (2023) From oceans to dinner plates: The impact of microplastics on human health. Heliyon, 26(9), e20440. .DOI: 10.1016/j.heliyon.2023.e20440

Ghosh, K., & Jones, B. H. (2021). Roadmap to Biodegradable Plastics—Current State and Research Needs. *ACS Sustainable Chemistry & Engineering*, 9(18), 6170–6187. DOI: 10.1021/acssuschemeng.1c00801

Ghosh, S., Sinha, J. K., Ghosh, S., Vashisth, K., Han, S., & Bhaskar, R. (2023). Microplastics as an Emerging Threat to the Global Environment and Human Health. *Sustainability (Basel)*, 15(14), 10821. DOI: 10.3390/su151410821

Habib, R. Z., Aldhanhani, J. A. K., Ali, A. H., Ghebremedhin, F., Elkashlan, M., Mesfun, M., Kittaneh, W., Al Kindi, R., & Thiemann, T. (2022). Trends of microplastic abundance in personal care products in the United Arab Emirates over the period of 3 years (2018-2020). *Environmental Science and Pollution Research International*, 29(59), 89614–89624. DOI: 10.1007/s11356-022-21773-y

Heléne Österlund. (2023) Microplastics in urban catchments: Review of sources, pathways, and entry into stormwater. Science of The Total Environment, 858(1), 159781. DOI: 10.1016/j.scitotenv.2022.159781

Issac, M. N., & Kandasubramanian, B. (2021). Effect of microplastics in water and aquatic systems. *Environmental Science and Pollution Research International*, 28(16), 19544–19562. DOI: 10.1007/s11356-021-13184-2

Júnior, R. (2024). Towards to Battery Digital Passport: Reviewing Regulations and Standards for Second-Life Batteries. *Batteries*, 10(4), 115. DOI: 10.3390/batteries10040115

Kole P.J. (2017). Wear and Tear of Tyres: A Stealthy Source of Microplastics in the Environment. Int J Environ Res Public Health, 20(14), 1265. DOI: 10.3390/ijerph14101265

Kumar, R., Verma, A., Shome, A., Sinha, R., Sinha, S., Jha, P. K., Kumar, R., Kumar, P., Shubham, , Das, S., Sharma, P., & Vara Prasad, P. V. (2021). Impacts of Plastic Pollution on Ecosystem Services, Sustainable Development Goals, and Need to Focus on Circular Economy and Policy Interventions. *Sustainability (Basel)*, 13(17), 9963. DOI: 10.3390/su13179963

Landrigan P.J. (2023) The Minderoo-Monaco Commission on Plastics and Human Health. Ann Glob Health, 21(89), 23. DOI: 10.5334/aogh.4056

Lapyote Prasittisopin. (2023) Microplastics in construction and built environment. Developments in the Built Environment, 15. DOI: 10.1016/j.dibe.2023.100188

Long B. (2023) Impact of plastic film mulching on microplastic in farmland soils in Guangdong province. China. Heliyon, 27(9), e16587. DOI: 10.1016/j.heliyon.2023.e16587

Lu, K., Zhan, D., Fang, Y., Li, L., Chen, G., Chen, S., & Wang, L. (2022, September 28). Microplastics, potential threat to patients with lung diseases. *Frontiers in Toxicology*, 4, 958414. DOI: 10.3389/ftox.2022.958414

Lwanga, E. H., Beriot, N., Corradini, F., Silva, V., Yang, X., Baartman, J., Rezaei, M., van Schaik, L., Riksen, M., & Geissen, V. (2022). Review of microplastic sources, transport pathways and correlations with other soil stressors: A journey from agricultural sites into the environment. *Chemical and Biological Technologies in Agriculture*, 9(1), 20. DOI: 10.1186/s40538-021-00278-9

Mihai, , Gündoğdu, S., Markley, L. A., Olivelli, A., Khan, F. R., Gwinnett, C., Gutberlet, J., Reyna-Bensusan, N., Llanquileo-Melgarejo, P., Meidiana, C., Elagroudy, S., Ishchenko, V., Penney, S., Lenkiewicz, Z., & Molinos-Senante, M. (2022). Plastic Pollution, Waste Management Issues, and Circular Economy Opportunities in Rural Communities. *Sustainability (Basel)*, 14(1), 20. DOI: 10.3390/su14010020

Mihai, F.-C., Gündoğdu, S., Markley, L. A., Olivelli, A., Khan, F. R., Gwinnett, C., Gutberlet, J., Reyna-Bensusan, N., Llanquileo-Melgarejo, P., Meidiana, C., Elagroudy, S., Ishchenko, V., Penney, S., Lenkiewicz, Z., & Molinos-Senante, M. (2022). Plastic Pollution, Waste Management Issues, and Circular Economy Opportunities in Rural Communities. *Sustainability (Basel)*, 14(1), 20. DOI: 10.3390/su14010020

Niccolai E. (2023) Adverse Effects of Micro- and Nanoplastics on Humans and the Environment. Int J Mol Sci., 31(24), 15822. .DOI: 10.3390/ijms242115822

Niedzielska. (2015) Oxidative Stress in Neurodegenerative Diseases. Mol Neurobiol., 53(6), 4094-4125. DOI: 10.1007/s12035-015-9337-5

Sangkham, S., Faikhaw, O., Munkong, N., Sakunkoo, P., Arunlertaree, C., Chavali, M., Mousazadeh, M., & Tiwari, A. (2022). A review on microplastics and nanoplastics in the environment: Their occurrence, exposure routes, toxic studies, and potential effects on human health. *Marine Pollution Bulletin*, 181, 113832. DOI: 10.1016/j.marpolbul.2022.113832

Schröder B. (2024). From the Automated Calculation of Potential Energy Surfaces to Accurate Infrared Spectra. J Phys Chem Lett., 21(15), 3159-3169. .DOI: 10.1021/acs.jpclett.4c00186

Todd, E. C. D. (2014). Foodborne Diseases: Overview of Biological Hazards and Foodborne Diseases. Encyclopedia of Food Safety, 221–42. DOI: 10.1016/B978-0-12-378612-8.00071-8

Wagner, M., Scherer, C., Alvarez-Muñoz, D., Brennholt, N., Bourrain, X., Buchinger, S., Fries, E., Grosbois, C., Klasmeier, J., Marti, T., Rodriguez-Mozaz, S., Urbatzka, R., Vethaak, A. D., Winther-Nielsen, M., & Reifferscheid, G. (2014). Microplastics in freshwater ecosystems: What we know and what we need to know. *Environmental Sciences Europe*, 26(1), 12. DOI: 10.1186/s12302-014-0012-7

Weis, J. S., & De Falco, F. (2022). Microfibers: Environmental Problems and Textile Solutions. *Microplastics*, 1(4), 626–639. DOI: 10.3390/microplastics1040043

Wright, S. L., Ulke, J., Font, A., Chan, K. L. A., & Kelly, F. J. (2019). Atmospheric microplastic deposition in an urban environment and an evaluation of transport. *Environment International*, 136, 105411. DOI: 10.1016/j.envint.2019.105411

Yee M.S. (2021) Impact of Microplastics and Nanoplastics on Human Health. Nanomaterials (Basel), 16(11), 496. .DOI: 10.3390/nano11020496

Yuan, D. (2023) Microplastics in the tropical Northwestern Pacific Ocean and the Indonesian seas. Journal of Sea Research, 194. DOI: 10.1016/j.seares.2023.102406

Zhang, Y., Wang, S., Olga, V., Xue, Y., Lv, S., Diao, X., Zhang, Y., Han, Q., & Zhou, H. (2021). The potential effects of microplastic pollution on human digestive tract cells. *Chemosphere*, 291(Pt 1), 132714. DOI: 10.1016/j.chemosphere.2021.132714

Chapter 3
Microplastics and Climate Change: A Mutually Reinforcing Relationship

Nazuk Bhasin
https://orcid.org/0009-0006-6510-8073
Banaras Hindu University, India

Sudhanshu Kumar
https://orcid.org/0000-0002-2859-9743
Banaras Hindu University, India

Anil Barla
Banaras Hindu University, India

Amit Kumar Tiwari
https://orcid.org/0000-0001-7667-8203
Banaras Hindu University, India

Gopal Shankar Singh
Banaras Hindu University, India

ABSTRACT

Plastics—macro, micro, or nano—have become persistent, pervasive, and potentially hazardous pollutants infiltrating the global environment. Microplastics (<5 mm) owing to their increased surface area as compared to their mass and small size are considered more harmful than larger plastics. The issue of their environmental presence has gained momentum due to their ability to act as sources and sinks for toxic substances, and also due to the intensification of climate change.

DOI: 10.4018/979-8-3693-3447-8.ch003

Climate change stimulates their deterioration, dispersal, and the interaction with the environmental compartments. In turn the plastic debris contributes directly or indirectly to greenhouse gas emissions during its life cycle. Plastics account for 3.3% of the global GHG emissions. Thus, microplastics and climate change share a mutually reinforcing relationship. For effective management of both these issues, it is imperative to understand the nature and dynamics of this complex relationship. This chapter aims to discuss the long-term ecological impacts of microplastics and climate change on each other.

1. INTRODUCTION

Plastics are produced from hydrocarbon-based resources (Pivato et al., 2022), which serve as feedstocks for plastic polymer manufacturing through techniques such as catalytic or steam cracking and various polymerization methods (Wong et al., 2015; Tanner, 1974). This conversion process has made plastic a versatile and ubiquitous material (Mah, 2022). However, plastic production and disposal pose significant environmental and sustainability challenges (Ayeleru et al., 2020; Hopewell et al., 2009) (Figure 1).

Figure 1. Global annual plastic waste generation (in tonnes, adapted from Ritchie, 2023)

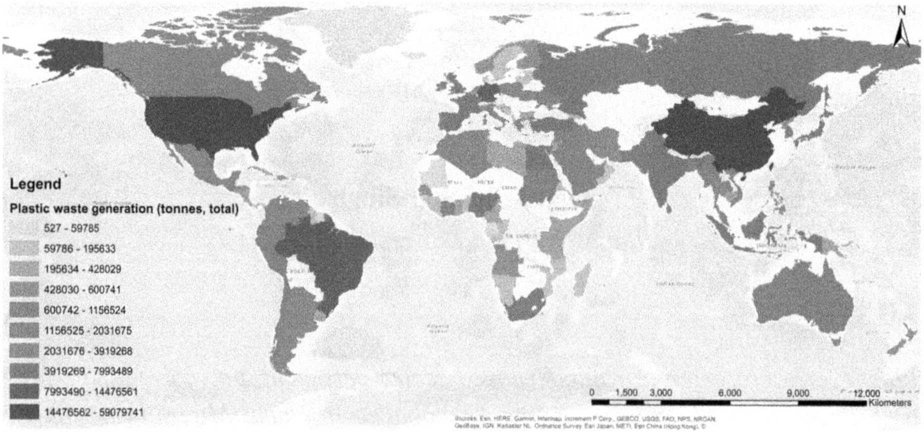

The infiltration of plastic as a non-biodegradable pollutant has led to pollution (Satti & Shah, 2020), habitat degradation (Kumar et al., 2021), and physical harm to wildlife (Puskic et al., 2020). Microplastics pose an even greater risk due to their small size (less than 5 mm in diameter), widespread distribution, increased bioavailability of toxins, and potential for bioaccumulation and biomagnification through the food web (Peng et al., 2017; Masura et al., 2015; Gouin, 2020; Carbery et al., 2018). Besides plastic pollution, climate change is another significant issue. Climate change refers to long-term alterations in Earth's climate, mainly caused by anthropogenic activities that emit greenhouse gases (Dore, 2005; Yoro & Daramola, 2020). It leads to global warming, melting ice caps, rising sea levels, ocean acidification, biodiversity loss, and socio-economic impacts (Upadhyay, 2020). Despite global efforts to mitigate these effects, the issue is worsening. NOAA predicts a 22% chance that 2024 will be the warmest year in 175 years, with a global surface temperature from January-March 2024 being 1.35°C higher than the 1901-2000 average (NOAA, 2024).

The challenges of microplastic pollution and climate change are interconnected (Patil et al., 2019), forming a complex and mutually reinforcing cycle. Climate change influences the lifecycle of microplastics, accelerating their transport, distribution, dispersal, and accumulation in the environment (Sharma et al., 2023; Hale et al., 2020). Conversely, microplastics can exacerbate climate change by increasing toxin bioavailability and facilitating bioaccumulation and biomagnification (Kumar et al., 2021). Feedback loops may exist, with microplastic degradation releasing greenhouse gases that contribute to global warming (Su et al., 2022; Shen et al., 2020). Additionally, microplastics in aquatic environments can alter local climate models by affecting radiation balance (Bernhard et al., 2020).

Addressing these issues requires an integrated approach that acknowledges their interconnected nature and seeks sustainable solutions to preserve ecosystems and environmental components (Kurniawan et al., 2023). However, a comprehensive understanding of the link between climate change and microplastics is lacking in current literature, particularly outside aquatic ecosystems. This chapter aims to bridge that gap and explore the association between microplastics and climate change.

2. RELATIONSHIP BETWEEN MICROPLASTICS AND CLIMATE CHANGE

The relationship between microplastics and climate change is intricate and multifaceted, governed by direct and indirect interactions (Sharma et al., 2023). Previous environmental studies on microplastics have primarily focused on their toxicity, occurrence, behavior, transport, and fate, with little attention to their interaction with the climate crisis.

Global mean temperatures have risen by 0.5-1.3°C between 1951 and 2010 due to increased greenhouse gas emissions (Bhatla et al., 2020). Projections suggest further increases of 0.3–0.7°C by 2035 and 0.3–4.8°C by 2100 compared to 1986-2005 (Moss et al., 2010). Climate change has been recognized as a global issue alongside microplastic pollution by UNEP in 2014. Efforts to mitigate climate change, such as the IPCC's recommendation to limit global temperature rise to 1.5°C, require reducing CO_2 emissions by 45% from 2010 levels and achieving net carbon neutrality by 2050 (Hoegh-Guldberg et al., 2019; Fadda, 2023). However, by 2050, cumulative GHG emissions from the lifecycle of plastics may consume 10-13% of the remaining carbon budget and exceed 56 gigatonnes (Hamilton et al., 2019). The adverse impacts of global warming, including ocean acidification, sea-level rise, and extreme weather events, are already evident, causing socio-economic and ecological harm (Ford et al., 2022).

Understanding the relationship between microplastics and climate change highlights the interconnectedness of environmental challenges and underscores the need for integrated approaches. These approaches should address plastic pollution, mitigate climate change impacts, and promote environmental sustainability to protect ecosystems and human well-being.

2.1 Influence of Microplastics on Climate Change

Plastics, including microplastics, significantly impact global climate throughout their lifecycle. Greenhouse gas emissions are not limited to production processes but also stem from plastic extraction, transportation, waste management, and environmental dispersion (Unuofin, 2020). The escalation of plastic manufacturing and waste generation exacerbates climate change (Cabernard et al., 2022). Plastics, derived from fossil fuels like crude oil and natural gas, undergo energy-intensive processes releasing CO_2 and other GHGs (Benavides et al., 2020). Fossil fuel extraction, processing, and transportation for plastic production further contribute to carbon emissions (Zheng & Suh, 2019).

Research indicates that in 2015, oil-well based energy production emitted 68 million tons CO_2eq GHG emissions between the well and refinery, with approximately 4% of crude oil used for plastic manufacture (Masnadi et al., 2018). Additionally, 72 plastic production units in the US emitted an estimated 17 million tons of CO_2eq in 2014 (Geyer et al., 2017). The adverse climate impacts of plastic emerge primarily after its use phase (Royer et al., 2018). Common methods for managing plastic waste include recycling, incineration, and landfilling. Incinerating plastic waste emitted 16 million tons of GHGs in 2015, projected to rise to 84 million tons by 2030 and 309 million tons by 2050 (PlasticsEurope, 2016). Additionally, plastic decomposition in the environment emits approximately 2,122 tons CO_2eq annually (Royer et al., 2018), though relatively minor compared to incineration, it remains a continuous process.

Microplastics act as carbon sinks, influencing carbon cycling and storage in ecosystems due to their increased surface area. They adsorb organic compounds, temporarily removing carbon from aquatic environments. Microplastics alter biogeochemical processes, impacting carbon dynamics and nutrient cycling. They provide surfaces for microbial colonization, affecting organic matter decomposition rates and microbial activity, influencing carbon mineralization, nutrient cycling, and GHG emissions from ecosystems.

The interaction of microplastics with climate change exacerbates negative impacts on ecosystem resilience and function. This alters ecosystem structure, disrupts nutrient cycling, and impairs ecosystem services like water purification and habitat provision. Such changes undermine resilience to climate change effects such as sea-level rise, ocean acidification, and habitat degradation, leading to cumulative environmental stress and biodiversity loss. Microplastic proliferation in marine environments negatively affects carbon fixation by disrupting phytoplankton communities, key for carbon fixation and oxygen production. Changes in phytoplankton dynamics cascade through marine food webs and biogeochemical cycles, potentially influencing climate feedback mechanisms (Nolte et al., 2017; Sjollema et al., 2016). Microplastic particulates also reduce metabolic rates, reproductive success, and survival of zooplankton, impacting carbon transfer to the deep sea (Galloway et al., 2017; Long et al., 2017), further affecting ecosystem productivity and habitat quality.

Light-colored microplastics in marine environments can alter surface albedo, impacting oceanic circulation, temperature gradients, and heat transfer processes, thereby affecting local and regional climate dynamics. These microplastic fragments decrease surface albedo, increasing solar radiation absorption and water heating. Climate change exacerbates environmental stressors and ecosystem vulnerabilities alongside microplastics. For instance, microplastics may adsorb and release chemical pollutants and GHGs, amplifying ocean acidification's effects on marine organisms and ecosystems. Extreme weather events like storms, hurricanes, and floods can mobilize and redistribute microplastics, increasing plastic pollution in affected areas. This movement may alter carbon and other element distribution patterns, impacting carbon fluxes and biogeochemical cycles locally and regionally.

The influence of microplastics on climate change involves uncertainties and knowledge gaps, such as the extent and spatial distribution of microplastic pollution, mechanisms of microplastic interactions with climate processes, and long-term environmental impacts on ecosystems and climate resilience. Interdisciplinary research efforts are needed to address these uncertainties and better understand the complex interactions between microplastics, climate change, and environmental sustainability.

2.2 Influence of Climate Change on Microplastics

Climate change-induced extreme weather events, such as alterations in oceanic circulation and acidity, rising sea levels, glacier retreat and melting, altered precipitation patterns leading to droughts and floods, and modifications of atmospheric circulations and wind patterns (Bindoff et al., 2019), exacerbate the microplastic problem by influencing the distribution and redistribution of microplastics as well as the microplastic sinks. Understanding the impact of climate change on the lifecycle of microplastics is crucial for comprehending their fate in the environment.

When glaciers thaw, trapped microplastics are released into the ocean (Obbard et al., 2014). These microplastics, with lower density than water, are carried by ocean currents over great distances, returning to aquatic environments as melted snow releases them. Additionally, in freshwater habitats, microplastics are more likely to settle due to the lower density of freshwater compared to saltwater. Melting sea ice accelerates ocean currents, impacting microplastic fate. Rising temperatures bring strong winds and increased thunderstorms, shifting surface water velocity and dispersing microplastics from their origin. These phenomena make understanding microplastic dispersion challenging. Strong winds not only move surface water horizontally but also vertically mix water, bringing submerged microplastics from the seafloor to the surface (Kukulka et al., 2012; Reisser et al., 2014).

Increased rainfall, along with sea level rise from retreating glaciers, will likely wash microplastics and other debris into water bodies from shorelines. This exposes plastic waste to additional weathering by ocean currents, potentially generating microplastics (Aragaw, 2020). This could lead to additional degradation or storage of microplastics in the aquatic ecosystem, accelerating their breakdown due to climate change.

Extreme weather events like storms and floods facilitate the spread of microplastics even into unpolluted areas. Drought conditions trap microplastic particles in the pedosphere (Lozano et al., 2021), emphasizing the need to understand how each climate change phenomenon affects microplastic fate.

Temperature: An increase in temperature expedites the decomposition of plastic waste during its transport. Sea ice melting released approximately one trillion microplastic particles into the Arctic Ocean over a decade (Obbard et al., 2014). Glacial lakes worldwide contain microplastics (Dong et al., 2021; Luoto et al., 2019), with global warming causing permafrost thawing, releasing microplastic debris into the pedosphere and hydrosphere (Chen et al., 2021). The melting and freezing cycle mechanically degrades microplastic particles, forming nanoplastics (Horton et al., 2017), accumulating in the environment (Chen et al., 2021).

Rivers originating from glaciers transport accumulated microplastics downstream. A Himalayan River flowing between China and Nepal contains microplastic contaminants ranging from 30 to 302 particles per unit volume (Yang et al., 2021).

Precipitation: Increased intensity and frequency of precipitation lead to surface water runoffs, transporting microplastics from the pedosphere to the hydrosphere (Napper et al., 2021). In Hong Kong, microplastic concentration in freshwater doubled that of shoreline marine water during a three-day rainfall event, dropping to a tenth after two hours, indicating runoff into the marine ecosystem (Cheung et al., 2019). Similarly, Xia et al. found a quadrupled microplastic concentration in surface waters following heavy precipitation (Xia et al., 2020). Precipitation also results in the wet deposition of atmospheric microplastic particles into ecosystems (Abbasi, 2021), with rainfall altering water movement mechanics and resuspending microplastics in waterbody sediments (Baldwin et al., 2016). Acid rain-induced coral bleaching releases stored microplastics into the environment (Haapkylä et al., 2011).

Drought: Drought conditions affect the vertical transport of microplastics in soil, as observed in laboratory studies (O'Connor et al., 2019), but field-level validation is necessary. While microplastic contamination negatively impacts biodiversity (Guo et al., 2020; Guzzetti et al., 2018; Rillig & Bonkowski, 2018; Wright et al., 2013), they have proven beneficial for flora under drought conditions, reducing nutrient loss by leaching (Lehmann et al., 2020; Lozano et al., 2021). Soil moisture content regulates microplastic movement; during droughts, microplastics accumulate in soil pores, enhancing water and nutrient retention (Lozano et al., 2021; Lozano & Rillig, 2020). Microfibers can promote root symbiotic microbial communities, benefiting onion shoots and arbuscular mycorrhiza (Lehmann et al., 2020). However, detrimental impacts on grass and wheat have also been observed (Lozano & Rillig, 2020; Boots et al., 2019), necessitating further research to understand plant-microplastic interactions under drought conditions.

Wind: Wind plays a crucial role in dispersing microplastics to various ecosystems, including remote locations like polar regions, the Pyrenees, and the Tibetan plateau (Brahney, 2020; Allen et al., 2019; Zhang et al., 2019; Bergmann et al., 2019). Hurricanes in China have intensified aquatic microplastic pollution (Wang et al., 2019), while Antarctic microplastic presence is attributed to wind transport (Aves et al., 2022). Atmospheric microplastics contribute to Earth's radiative forcing and regional temperature variations (Revell et al., 2021; Evangeliou et al., 2020). Wind velocity affects microplastic transport and accumulation in surface water (Jong et al., 2022).

Microplastics may accumulate in the hydrosphere, depending on their nature and environmental conditions (Haque & Fan, 2023). Global warming-induced rainfall and wind enhance microplastic transport to distant aquatic habitats like the Arctic Sea (Gupta et al., 2023). Elevated temperatures aid in removing microplastics from melting snow caps (Ásmundsdóttir & Schulz, 2020). Inundation and heatwaves influence terrestrial microplastic fate, causing movement between marine and terrestrial ecosystems or retention in the pedosphere (Cheung & Not, 2023).

3. IMPLICATIONS

The significant and enduring changes in Earth's climate due to global warming and climate change have led to various environmental, social, and economic impacts. While natural climate variability occurs over geological time scales, contemporary climate change is primarily driven by human activities, increasing greenhouse gas (GHG) concentrations in the atmosphere. This has resulted in adverse effects on meteorological parameters such as temperature, precipitation, wind patterns, sea levels, glacial retreats, ocean acidification, and extreme weather events. In remote, delicate ecosystems like polar regions (Vethaak and Leslie, 2016), these impacts have been exacerbated by the growing presence of microplastics, affecting ecosystem structure and function. Factors influencing microplastic incidence in these environments include shape, size, wind patterns, local atmospheric instabilities, precipitation patterns, and deposition rates (Enyoh et al., 2019).

For example, microplastics are deposited in snowfall, with 1760 particles per unit volume found in snow samples from Greenland and Svalbard (Carrington, 2019). Similarly, glacier samples from Europe contained 24,600 microplastic particles per unit volume. The vulnerability of these ecosystems is attributed to low genetic variability (Rowlands et al., 2021), with microplastics posing a threat to their fragility. Microplastic waste is assimilated by biotic components, accumulating throughout the food chain, impacting organisms at all levels. Sediments and sea ice contain significant microplastic amounts, ingested by seabirds and other fauna (Amélineau et al., 2016; Munari et al., 2017). Lusher et al. (2013) found comparable concentrations of microplastic fragments in fish and sea ice samples. In the Arctic, declining sea ice reduces ice algae levels, affecting zooplankton, Arctic cod, seals, and polar bears. This decline in ice algae contributes to the decline in higher trophic levels (Melillo et al., 2014).

Various aquatic organisms, including marine species, freshwater fish, crustaceans, and plankton, have been found contaminated with microplastics (Ferreira et al., 2016; Green, 2016; Setälä et al., 2018; Devriese et al., 2015). Through feeding, filter-feeding, and respiration, they can be ingested by organisms, leading to physiological and ecological consequences (Setälä et al., 2016; Teuten et al., 2009). Aquatic creatures are vulnerable to physical injury, digestive blockages, and toxicological effects due to microplastic accumulation in their tissues, organs, and digestive tracts (Setälä et al., 2016).

Furthermore, the persistence of microplastics poses an ongoing threat to biota, with significant increases in microplastic concentrations observed in pelagic species. Research indicates that pelagic animals exhibit higher intestinal concentrations of plastic debris, with open-ocean pelagic animals particularly susceptible to plastic ingestion compared to benthic animals (Pereira et al., 2020). Additionally, alongside genetic changes induced by climate change, the rise of plastics is expected to contribute to various morphological alterations in aquatic invertebrates, impacting photoperiodic deviations and resulting in phenological and thermal adaptations (Stoks et al., 2014).

Microplastics transfer through trophic levels, reaching aquatic food webs (Setälä et al., 2016; Nobre et al., 2015), posing risks to apex predators due to biomagnification and bioaccumulation (Eerkes-Medrano et al., 2015). In terrestrial ecosystems, the pedosphere supports nutrient cycles and food production (Sharma et al., 2023). Plastic waste sources, including compost and plastic mulch, alter soil properties, emitting GHGs and affecting fertility. Plastic residues change decomposition rates, soil temperature, impacting soil ecosystem services (Kumar et al., 2021). Soil temperature affects species hatching and gender determination, while microbial populations and forest health suffer from soil changes, leading to erosion, biodiversity loss, and increased forest fire risk.

Numerous studies have explored the ingestion of microplastics by terrestrial organisms (Lwanga et al., 2016, 2017; Zhao et al., 2016). Direct consumption and leaching of toxic additives from weathered microplastics pose significant toxicity risks to terrestrial fauna (Forschungsverbund, 2018). Terrestrial microplastics can biomagnify in trophic levels, similar to freshwater microplastics (Ašmonaitė and Almroth, 2019). Lwanga et al. (2017) investigated microplastic movement in terrestrial food chains, finding LDPE microplastics in chicken excreta from earthworms exposed to microplastics. Plastic debris was found in the digestive tracts of 27% of 230 goats and 50% of 185 sheep (Omidi et al., 2012). Zhao et al. (2016) observed 94% of 17 terrestrial birds consuming 364 microplastics. The estrogenic effects of leached chemicals from microplastic additives disrupt vertebrate endocrine systems, including phthalates and bisphenol A (de Souza Machado et al., 2018). Soil biotas, such as springtails and earthworms, can transport microplastics like polyethylene beads as deep as 10 cm within 21 days (Fahrenkamp-Uppenbrink, 2016).

Microplastics in the atmosphere disrupt pollination by obstructing pollen grains, hindering the process (Zang et al., 2020). This threatens the diverse plant species in tropical regions, particularly endemic species with limited seed banks, leading to population declines. Airborne microplastics also pose health risks, causing ocular infections, lung congestion, ulcers, and other issues due to their toxicity.

Concerns over plastic's impact on animal and human health are growing as microplastics accumulate in food chains. Livestock products like meat, milk, and blood contain microplastics, as do globally consumed foods like salt, fish, and shellfish. Research on Tuticorin salt samples found an average ingestion of 216 microplastic particles per year for every 5 kg of sea salt and 48 microplastic particles annually for every 5 kg of bore-well salt (Sathish et al., 2020). Microplastic fragments, when inhaled or consumed by humans, can lead to various health issues, including apoptosis, cardiovascular disease, autoimmune disorders, and degenerative neural diseases (Azoulay et al., 2019). They can also cause tissue injury, particularly in the gut and lungs, with particles as small as 10 μm detected in maternal and foetal placentas (Carrington, 2019). Additionally, microplastics can induce inflammation, cell damage, and oxidative stress (Vethaak and Leslie, 2016).

The toxicity of plastics is exacerbated by synthetic additives like phthalates, bisphenol A (BPA), lead, and brominated compounds, added during manufacturing (Hahladakis et al., 2018). BPA exposure during developmental stages in mice results in insulin resistance, sperm count reduction, and disruption of brain connections (Talsness et al., 2009). Phthalate exposure, particularly in males, can lead to reproductive abnormalities in guinea pigs and mice (Sathyanarayana et al., 2008; Swan, 2008). Moreover, phthalates can disrupt endocrine functions in aquatic organisms and humans (Wang and Qian, 2021). Lead, used in thermal stabilizers, poses long-term organ damage risks (Jubsilp et al., 2021).

Apart from toxic chemical additives, microplastics and plastic fragments can transport harmful substances and microorganisms into ecosystems, disrupting their balance (de Souza Machado et al., 2018). Reports have identified *Escherichia coli*, *Stenotrophomonas maltophilia*, and *Bacillus cereus* in plastic waste along the Belgian shore (Vethaak and Leslie, 2016). These bacteria thrive on microplastic surfaces, forming biofilms that enter freshwater, increasing disease risks for humans. Plastic debris in damp environments can breed mosquito larvae, aiding in the transmission of diseases like dengue and Zika virus (Krystosik et al., 2020).

To achieve environmental sustainability, comprehensive research is crucial to understand exposure levels, concentration, and pathways through which plastic waste, particularly microplastics and nanoparticles, affect ecosystem health. Recent studies highlight plastic pollution as a growing concern for biota, including humans, potentially straining healthcare infrastructure and lowering the Human Development Index (HDI) (Kumar et al., 2021).

4. CONCLUSION

While majority of the studies associated with this topic are primarily either in the field of marine plastic pollution or related to the behaviour, fate and toxicology of microplastics in the environment. This chapter approached the understanding of prevalence of microplastics and their association to climate change.

Microplastics disperse widely from their sources and can originate from the breakdown of larger plastic items in the environment. They are embedded in soils and sediments and cycle through various environmental realms influenced by meteorological conditions and hydrodynamic forces. Properties like size, shape, and density determine their behavior, while biological, chemical, and physical interactions also influence their transport dynamics. Plastic pollutants, including micro and nano plastics, have created a synthetic ecosystem known as the plastisphere (Sharma et al., 2023).

Attaining a comprehensive understanding of microplastic pollution necessitates a thorough examination of the intricate interplay between microplastics and the environment across a spectrum of matrices, encompassing air, water, soil, and sediment, with special focus on the interplay between microplastics and climate change. Such investigations are crucial for discerning the combined ecological ramifications of microplastics and climate change at a systemic level and to formulate targeted, comprehensive, and holistic approaches in mitigating and adapting to the combined impacts of both.

REFERENCES

Abbasi, S. (2021). Microplastics washout from the atmosphere during a monsoon rain event. *Journal of Hazardous Materials Advances*, 4, 100035. DOI: 10.1016/j.hazadv.2021.100035

Allen, S., Allen, D., Phoenix, V. R., Le Roux, G., Durántez Jiménez, P., Simonneau, A., Binet, S., & Galop, D. (2019). Atmospheric transport and deposition of microplastics in a remote mountain catchment. *Nature Geoscience*, 12(5), 339–344. DOI: 10.1038/s41561-019-0335-5

Amélineau, F., Bonnet, D., Heitz, O., Mortreux, V., Harding, A. M., Karnovsky, N., Walkusz, W., Fort, J., & Grémillet, D. (2016). Microplastic pollution in the Greenland Sea: Background levels and selective contamination of planktivorous diving seabirds. *Environmental Pollution*, 219, 1131–1139. DOI: 10.1016/j.envpol.2016.09.017

Aragaw, T. A. (2020). Surgical face masks as a potential source for microplastic pollution in the COVID-19 scenario. *Marine Pollution Bulletin*, 159, 111517. DOI: 10.1016/j.marpolbul.2020.111517

Ašmonaitė, G., & Almroth, B. C. (2019). *Effects of microplastics on organisms and impacts on the environment: Balancing the known and unknown. Department of Biological and Environmental Sciences*. University of Gothenburg.

Ásmundsdóttir, Á. M., & Schulz, B. (2020). Effects of Microplastics in the Cryosphere. In Rocha-Santos, T., Costa, M. F., & Mouneyrac, C. (Eds.), *Handbook of Microplastics in the Environment*. Springer. DOI: 10.1007/978-3-030-39041-9_47

Aves, A. R., Revell, L. E., Gaw, S., Ruffell, H., Schuddeboom, A., Wotherspoon, N. E., LaRue, M., & McDonald, A. J. (2022). First evidence of microplastics in Antarctic snow. *The Cryosphere*, 16(6), 2127–2145. DOI: 10.5194/tc-16-2127-2022

Ayeleru, O. O., Dlova, S., Akinribide, O. J., Ntuli, F., Kupolati, W. K., Marina, P. F., Blencowe, A., & Olubambi, P. A. (2020). Challenges of plastic waste generation and management in Sub-Saharan Africa: A review. *Waste Management (New York, N.Y.)*, 110, 24–42. DOI: 10.1016/j.wasman.2020.04.017

Azoulay, D., Villa, P., Arellano, Y., Gordon, M. F., Moon, D., Miller, K. A., & Thompson, K. (2019). Plastic & health: the hidden costs of a plastic planet. *Center for International Environmental Law (CIEL)*. Available at https://www.ciel.org/plasticandhealth

Baldwin, A. K., Corsi, S. R., & Mason, S. A. (2016). Plastic debris in 29 great lakes tributaries: Relations to watershed attributes and hydrology. *Environmental Science & Technology*, 50(19), 10377–10385. DOI: 10.1021/acs.est.6b02917

Benavides, P. T., Lee, U., & Zarè-Mehrjerdi, O. (2020). Life cycle greenhouse gas emissions and energy use of polylactic acid, bio-derived polyethylene, and fossil-derived polyethylene. *Journal of Cleaner Production*, 277, 124010. DOI: 10.1016/j.jclepro.2020.124010

Bergmann, M., Mützel, S., Primpke, S., Tekman, M. B., Trachsel, J., & Gerdts, G. (2019). White and wonderful? Microplastics prevail in snow from the Alps to the Arctic. *Science Advances*, 5(8), eaax1157. DOI: 10.1126/sciadv.aax1157

Bernhard, G. H., Neale, R. E., Barnes, P. W., Neale, P. J., Zepp, R. G., Wilson, S. R., Andrady, A. L., Bais, A. F., McKenzie, R. L., Aucamp, P. J., Young, P. J., Liley, J. B., Lucas, R. M., Yazar, S., Rhodes, L. E., Byrne, S. N., Hollestein, L. M., Olsen, C. M., Young, A. R., & White, C. C. (2020). Environmental effects of stratospheric ozone depletion, UV radiation and interactions with climate change: UNEP Environmental Effects Assessment Panel, update 2019. *Photochemical & Photobiological Sciences*, 19(5), 542–584. DOI: 10.1039/d0pp90011g

Bhatla, R., Verma, S., Ghosh, S., & Gupta, A. (2020). Abrupt changes in mean temperature over India during 1901–2010. *Journal of Earth System Science*, 129(1), 1–11. DOI: 10.1007/s12040-020-01421-0

Bindoff, N. L., Cheung, W. W., Kairo, J. G., Arístegui, J., Guinder, V. A., Hallberg, R., ... & Williamson, P. (2019). Changing ocean, marine ecosystems, and dependent communities. *IPCC Special Report on the Ocean and Cryosphere in a Changing Climate,* 477-587.

Boots, B., Russell, C. W., & Green, D. S. (2019). Effects of microplastics in soil ecosystems: Above and below ground. *Environmental Science & Technology*, 53(19), 11496–11506. DOI: 10.1021/acs.est.9b03304

Brahney, J. (2020). Wet and dry plastic deposition data for western US National Atmospheric Deposition Program sites (2017-2019). Constraining physical understanding of aerosol loading, biogeochemistry, and snowmelt hydrology from hillslope to watershed scale in the east river scientific focus area. *ESS-DIVE repository.* DOI: 10.15485/1773176

Cabernard, L., Pfister, S., Oberschelp, C., & Hellweg, S. (2022). Growing environmental footprint of plastics driven by coal combustion. *Nature Sustainability*, 5(2), 139–148. DOI: 10.1038/s41893-021-00807-2

Carbery, M., O'Connor, W., & Palanisami, T. (2018). Trophic transfer of microplastics and mixed contaminants in the marine food web and implications for human health. *Environment International*, 115, 400–409. DOI: 10.1016/j.envint.2018.03.007

Carrington, D. 2019. Microplastics 'significantly contaminating the air', scientists warn. *The Guardian.* Retrieved 05 April 2024, from https://www.theguardian.com/environment/2019/aug/14/microplastics-found-at-profuse-levels-in-snow-from-arctic-to-alps-contamination

Chen, X., Huang, G., Gao, S., & Wu, Y. (2021). Effects of permafrost degradation on global microplastic cycling under climate change. *Journal of Environmental Chemical Engineering*, 9(5), 106000. DOI: 10.1016/j.jece.2021.106000

Cheung, C. K. H., & Not, C. (2023). Impacts of extreme weather events on microplastic distribution in coastal environments. *The Science of the Total Environment*, 904, 166723. DOI: 10.1016/j.scitotenv.2023.166723

Cheung, P. K., Hung, P. L., & Fok, L. (2019). River microplastic contamination and dynamics upon a rainfall event in Hong Kong, China. *Environmental Processes*, 6(1), 253–264. DOI: 10.1007/s40710-018-0345-0

de Souza Machado, A. A., Kloas, W., Zarfl, C., Hempel, S., & Rillig, M. C. (2018). Microplastics as an emerging threat to terrestrial ecosystems. *Global Change Biology*, 24(4), 1405–1416. DOI: 10.1111/gcb.14020

Devriese, L. I., Van der Meulen, M. D., Maes, T., Bekaert, K., Paul-Pont, I., Frère, L., Robbens, J., & Vethaak, A. D. (2015). Microplastic contamination in brown shrimp (*Crangon crangon*, Linnaeus 1758) from coastal waters of the Southern North Sea and Channel area. *Marine Pollution Bulletin*, 98(1-2), 179–187. DOI: 10.1016/j.marpolbul.2015.06.051

Dong, H., Wang, L., Wang, X., Xu, L., Chen, M., Gong, P., & Wang, C. (2021). Microplastics in a remote lake basin of the Tibetan Plateau: Impacts of atmospheric transport and glacial melting. *Environmental Science & Technology*, 55(19), 12951–12960. DOI: 10.1021/acs.est.1c03227

Dore, M. H. (2005). Climate change and changes in global precipitation patterns: What do we know? *Environment International*, 31(8), 1167–1181. DOI: 10.1016/j.envint.2005.03.004

Eerkes-Medrano, D., Thompson, R. C., & Aldridge, D. C. (2015). Microplastics in freshwater systems: A review of the emerging threats, identification of knowledge gaps and prioritisation of research needs. *Water Research*, 75, 63–82. DOI: 10.1016/j.watres.2015.02.012

Enyoh, C. E., Verla, A. W., Verla, E. N., Ibe, F. C., & Amaobi, C. E. (2019). Airborne microplastics: A review study on method for analysis, occurrence, movement and risks. *Environmental Monitoring and Assessment*, 191(11), 1–17. DOI: 10.1007/s10661-019-7842-0

Evangeliou, N., Grythe, H., Klimont, Z., Heyes, C., Eckhardt, S., Lopez-Aparicio, S., & Stohl, A. (2020). Atmospheric transport is a major pathway of microplastics to remote regions. *Nature Communications*, 11(1), 3381. DOI: 10.1038/s41467-020-17201-9

Fadda, G. (2023). CO_2 Emissions and underground storage analysis: towards achieving net-zero targets by 2030 and 2050 (*Masters' Dissertation, Politecnico di Torino*) https://webthesis.biblio.polito.it/id/eprint/29166

Fahrenkamp-Uppenbrink, J. (2016). Earthworms on a microplastics diet. *Science*, 351(6277), 1039–1039. DOI: 10.1126/science.351.6277.1039-a

Ferreira, P., Fonte, E., Soares, M. E., Carvalho, F., & Guilhermino, L. (2016). Effects of multi-stressors on juveniles of the marine fish *Pomatoschistus microps*: Gold nanoparticles, microplastics and temperature. *Aquatic Toxicology (Amsterdam, Netherlands)*, 170, 89–103. DOI: 10.1016/j.aquatox.2015.11.011

Ford, H. V., Jones, N. H., Davies, A. J., Godley, B. J., Jambeck, J. R., Napper, I. E., Suckling, C. C., Williams, G. J., Woodall, L. C., & Koldewey, H. J. (2022). The fundamental links between climate change and marine plastic pollution. *The Science of the Total Environment*, 806, 150392. DOI: 10.1016/j.scitotenv.2021.150392

Forschungsverbund, B. (2018). An underestimated threat: Land-based pollution with microplastics. Retrieved 5 April 2024, from https://www.sciencedaily.com/releases/2018/02/180205125728.htm

Galloway, T. S., Cole, M., & Lewis, C. (2017). Interactions of microplastic debris throughout the marine ecosystem. *Nature Ecology & Evolution, 1*(5), 0116.

Geyer, R., Jambeck, J. R., & Law, K. L. (2017). Production, use, and fate of all plastics ever made. *Science Advances*, 3(7), e1700782. DOI: 10.1126/sciadv.1700782

Gouin, T. (2020). Toward an improved understanding of the ingestion and trophic transfer of microplastic particles: Critical review and implications for future research. *Environmental Toxicology and Chemistry*, 39(6), 1119–1137. DOI: 10.1002/etc.4718

Green, D. S. (2016). Effects of microplastics on European flat oysters, Ostrea edulis and their associated benthic communities. *Environmental Pollution*, 216, 95–103. DOI: 10.1016/j.envpol.2016.05.043

Guo, J. J., Huang, X. P., Xiang, L., Wang, Y. Z., Li, Y. W., Li, H., Cai, Q.-Y., Mo, C.-H., & Wong, M. H. (2020). Source, migration and toxicology of microplastics in soil. *Environment International*, 137, 105263. DOI: 10.1016/j.envint.2019.105263

Gupta, S., Kumar, R., Rajput, A., Gorka, R., Gupta, A., Bhasin, N., Yadav, S., Verma, A., Ram, K., & Bhagat, M. (2023). Atmospheric microplastics: Perspectives on origin, abundances, ecological and health risks. *Environmental Science and Pollution Research International*, 30(49), 107435–107464. DOI: 10.1007/s11356-023-28422-y

Guzzetti, E., Sureda, A., Tejada, S., & Faggio, C. (2018). Microplastic in marine organism: Environmental and toxicological effects. *Environmental Toxicology and Pharmacology*, 64, 164–171. DOI: 10.1016/j.etap.2018.10.009

Haapkylä, J., Unsworth, R. K., Flavell, M., Bourne, D. G., Schaffelke, B., & Willis, B. L. (2011). Seasonal rainfall and runoff promote coral disease on an inshore reef. *PLoS One*, 6(2), e16893. DOI: 10.1371/journal.pone.0016893

Hahladakis, J. N., Velis, C. A., Weber, R., Iacovidou, E., & Purnell, P. (2018). An overview of chemical additives present in plastics: Migration, release, fate and environmental impact during their use, disposal and recycling. *Journal of Hazardous Materials*, 344, 179–199. DOI: 10.1016/j.jhazmat.2017.10.014

Hale, R. C., Seeley, M. E., La Guardia, M. J., Mai, L., & Zeng, E. Y. (2020). A global perspective on microplastics. *Journal of Geophysical Research: Oceans, 125*(1), e2018JC014719.

Hamilton, L. A., Feit, S., Muffett, C., Kelso, M., Rubright, S. M., Bernhardt, C., . . . Labbé-Bellas, R. (2019). Plastic and Climate: The hidden costs of a plastic planet. *Center for International Environmental Law (CIEL).* Available online at www.ciel .org/plasticandclimate

Haque, F., & Fan, C. (2023). Fate and impacts of microplastics in the environment: Hydrosphere, pedosphere, and atmosphere. *Environments (Basel, Switzerland)*, 10(5), 70. DOI: 10.3390/environments10050070

Hoegh-Guldberg, O., Jacob, D., Taylor, M., Guillén Bolaños, T., Bindi, M., Brown, S., Camilloni, I. A., Diedhiou, A., Djalante, R., Ebi, K., Engelbrecht, F., Guiot, J., Hijioka, Y., Mehrotra, S., Hope, C. W., Payne, A. J., Pörtner, H.-O., Seneviratne, S. I., Thomas, A., & Zhou, G. (2019). The human imperative of stabilizing global climate change at 1.5°C. *Science*, 365(6459), eaaw6974. DOI: 10.1126/science.aaw6974

Hopewell, J., Dvorak, R., & Kosior, E. (2009). Plastics recycling: Challenges and opportunities. *Philosophical Transactions of the Royal Society of London. Series B, Biological Sciences*, 364(1526), 2115–2126. DOI: 10.1098/rstb.2008.0311

Horton, A. A., Walton, A., Spurgeon, D. J., Lahive, E., & Svendsen, C. (2017). Microplastics in freshwater and terrestrial environments: Evaluating the current understanding to identify the knowledge gaps and future research priorities. *The Science of the Total Environment*, 586, 127–141. DOI: 10.1016/j.scitotenv.2017.01.190

Jong, M. C., Tong, X., Li, J., Xu, Z., Chng, S. H. Q., He, Y., & Gin, K. Y. H. (2022). Microplastics in equatorial coasts: Pollution hotspots and spatiotemporal variations associated with tropical monsoons. *Journal of Hazardous Materials*, 424, 127626. DOI: 10.1016/j.jhazmat.2021.127626

Jubsilp, C., Asawakosinchai, A., Mora, P., Saramas, D., & Rimdusit, S. (2021). Effects of organic based heat stabilizer on properties of polyvinyl chloride for pipe applications: A comparative study with Pb and CaZn systems. *Polymers*, 14(1), 133. DOI: 10.3390/polym14010133

Krystosik, A., Njoroge, G., Odhiambo, L., Forsyth, J. E., Mutuku, F., & LaBeaud, A. D. (2020). Solid wastes provide breeding sites, burrows, and food for biological disease vectors, and urban zoonotic reservoirs: A call to action for solutions-based research. *Frontiers in Public Health*, 7, 405. DOI: 10.3389/fpubh.2019.00405

Kukulka, T., Proskurowski, G., Morét-Ferguson, S., Meyer, D. W., & Law, K. L. (2012). The effect of wind mixing on the vertical distribution of buoyant plastic debris. *Geophysical Research Letters*, 39(7), 2012GL051116. DOI: 10.1029/2012GL051116

Kumar, R., Verma, A., Shome, A., Sinha, R., Sinha, S., Jha, P. K., Kumar, R., Kumar, P., Shubham, , Das, S., Sharma, P., & Vara Prasad, P. V. (2021). Impacts of plastic pollution on ecosystem services, sustainable development goals, and need to focus on circular economy and policy interventions. *Sustainability (Basel)*, 13(17), 9963. DOI: 10.3390/su13179963

Kurniawan, T. A., Haider, A., Mohyuddin, A., Fatima, R., Salman, M., Shaheen, A., Ahmad, H. M., Al-Hazmi, H. E., Othman, M. H. D., Aziz, F., Anouzla, A., & Ali, I. (2023). Tackling microplastics pollution in global environment through integration of applied technology, policy instruments, and legislation. *Journal of Environmental Management*, 346, 118971. DOI: 10.1016/j.jenvman.2023.118971

Lehmann, A., Leifheit, E. F., Feng, L., Bergmann, J., Wulf, A., & Rillig, M. C. (2020). Microplastic fiber and drought effects on plants and soil are only slightly modified by arbuscular mycorrhizal fungi. *Soil Ecology Letters*, 1–13.

Long, M., Paul-Pont, I., Hegaret, H., Moriceau, B., Lambert, C., Huvet, A., & Soudant, P. (2017). Interactions between polystyrene microplastics and marine phytoplankton lead to species-specific hetero-aggregation. *Environmental Pollution*, 228, 454–463. DOI: 10.1016/j.envpol.2017.05.047

Lozano, Y. M., Aguilar-Trigueros, C. A., Onandia, G., Maaß, S., Zhao, T., & Rillig, M. C. (2021). Effects of microplastics and drought on soil ecosystem functions and multifunctionality. *Journal of Applied Ecology*, 58(5), 988–996. DOI: 10.1111/1365-2664.13839

Lozano, Y. M., & Rillig, M. C. (2020). Effects of microplastic fibers and drought on plant communities. *Environmental Science & Technology*, 54(10), 6166–6173. DOI: 10.1021/acs.est.0c01051

Luoto, T. P., Rantala, M. V., Kivilä, E. H., Nevalainen, L., & Ojala, A. E. (2019). Biogeochemical cycling and ecological thresholds in a High Arctic lake (Svalbard). *Aquatic Sciences*, 81(2), 1–16. DOI: 10.1007/s00027-019-0630-7

Lusher, A. L., Mchugh, M., & Thompson, R. C. (2013). Occurrence of microplastics in the gastrointestinal tract of pelagic and demersal fish from the English Channel. *Marine Pollution Bulletin*, 67(1-2), 94–99. DOI: 10.1016/j.marpolbul.2012.11.028

Lwanga, E. H., Gertsen, H., Gooren, H., Peters, P., Salánki, T., Van Der Ploeg, M., & Geissen, V. (2016). Microplastics in the terrestrial ecosystem: Implications for *Lumbricus terrestris* (Oligochaeta, Lumbricidae).*Environmental Science & Technology*, 50(5), 2685–2691. DOI: 10.1021/acs.est.5b05478

Lwanga, E. H., Gertsen, H., Gooren, H., Peters, P., Salánki, T., van der Ploeg, M., & Geissen, V. (2017). Incorporation of microplastics from litter into burrows of *Lumbricus terrestris.Environmental Pollution*, 220, 523–531. DOI: 10.1016/j.envpol.2016.09.096

Mah, A. (2022). *Plastic unlimited: How corporations are fuelling the ecological crisis and what we can do about it*. John Wiley & Sons.

Masnadi, M. S., El-Houjeiri, H. M., Schunack, D., Li, Y., Englander, J. G., Badahdah, A., Monfort, J.-C., Anderson, J. E., Wallington, T. J., Bergerson, J. A., Gordon, D., Koomey, J., Przesmitzki, S., Azevedo, I. L., Bi, X. T., Duffy, J. E., Heath, G. A., Keoleian, G. A., McGlade, C., & Brandt, A. R. (2018). Global carbon intensity of crude oil production. *Science*, 361(6405), 851–853. DOI: 10.1126/science.aar6859

Masura, J., Baker, J. E., Foster, G. D., Arthur, C., & Herring, C. (2015). Laboratory methods for the analysis of microplastics in the marine environment: recommendations for quantifying synthetic particles in waters and sediments. *NOAA Technical Memorandum NOS-OR&R-48.*

Melillo, J. M., Richmond, T. T., & Yohe, G. W. (Eds.). (2014). *Climate change impacts in the United States: Third National Climate Assessment*. U.S. Global Change Research Program. DOI: 10.7930/J0Z31WJ2

Moss, R. H., Edmonds, J. A., Hibbard, K. A., Manning, M. R., Rose, S. K., Van Vuuren, D. P., Carter, T. R., Emori, S., Kainuma, M., Kram, T., Meehl, G. A., Mitchell, J. F. B., Nakicenovic, N., Riahi, K., Smith, S. J., Stouffer, R. J., Thomson, A. M., Weyant, J. P., & Wilbanks, T. J. (2010). The next generation of scenarios for climate change research and assessment. *Nature*, 463(7282), 747–756. DOI: 10.1038/nature08823

Munari, C., Infantini, V., Scoponi, M., Rastelli, E., Corinaldesi, C., & Mistri, M. (2017). Microplastics in the sediments of Terra Nova bay (Ross sea, Antarctica). *Marine Pollution Bulletin*, 122(1-2), 161–165. DOI: 10.1016/j.marpolbul.2017.06.039

Napper, I. E., Baroth, A., Barrett, A. C., Bhola, S., Chowdhury, G. W., Davies, B. F., Duncan, E. M., Kumar, S., Nelms, S. E., Hasan Niloy, M. N., Nishat, B., Maddalene, T., Thompson, R. C., & Koldewey, H. (2021). The abundance and characteristics of microplastics in surface water in the transboundary Ganges River. *Environmental Pollution*, 274, 116348. DOI: 10.1016/j.envpol.2020.116348

NOAA. (2024). National Centers for Environmental Information, Monthly Global Climate Report for March 2024. Retrieved 03 April 2024, from https://www.ncei.noaa.gov/access/monitoring/monthlyreport/global/202403/supplemental/page-1

Nobre, C. R., Santana, M. F. M., Maluf, A., Cortez, F. S., Cesar, A., Pereira, C. D. S., & Turra, A. (2015). Assessment of microplastic toxicity to embryonic development of the sea urchin *Lytechinus variegatus* (Echinodermata: Echinoidea). *Marine Pollution Bulletin*, 92(1-2), 99–104. DOI: 10.1016/j.marpolbul.2014.12.050

Nolte, T. M., Hartmann, N. B., Kleijn, J. M., Garnæs, J., Van De Meent, D., Hendriks, A. J., & Baun, A. (2017). The toxicity of plastic nanoparticles to green algae as influenced by surface modification, medium hardness and cellular adsorption. *Aquatic Toxicology (Amsterdam, Netherlands)*, 183, 11–20. DOI: 10.1016/j.aquatox.2016.12.005

O'Connor, D., Pan, S., Shen, Z., Song, Y., Jin, Y., Wu, W. M., & Hou, D. (2019). Microplastics undergo accelerated vertical migration in sand soil due to small size and wet-dry cycles. *Environmental Pollution*, 249, 527–534. DOI: 10.1016/j.envpol.2019.03.092

Obbard, R. W., Sadri, S., Wong, Y. Q., Khitun, A. A., Baker, I., & Thompson, R. C. (2014). Global warming releases microplastic legacy frozen in Arctic Sea ice. *Earth's Future*, 2(6), 315–320. DOI: 10.1002/2014EF000240

Omidi, A., Naeemipoor, H., & Hosseini, M. (2012). Plastic debris in the digestive tract of sheep and goats: An increasing environmental contamination in Birjand, Iran. *Bulletin of Environmental Contamination and Toxicology*, 88(5), 691–694. DOI: 10.1007/s00128-012-0587-x

Patil, S. S., Bhagwat, R. V., Kumar, V., & Durugkar, T. (2019). Megaplastics to nanoplastics: emerging environmental pollutants and their environmental impacts. In *Environmental Contaminants: Ecological Implications and Management,* (pp. 205-235). Springer. DOI: 10.1007/978-981-13-7904-8_10

Peng, J., Wang, J., & Cai, L. (2017). Current understanding of microplastics in the environment: Occurrence, fate, risks, and what we should do. *Integrated Environmental Assessment and Management*, 13(3), 476–482. DOI: 10.1002/ieam.1912

Pereira, J. M., Rodríguez, Y., Blasco-Monleon, S., Porter, A., Lewis, C., & Pham, C. K. (2020). Microplastic in the stomachs of open-ocean and deep-sea fishes of the North-East Atlantic. *Environmental Pollution*, 265, 115060. DOI: 10.1016/j.envpol.2020.115060

Pivato, A. F., Miranda, G. M., Prichula, J., Lima, J. E., Ligabue, R. A., Seixas, A., & Trentin, D. S. (2022). Hydrocarbon-based plastics: Progress and perspectives on consumption and biodegradation by insect larvae. *Chemosphere*, 293, 133600. DOI: 10.1016/j.chemosphere.2022.133600

PlasticsEurope. (2016). *Plastics – the facts 2016. An analysis of European plastics production, demand and waste data.* PlasticsEurope.

Puskic, P. S., Lavers, J. L., & Bond, A. L. (2020). A critical review of harm associated with plastic ingestion on vertebrates. *The Science of the Total Environment*, 743, 140666. DOI: 10.1016/j.scitotenv.2020.140666

Reisser, J., Shaw, J., Hallegraeff, G., Proietti, M., Barnes, D. K., Thums, M., Wilcox, C., Hardesty, B. D., & Pattiaratchi, C. (2014). Millimeter-sized marine plastics: A new pelagic habitat for microorganisms and invertebrates. *PLoS One*, 9(6), e100289. DOI: 10.1371/journal.pone.0100289

Revell, L. E., Kuma, P., Le Ru, E. C., Somerville, W. R., & Gaw, S. (2021). Direct radiative effects of airborne microplastics. *Nature*, 598(7881), 462–467. DOI: 10.1038/s41586-021-03864-x

Rillig, M. C., & Bonkowski, M. (2018). Microplastic and soil protists: A call for research. *Environmental Pollution*, 241, 1128–1131. DOI: 10.1016/j.envpol.2018.04.147

Ritchie, H. (2023). How much of global greenhouse gas emissions come from plastics? Retrieved 03 April 2024, from: https://ourworldindata.org/ghg-emissions-plastics

Rowlands, E., Galloway, T., Cole, M., Lewis, C., Peck, V., Thorpe, S., & Manno, C. (2021). The effects of combined ocean acidification and nanoplastic exposures on the embryonic development of Antarctic krill. *Frontiers in Marine Science*, 8, 709763. DOI: 10.3389/fmars.2021.709763

Royer, S. J., Ferrón, S., Wilson, S. T., & Karl, D. M. (2018). Production of methane and ethylene from plastic in the environment. *PLoS One*, 13(8), e0200574. DOI: 10.1371/journal.pone.0200574

Sathish, M. N., Jeyasanta, I., & Patterson, J. (2020). Occurrence of microplastics in epipelagic and mesopelagic fishes from Tuticorin, Southeast coast of India. *The Science of the Total Environment*, 720, 137614. DOI: 10.1016/j.scitotenv.2020.137614

Sathyanarayana, S., Karr, C. J., Lozano, P., Brown, E., Calafat, A. M., Liu, F., & Swan, S. H. (2008). Baby care products: Possible sources of infant phthalate exposure. *Pediatrics*, 121(2), e260–e268. DOI: 10.1542/peds.2006-3766

Satti, S. M., & Shah, A. A. (2020). Polyester-based biodegradable plastics: An approach towards sustainable development. *Letters in Applied Microbiology*, 70(6), 413–430. DOI: 10.1111/lam.13287

Setälä, O., Lehtiniemi, M., Coppock, R., & Cole, M. (2018). Microplastics in marine food webs. In Zeng, E. Y. (Ed.), *Microplastic Contamination in Aquatic Environments: An Emerging Matter of Environmental Urgency* (pp. 339–363). Elsevier. DOI: 10.1016/B978-0-12-813747-5.00011-4

Setälä, O., Norkko, J., & Lehtiniemi, M. (2016). Feeding type affects microplastic ingestion in a coastal invertebrate community. *Marine Pollution Bulletin*, 102(1), 95–101. DOI: 10.1016/j.marpolbul.2015.11.053

Sharma, S., Sharma, V., & Chatterjee, S. (2023). Contribution of plastic and microplastic to global climate change and their conjoining impacts on the environment: A review. *The Science of the Total Environment*, 875, 162627. DOI: 10.1016/j.scitotenv.2023.162627

Shen, M., Huang, W., Chen, M., Song, B., Zeng, G., & Zhang, Y. (2020). (Micro) plastic crisis: Un-ignorable contribution to global greenhouse gas emissions and climate change. *Journal of Cleaner Production*, 254, 120138. DOI: 10.1016/j.jclepro.2020.120138

Sjollema, S. B., Redondo-Hasselerharm, P., Leslie, H. A., Kraak, M. H., & Vethaak, A. D. (2016). Do plastic particles affect microalgal photosynthesis and growth? *Aquatic Toxicology (Amsterdam, Netherlands)*, 170, 259–261. DOI: 10.1016/j.aquatox.2015.12.002

Stoks, R., Geerts, A. N., & De Meester, L. (2014). Evolutionary and plastic responses of freshwater invertebrates to climate change: Realized patterns and future potential. *Evolutionary Applications*, 7(1), 42–55. DOI: 10.1111/eva.12108

Su, L., Xiong, X., Zhang, Y., Wu, C., Xu, X., Sun, C., & Shi, H. (2022). Global transportation of plastics and microplastics: A critical review of pathways and influences. *The Science of the Total Environment*, 831, 154884. DOI: 10.1016/j.scitotenv.2022.154884

Swan, S. H. (2008). Environmental phthalate exposure in relation to reproductive outcomes and other health endpoints in humans. *Environmental Research*, 108(2), 177–184. DOI: 10.1016/j.envres.2008.08.007

Talsness, C. E., Andrade, A. J., Kuriyama, S. N., Taylor, J. A., & Vom Saal, F. S. (2009). Components of plastic: Experimental studies in animals and relevance for human health. *Philosophical Transactions of the Royal Society of London. Series B, Biological Sciences*, 364(1526), 2079–2096. DOI: 10.1098/rstb.2008.0281

Tanner, D. J. (1974). *Plastics waste: A technological and economic study*. University of Surrey.

Teuten, E. L., Saquing, J. M., Knappe, D. R., Barlaz, M. A., Jonsson, S., Björn, A., Rowland, S. J., Thompson, R. C., Galloway, T. S., Yamashita, R., Ochi, D., Watanuki, Y., Moore, C., Viet, P. H., Tana, T. S., Prudente, M., Boonyatumanond, R., Zakaria, M. P., Akkhavong, K., & Takada, H. (2009). Transport and release of chemicals from plastics to the environment and to wildlife. *Philosophical Transactions of the Royal Society of London. Series B, Biological Sciences*, 364(1526), 2027–2045. DOI: 10.1098/rstb.2008.0284

Unuofin, J. O. (2020). Garbage in garbage out: The contribution of our industrial advancement to wastewater degeneration. *Environmental Science and Pollution Research International*, 27(18), 22319–22335. DOI: 10.1007/s11356-020-08944-5

Upadhyay, R. K. (2020). Markers for global climate change and its impact on social, biological and ecological systems: A review. *American Journal of Climate Change*, 9(03), 159–203. DOI: 10.4236/ajcc.2020.93012

Vethaak, A. D., & Leslie, H. A. (2016). Plastic debris is a human health issue. *Environmental Science & Technology*, 50(13), 6825–6826. DOI: 10.1021/acs.est.6b02569

Wang, J., Lu, L., Wang, M., Jiang, T., Liu, X., & Ru, S. (2019). Typhoons increase the abundance of microplastics in the marine environment and cultured organisms: A case study in Sanggou Bay, China. *The Science of the Total Environment*, 667, 1–8. DOI: 10.1016/j.scitotenv.2019.02.367

Wang, Y., & Qian, H. (2021). Phthalates and their impacts on human health. *Healthcare (Basel)*, 9(5), 603. DOI: 10.3390/healthcare9050603

Wong, S. L., Ngadi, N., Abdullah, T. A. T., & Inuwa, I. M. (2015). Current state and future prospects of plastic waste as source of fuel: A review. *Renewable & Sustainable Energy Reviews*, 50, 1167–1180. DOI: 10.1016/j.rser.2015.04.063

Wright, S. L., Thompson, R. C., & Galloway, T. S. (2013). The physical impacts of microplastics on marine organisms: A review. *Environmental Pollution*, 178, 483–492. DOI: 10.1016/j.envpol.2013.02.031

Xia, W., Rao, Q., Deng, X., Chen, J., & Xie, P. (2020). Rainfall is a significant environmental factor of microplastic pollution in inland waters. *The Science of the Total Environment*, 732, 139065. DOI: 10.1016/j.scitotenv.2020.139065

Yang, L., Luo, W., Zhao, P., Zhang, Y., Kang, S., Giesy, J. P., & Zhang, F. (2021). Microplastics in the Koshi River, a remote alpine river crossing the Himalayas from China to Nepal. *Environmental Pollution*, 290, 118121. DOI: 10.1016/j.envpol.2021.118121

Yoro, K. O., & Daramola, M. O. (2020). CO_2 emission sources, greenhouse gases, and the global warming effect. In *Advances in Carbon Capture* (pp. 3-28). Woodhead Publishing. DOI: 10.1016/B978-0-12-819657-1.00001-3

Zang, H., Zhou, J., Marshall, M. R., Chadwick, D. R., Wen, Y., & Jones, D. L. (2020). Microplastics in the agroecosystem: Are they an emerging threat to the plant-soil system? *Soil Biology & Biochemistry*, 148, 107926. DOI: 10.1016/j.soilbio.2020.107926

Zhang, Y., Gao, T., Kang, S., & Sillanpää, M. (2019). Importance of atmospheric transport for microplastics deposited in remote areas. *Environmental Pollution*, 254, 112953. DOI: 10.1016/j.envpol.2019.07.121

Zhao, S., Zhu, L., & Li, D. (2016). Microscopic anthropogenic litter in terrestrial birds from Shanghai, China: Not only plastics but also natural fibers. *The Science of the Total Environment*, 550, 1110–1115. DOI: 10.1016/j.scitotenv.2016.01.112

Zheng, J., & Suh, S. (2019). Strategies to reduce the global carbon footprint of plastics. *Nature Climate Change*, 9(5), 374–378. DOI: 10.1038/s41558-019-0459-z

KEY TERMS AND DEFINITIONS

Bioaccumulation: Bioaccumulation is the process by which substances, typically chemicals or pollutants, accumulate within living organisms over time, often reaching higher concentrations than those found in the surrounding environment. This phenomenon occurs as organisms ingest, absorb, or accumulate substances from their food, water, or surroundings at a rate faster than they can metabolize or excrete them.

Biomagnification: Biomagnification is the natural process by which the concentration of a substance, typically a pollutant or toxin, increases as it moves up the food chain. Organisms higher in the food chain tend to have higher concentrations of the toxin than those at lower levels, potentially leading to harmful effects on top predators and overall ecosystem health.

Ecosystem: An ecosystem is a complex community of living organisms, their physical environment, and the intricate interactions between them. It encompasses all biotic (living) organisms and abiotic (non-living) components within a designated area, such as a forest, a coral reef, or a freshwater pond. Ecosystems can vary in size and complexity, but they all share the fundamental characteristics of interconnectedness and interdependence among the organisms and their environment.

Environment: The environment encompasses the entirety of the natural world surrounding us, including the air, water, land, and all living organisms. It encompasses both the physical and biological components of Earth, along with the interactions and relationships between them.

Global Warming: Global warming refers to the long-term increase in Earth's average surface temperature, primarily attributed to human activities that release greenhouse gases into the atmosphere. These gases, such as carbon dioxide (CO_2), methane (CH_4), and nitrous oxide (N_2O), trap heat from the sun, preventing it from escaping back into space and leading to the gradual warming of the planet. Global warming is a significant aspect of climate change and is associated with various impacts, including melting polar ice caps, rising sea levels, more frequent and intense heatwaves, altered precipitation patterns, and shifts in ecosystems and biodiversity.

Greenhouse Gases: Greenhouse gases are gases in the atmosphere that can trap heat and contribute to the greenhouse effect, leading to the warming of the planet. These gases absorb and re-emit infrared radiation emitted by the Earth's surface, preventing some of it from escaping back into space. The main greenhouse gases include carbon dioxide (CO_2), methane (CH_4), nitrous oxide (N_2O), ozone (O_3), water vapor (H_2O), and fluorinated gases.

Nanoplastics: Nanoplastics refer to tiny plastic particles with dimensions typically ranging from 1 to 1000 nanometers (nm), which are smaller than conventional microplastics. These particles can originate from the degradation of larger plastic items or be intentionally manufactured at the nanoscale for various purposes.

Trophic Level: A trophic level is the arrangement of positions in a food chain, or an ecological pyramid occupied by organisms with similar feeding habits and nutritional roles. It represents the hierarchical level of an organism's position in the energy flow within an ecosystem based on its feeding relationships.

Chapter 4
Detection, Monitoring, and Absorption of Micro/Nano-Plastics in Soil System

Rajeev Kumar
https://orcid.org/0000-0002-0820-5970
Manav Rachna International Institute of Research and Studies, India

Jyoti Chawla
Manav Rachna International Institute of Research and Studies, India

Jyoti Syal
Maharishi Markandeshwar Engineering College (Deemed), India

ABSTRACT

Micro/nano plastics (MNPs) pollution in soil system is a big threat to all living organisms. Plastic may alter the physiochemical and biological properties of soil and also has long term effects on biodiversity. Plastic on land is disintegrated into smaller particles having size less than 5 milimeters (micro) or 0.1 micrometer (nano). The interaction between MNPs present in soil and plants affects physiology and morphology of plants. Thus, regular monitoring, detection, and their removal from the soil system are important during plant growth. MNPs have been typically monitored and detected by different analytical methods. FTIR, Raman spectroscopy, pyrolysis coupled with GCMS, UV-visible spectroscopy, scanning electron microscopy, and atomic force microscopy have been widely applied detection and monitoring techniques for MNPs. MNPs have the capability to adsorb various types of contaminants from environment due to hydrophobicity and high surface area

DOI: 10.4018/979-8-3693-3447-8.ch004

to volume ratio. In this chapter, various techniques for monitoring, detection, and adsorption behavior of MNPs have been discussed.

1. INTRODUCTION

Plastics are polymer having some characteristics properties which preferred them in various applications then the other materials such as metal, glass and ceramics. Durability, flexibility and affordable cost make the plastics more popular than others in different application. Most of the products in market are replaced by plastic materials because of these properties. High demand of plastics in market is increasing the risk of plastic pollution in the environment. There are mainly two main sources of micro/nanoplastics in the environment. Primary and secondary sources are responsible for micro/nanoplastic pollution in the environment. Primary sources of MNPs are intentionally prepared for various applications in different fields. However secondary sources of MNPs are derived from chemical, biological or physical breakdown of the large size of plastic materials dumped into different areas. Landfill dumping, sewage sludge, agricultural practices, and uncollected waste are the basic sources of both primary and secondary MNPs in soil. Both primary and secondary MNPs are found in various forms in the soil such as fibre, pellets, films etc. (Fraser et al 2020, Mehdinia et al, 2020, Fan et al 2022).

Plastic after use degrades on terrestrial habitats by variety of processes, such as photodecomposition or hydrolysis. These two processes break the polymers into smaller fragments such as micro/nano-plastics (MNPs) having the size from few micro to nanometers (Gupta et al 2024, Matthews et al 2021, Kumar et al 2023). According to the European Food Safety Authority (EFSA), plastics classified as MPs have a diameter between 0.1 and 5000 micrometers, while NPs have a diameter between 0.001 and 0.1. Size, shape, chemical characteristics, colour and morphology of these micro/nanoplastics play a crucial role towards the environment (Rathod et al 2023, Malinowska et al 2022). Significant impacts have been observed in different ecosystem due to presence of these micro/nanoplastics in aquatic, terrestrial and air medium (Li et al 2023). According to recent studies, it has been observed that MNPs can enter into soil and other terrestrial ecosystems through a variety of routes, including sewage sludge, irrigation, littering, and other means.

Presence of high concentration of MNPs in the soil can potentially permeate all parts of the soil ecosystem which has potential risks to the soil ecosystem's structure and functionality. Because of small size of MNPs, they are readily consumed by a variety of producers and consumers and build up in the existing food chain (Lia et al 2023, Sohail et al 2023). The vast majority of experimental data suggests that MNPs may have a variety of detrimental consequences on the chemical, physical, and biological characteristics of soils. Chemical consequences are leached plastic additives or adsorb/desorb chemicals. However, physical consequences are affected the stability of soil aggregates, density, porosity, and water-bearing capacity. Biological consequences are affected living things, such as microbial populations. Geno-toxicity, oxidative stress, suppression of growth and development, reduced feeding activity, immunological and neurotransmission dysfunction, disturbance of energy metabolism, and endocrine disruption are the various detrimental effects due to accumulation of MNPs in organisms (Guidelli et al 2011, Asefnejad et al 2011).

In order to provide essential supports that meet the needs of both humans and ecosystems, soil is necessary. The production of biomass, maintaining the temperature of earth, environmental purification, and biodiversity protection & preservation, all depend heavily on soil. Soil biodiversity refers to the diversity of subsurface organisms, as defined by the Food and Agriculture Organization of the United Nations (FAO) (Usman et al 2016, Pérez-Reverón et al 2023). This includes genes, species, and their constituent communities, as well as the ecological complexes to which they belong and contribute. Plastic's presence in terrestrial habitat, such as soils, is also very concerning (Renner et al 2018). As a matter of fact, mishandled plastic trash is thought to be anywhere between 4-23 times more than that which is documented for marine habitats (Valiyaveettil et al 2024, Agboola et al 2021,). A staggering quantity of plastic materials is used in conventional farming, including as water pipes, mulching, packing, greenhouse shedding, and seedbeds. When it comes to MPs and NPs, treated wastewater irrigation or the application of compost or sewage sludge are particularly significant. The current chapter objectives are to address the current level of MNPs in soil system, their detection by various technologies, monitoring and adsorption from the system.

2. MICRO/NANO PLASTICS AND DETECTION TECHNIQUES

Monitoring and analysis of micro/nano plastics in environmental matrices is a complex task due to diversity in composition, shapes and size etc. There is a need to integrate the results obtained from specialized techniques for complete physiochemical analysis that depends on the nature of micro/nano plastics. In this section, most suitable techniques available till date for the monitoring and analysis of micro/nano plastics in environmental samples are being discussed along with the associated advantages and challenges for each technique. Commonly preferred techniques include UV-Visible spectroscopy (UVS), Raman spectroscopy (RS), pyrolysis coupled with gas chromatography mass spectrometry (Pyro-GCMS), scanning electron microscopy (SEM), and atomic force microscopy (AFM) (Figure

1). Different techniques used for detection of various MNPs in different samples are summarized in Table 1.

Figure 1. Detection Techniques for MNPs

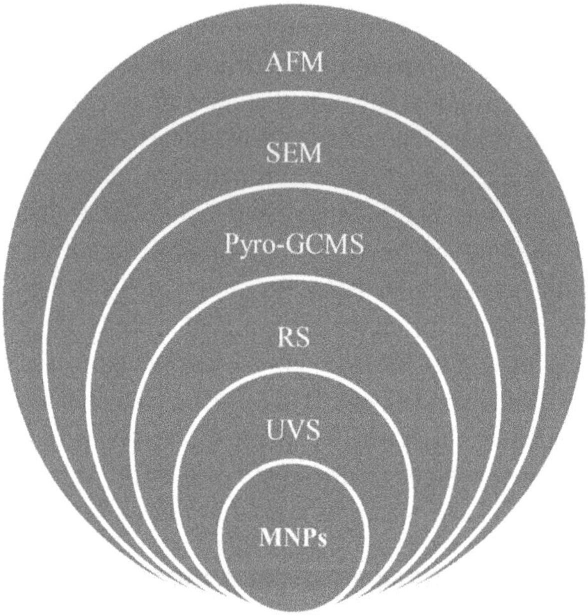

2.1. Fourier Transform Infrared (FTIR)

FTIR spectroscopy is a simple, efficient, and non-destructive approach to identify different types of plastic polymers (Asensio et al 2009, Guidelli et al 2011, Asefnejad et al 2011) and it is mostly used for identification of polymer in soil/marine debris. Microplastics mostly contain polyethylene, polypropylene, polystyrene, and polyvinyl chloride, polyester, cellulose acetate. The FTIR peaks correspond to O–H, C=O & C=C stretching are observed in 700 cm^{-1} - 3700 cm^{-1} (Li et al 2020). FTIR can operate in different modes including reflection, transmission and attenuated total reflection (ATR). Particles with size more than 200 μm can be analysed by attenuated total reflectance Fourier transform infrared spectroscopy (ATR-FTIR). However, in ATR due to applied pressure particles may be affected. ATR-FTIR permits direct and rapid testing of microplastics larger than 200 μm (Fuller and Gautam, 2016). As morphology of microplastics cannot be analysed through FTIR spectra, other tools like scanning electron microscope or atomic force microscope are required for same. Atomic force microscope can also help to analyse nanoplastics (Shim et al 2016). During a study on soil by Helcoski et al 2020 wetland soil samples treated with 30% H_2O_2 &0.500 M Fe(II) were analysed and polystyrene, polyethylene and polypropylene were detected though ATR-FTIR. Scanning electron microscopy was also used for further analysis of soil samples. Wang et al., 2021 also detected microplastics including polystyrene, polyethylene, polyethersulfone, polyamide and polypropylene in soil samples followed by oxidative digestion with 30% H_2O_2 through micro FTIR. Corradini et al., 2019 utilized vis- near infrared spectroscopy in 50-2500 nm range to quantify the microplastics in artificially polluted soil samples. One polymer (low density polyethylene or polyethylene terephthalateor polyvinyl chloride) was added in three different soil samples and one another soil sample was prepared with all three polymers. It was justified that microplastics in soil samples on pollution sites can be detected using this method and no extraction steps were required. The detection limit was found to 15 g kg^{-1}. Jung et al 2018 validated ATR-FTIR for identification of type of plastic in 828 plastic pieces ingested by 50 pacific sea turtles. Particles (96%) were identified as high-density polyethylene, low density polyethylene, unknown poly ethylene, polypropylene, mixture of poly ethylene and poly propylene, nylon, polystyrene, and polyvinyl chloride. Mecozzi et al 2016 used FTIR along with independent component analysis database and Mahalanobis distance to find the type of plastics ingested by loggerhead sea turtles in the Mediterranean Sea. 190 FTIR spectra of plastic samples in a digital database were considered and submitted to Independent Component Analysis and it was found that plastic types were polypropylene, high density polyethylene, low density polyethylene, high density polyethylene terephthalate, low density polyethylene terephthalate, polystyrene, Nylon, polyethylene oxide, and Teflon. It has been further confirmed (Asensio et al 2009, Nishikida and Coates 2003) that low density polyethylene shows a small unique band at. However, no band is seen at 1377 cm^{-1} for high density polyethylene. Also, there is an effect of environmental conditions, aging etc, that may pose certain challenges to identify the type of plastics. During a study by Brandon et al 2016 polypropylene, low density polyethylene, and high-density polyethylene pre-production plastic pellets were weathered for 3 years in different experimental and observed changes with weathering using FTIR spectroscopy. Results obtained experimentally weathered plastics were also compared to plastics collected from oceanic surface waters. Campanale et al 2023 also compared environmental particles and weathered artificial microplastics in controlled light/temperature conditions through FTIR. The degree of ageing was assessed in terms of Carbonyl, Hydroxyl, and Carbon Oxygen Index and confirmed the changes with respected to peaks for hydroxyl groups (3100 to 3700 cm^{-1}), alkenes or carbon double bonds (1600 and 1680 cm^{-1}), and carbonyl groups (1690 and 1810 cm^{-1}) in artificial and natural weathered particles compared to the pristine ones. Andrade et al., 2019 confirmed that FTIR can be used to monitor changes in polyamide, polypropylene and polystyrene microplastics during oceanic aging. In polyamide particles, intensity of 1150 cm^{-1} band increases in FTIR spectra due to oxidation. FTIR spectra of polypropylene changes with weathering and it is confirmed by bands appearing at 3370 and 3240 cm^{-1} (-OH group), 1640 cm^{-1} (>C=O), 1530 cm^{-1} (>C=O), 1440 cm^{-1} (-COOH), 1140 cm^{-1} (alkanes), 1100 cm^{-1} (C–O bond), and 720 cm^{-1} (CH_2) (Fernandez-Gonzalezet al 2019; Tang et al., 2019). The FTIR spectra of weathered polystyrene microplastic showed the appearance of bands at 3360–3240 cm^{-1} (-OH), 1640 cm^{-1} (>C=O), and 1100 cm^{-1} (C-O bond). Small particles (less than 500 nm) can be analysed with the help of micro FTIR in ATR /reflectance mode. It has been further seen during a study of finding plastic polymer types in sea turtle by Jung et al 2018 that ATR needs infrared transparent substrate and reflectance mode is appropriate for thick samples. It was also discussed that in case of uneven particle surface some interference may arise because of refractive error. It has also been emphasized that the method of sample collection affects the micro FTIR analysis. Laser direct infrared is latest tool that uses quantum cascade laser technology and fast scanning optics. It can also be employed for detection of microplastics (Tian et al 2022). Jia et al 2022 used laser direct infrared and Fourier-transform infrared to examine the microplastics in farmland and detected 26 types of polymers microplastics including polyethylene, polypropylene, polyvinyl chloride, polyamide, and polytetrafluoroethylene in more percentage (fibres, pellets or films). Primpke et al 2020 also identified and quantified microplastics in the environment by quantum cascade laser based hyper spectral IR and compared the results with FTIR. Reported studies confirmed that FTIR can be utilized to find the composition of microplastics accurately. Small plastic particles can also be analysis with the help of micro FTIR. However, it's an expensive and little time-consuming tool but provides non-destructive analysis. Refractive error may also affect the accuracy. Other techniques can be combined with FTIR to get more reliable results.

2.2. UV-Visible Spectroscopy

Using UV-VIS-NIR spectroscopy, microplastics including high density polyethylene, polypropylene polystyrene, and polyvinyl chloride can be analysed (Pieniazek et al 2023). Jang et al 2021 measured the maximum wavelength for poly propylene, polyethylene, polyethylene terephthalate and poly methyl methacrylate with UV/Vis/NIR spectrometer in the range of 800-1000 nm. The wavelengths were 920 and 927 nm for poly propylene, 873 and 914 nm for the polyethylene terephthalate, 903 nm for the poly methyl methacrylate, and 933 nm for the polyethylene. Garaba et al 2018 during a study on marine harvested microplastics showed that absorption peak wavelengths of the poly propylene and polyethylene are almost same, and those of the polyethylene terephthalate and poly methyl methacrylate have similar results. Fan et al 2022 used UV-Visible spectrophotometry to develop quantification procedure for analysing the polyethylene terephthalate in the environmental samples.

2.3. Raman Spectroscopy

Raman spectroscopy is a reliable technique based on inelastic light scattering that use laser for excitation for structural analysis of samples under investigation. Raman spectroscopic studies related to compositional analysis of plastic microparticles have indicated the presence of polystyrene, polyethylene terephthalate, polyethylene, and polypropylene in environmental water samples. Characteristic peaks unique to specific molecules are obtained in Raman spectra. Micro Raman spectroscopy use Raman spectrometer along with an optical microscope to analyse the samples sub-micron scales. Raman Spectroscopy can provide resolution (up to 1μm) with wide spectral range and thinner bands in comparison to FTIR spectroscopy. Raman Spectroscopic analysis may lead to increase in temperature of polymer sample under investigation and its degradation as it involves the use of laser ((Ivleva et al, 2017, Silva et al, 2018,). Liu et al 2024 confirmed during Raman spectroscopic studies that different types of plastics shows unique peaks. Peaks for polythene at 1067, 1132 and 1308 cm^{-1} indicated C-C stretching and CH_2 twisting whereas for polypropylene peak at 808 cm^{-1} was there for C-C stretching and CH_2 rocking. Polytetrafluoroethylene showed bands at 732 and 1380 cm^{-1} which indicated CF_2 and CF stretching. Similarly, unique bands were reported for polystyrene and polyvinyl chloride. Ruan et al 2024 introduced a sol-based surface enhanced Raman spectroscopic approach for the detection of micro and Nano plastics in aqueous samples. It has been highlighted that particles of size less than 1 μm cannot be analysed accurately and the presence of organic materials on the surface of microparticles also interfere with the result. Feng et al 2020 also confirmed the presence of microplastics in soil using Raman spectroscopy in 800 - 2900 cm^{-1} range through peaks for C–O/ C=O stretching and CH_2 bending. Xie et al 2023 identified microplastics using surface enhanced Raman spectroscopy. It has been highlighted that the Raman spectroscopy cannot identify trace substances reliably and presence of non-plastic compounds also interfere with the results. This tool can be suitable for analysing the nonpolar bond vibration of the same atom where as for the polar bond vibration of different atoms infrared is suitable (Phan et al 2022) and both Raman and infrared spectroscopy can be uses together for better analysis.

2.4 Pyrolysis Coupled With Gas Chromatography Mass Spectrometry (Pyro-GCMS)

Pyrolysis coupled with gas chromatography-mass spectrometry (Pyro-GCMS) has great potential for analysis of various complicated compounds and environmental particles. The primary use of Pyro-GCMS is the chemical identification of macromolecules that are not possible by LC and GC techniques because of their size. Macromolecules or polymers are reduced to smaller molecules through pyrolysis, a controlled thermal decomposition process, in the absence of oxygen, combining gas chromatography and mass spectrometric (GC/MS) technologies to produce volatile and semi-volatile compounds that can be studied quickly (Pico et al 2020). Pyrolysis, in turn, makes use of mass spectrometry's identification capabilities and gas chromatography's resolving power for non-volatile macromolecules or polymers.

The sample is heated to a relatively low pressure in the pyrolyzer using GC carrier gas, which is normally an inert gas (Helium) (Li et al 2024). The polymer or macromolecules undergo increasing bond breakdown as a result of the high temperatures, which ranges from ranked from worst to strongest based on temperature. The primary reactions that take place are random excision, de-polymerization, and side group removal. Minor but conceivable reactions include oxidation, isomerization, hydrogenation, cyclization, cross-linking, and chard formation (Pico et al 2020). The capacity of the free radicals produced to produce stable products and the relative strengths of the molecular bonds determine how molecules fragment during pyrolysis and the products that are produced. Currently, various Pyro-GCMS operating modes have been described based on the analysis's goal (Figure 2).

Figure 2. Pyro-GCMS various operating modes

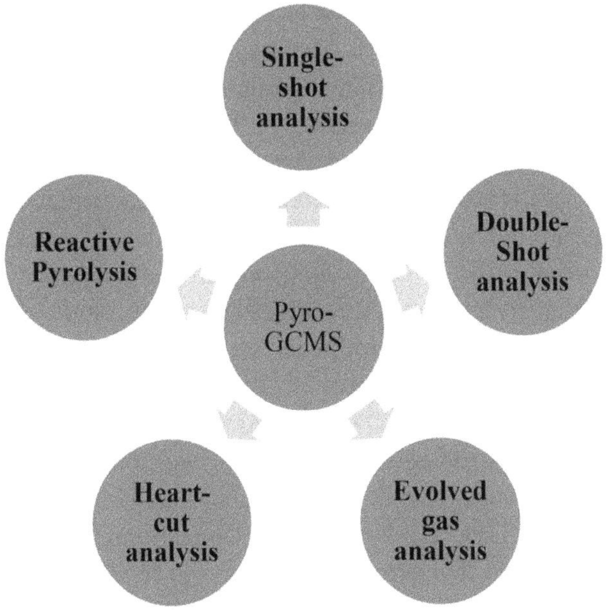

Double-shot analysis involved pyrolysis at two distinct temperatures, whereas single-shot analysis involved pyrolysis at a single temperature (>500°C) (Mattonai et al 2020). EGA was used to obtain a thermogram for heart-cut analysis, which involved selectively introducing evaporated components to a GC column where they were temporarily trapped at the beginning of the column before being analyzed by GC-MS. Reactive pyrolysis involved subjecting macromolecules to a chemical derivatization reaction in the pyrolysis chamber. At low and high temperatures, evolved gas analysis involved separation of degradation products from macromolecules according to the temperature at which they are formed rather than according to their volatilization temperature. After further examination, it was discovered that soil contained styrene butadiene rubber, polyvinyl acetate, and poly(acrylonitrile butadiene styrene) (Corradini et al 2019). Only microplastics derived from polystyrene derivative products may be identified using this approach. To create molecular identifiers for a range of polymers, more research is required (Watteau et al 2018).

2.5 Scanning Electron Microscopy (SEM)

Scanning electron microscopy is an analytical method frequently used to examine micro-nanoplastic. It provides sharp, multi-magnified images even for minuscule particles. High-resolution image show that organic particles and microplastics can be easily distinguished by their different particle surface textures. Examining the shape, composition, and surface properties of these particles are its primary goal. The elemental composition is determined by additional investigation using energy-dispersive X-ray spectroscopy (EDX). Additive components like Ca, Na, Al, Mg, and Si are commonly collected as markers in order to identify microplastics using EDX (Watteau et al 2018). It is possible to distinguish between inorganic and carbon-rich polymers by looking at the surface elemental composition of the particle. The time and high cost of the equipment limit the number of samples that can be processed by SEM. Pico and Barcelo 2022, have been analyzed MNPs in green and sustainable ways. SEM with EDX was used to analyze the shape, composition, and surface properties of the present polymers in the soil. The size of polymers was determined by SEM. Qi et al 2024 have been studied Single particle detection of MNPs using SEM. Silva et al 2022 have been analyzed MNPs using SEM for identification the size and characteristics properties of MNPs. Naji et al. 2019 have applied new technology for detection of MNPs. FESEM, which operates at low voltage and enables the acquisition of high-quality, high-magnification pictures of MNPs samples without the need for sample preparation prior to observation. This method is rapid and simple since the MNPs pieces are deposited directly into the carbon tape on the aluminum stub, eliminating the need to cover the sample with metal or carbon. However, in this instance, an EDX analysis cannot be made in addition to utilizing the low voltage used.

2.6 Atomic Force Microscopy (AFM)

Atomic force microscopy (AFM) is widely used techniques for characterization of MNPs in different samples. It is widely applied as detection techniques for non-conductive materials. A tiny, extremely rigid conductive material tip is attached to the end of a rod or micro-lever (cantilever) in AFM, and the tip presses down on the sample while the measurement is being taken. A piezoelectric mechanism causes the cantilever to move on a sample's surface as it moves along the three Cartesian axes. This technique provides high-resolution three-dimensional image very quickly. AFM is better than SEM for characterization of specific materials. Because AFM only requires basic sample preparation, it maintains the sample surface. AFM is a useful tool for studying the surface of non-conductive polymers such as MNPs. Direct three-dimensional images of the polymer's surface structure are provided using

AFM. AFM has been used extensively in the characterization and identification of size and surface characteristics of various MNPs in different samples (Sohail et al 2023, Enfrin et al 2021, Fan et al 2023, Thaiba al 2023). Many researchers have been applied SEM, AFM and TEM techniques for surface morphology, size of particles and nature of the particles. These techniques are very helpful for determination of size and surface morphology. The type of elements and their abundance have also been identified using EDX.

3. MONITORING OF MICRO/NANO-PLASTICS

On the side of the road, in apartment buildings, and in park different types of plastics in different forms from domestic wastes were found in the different analysis. Plastic bags, rubber, PVC materials, bottles, poly-package, polyethylene, etc are more common plastics on roadside and agricultural land and are detected more frequently than in parks and forest areas. Their findings verified that one of the main factors influencing plastic pollution was the frequency of human and agricultural activities. While the levels of microplastics found in parks and residential areas were in the mid-range, those found in agricultural land and roadside soils displayed statistically significant variations from those found in forest soils. Roadside and agricultural area soils had the largest percentages of microplastics greater than 1 mm (20% and 17%, respectively), whereas forest soils had the lowest percentage (1%) (Yoon et al 2024). Many research studies had found the largest proportion of microplastic in agricultural land, roadside and residential areas (Wang et al 2021 and Choi et al 2021). Physical forces by the living beings are the prime force to break these plastics into microplastics in residential, agricultural and road side areas. Plastics larger than 1 mm were found in modest quantities in forest soils places that are rarely visited by humans due to low trash input. Despite the large percentage of microscopic particles, the plastic came from either plastic fallout or plastic that was already there and took a long time to disintegrate. MNPs are dispersed by wind to aquatic ecosystems or deposited on agricultural surfaces after being carried to remote locations by atmospheric circulation. Rainwater runoff and air dust fall are two ways that MNPs can get into the soil. MNP introduction into soils is further enhanced by practices such as sewage discharge, irrigation, and agricultural activity (Li et al 2023).

Micro/Nano particles in different forms including fragments/fibres/films/pellets have been detected in water, soil, and air. The Microplastics mostly contains polyethylene, polypropylene, polystyrene, and polyvinyl chloride, polyester, cellulose acetate, polyethylene terephthalate, poly vinyl alcohol and polyethersulfone. Feng et al 2020 conducted a study related to monitoring of microplastics in soil and water in a remote region of Tibetan Plateau (China) simultaneously and detected microplastics in surface water, sediment & soil in range of 66.6-733.3 number/ m^3, 20 -160 items/kg, and 20 -110 items/kg, respectively in terms of abundance. In surface water fibres were significantly observes whereas film was observed in soil. Compositional studies of microplastics in water and soil samples confirmed the presence of polypropylene and polyethylene. Mason et al., 2018 during a study related to presence of microplastic in drinking water (Bottled water of 11 brands from 9 countries) confirmed that 93% of samples contain microplastics. Fifty percent of the particles were found to be polymeric as confirmed through FTIR spectroscopy and polypropylene was the most common (54%). Polypropylene is used in bottle caps. 95% of particles were having 6.5- 100 μm in size. Figure 3 represents the ways through which micro/nano plastics enter the different environment segments. Jia et al 2022 used laser direct infrared and Fourier-transform infrared to examine the microplastics in farmland and showed the total abundance of microplastics as $1.98 \pm 0.41 \times 10^5$, $1.57 \pm 0.28 \times 10^5$, $1.78 \pm 0.27 \times 10^5$, and $3.20 \pm 0.41 \times 10^5$ particles/kg soil in cotton fields with film mulching of 5, 10, 20, and ore than30 years, respectively. Results indicated that microplastics of size 10 - 500 μm are of 96.5-99.9% of the total microplastics present in the soil.

Figure 3. Routes of entering MNPs in different environment segments

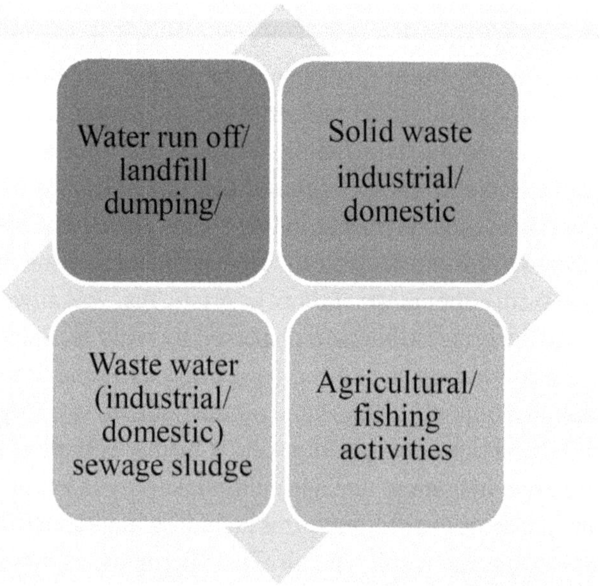

4. ADSORPTION BEHAVIOR OF MNPS

MNPs are widely found in a wide range of terrestrial ecosystem i.e. food chains and food web. Soil characteristics can be impacted by the presence of various types of MNPs (Gupta et al 2023). Through direct adsorption, it influences the enzymatic activity of the soil. MNPs have been found in a vast variety of flora and fauna. MNPs are found in flora and fauna through mechanical transport, and their size and densities are comparable to those of the minerals absorbed by plants (Astner et al 2023). Living organisms are susceptible to a range of metabolic and reproductive problems due to consumption of MNPs (Matthews et al 2021, Russo et al 2023, Giri et al 2024). Various contaminants, heavy metals, pharmaceutical materials, organic and inorganic compounds are easily adsorbed at MNPs and entered into ecosystem. Size and shape of MNPs play a crucial role for adsorption of these materials on MNPs. Environmental factor, soil pH, salinity of soil, presence of other materials in the soil are responsible for adsorption of these materials on MNPs (Kumar et al 2023) (Figure 4). MNPs also act as carrier for travel of microorganism in food chain. MNPs have ability to adsorb and desorb the various materials. These properties make them for bioaccumulation, decomposition and migration of various contaminants.

Wang et al 2024 have reported that THP-1 cells had the ability to ingest 100 nm polystyrene nanoplastics and selectively adsorbed intracellular proteins. Using a proteomics technique, 773 proteins were found to be adsorbed onto NPs with great reliability. These proteins were then evaluated using bioinformatics to predict the distribution and route of NPs after they have been internalized into cells. In order to better understand protein adsorption onto NPs and its biological implications, more research was done on the representative proteins found through Kyoto Encyclopedia of Genes and Genomes pathway analysis. The results of the analysis showed that NPs primarily internalize into cells through clathrin-mediated endocytosis with concurrent clathrin heavy chain adsorption, influence glycolysis through pyruvate kinase M (PKM) adsorption, and initiate the unfolded protein response through the adsorption of ribophorin 1 (RPN1) and heat shock 70 protein 8 (HSPA8). Liao et al 2023 have reported that polystyrene nanoplastics and di(2-ethylhexyl) phthalate trophically transferred among algae-crustacean-fish species such as Chlorella pyrenoidosa, Daphnia magna, and Micropterus salmoides; fish lipid metabolism at a higher trophic level. MNPs containing adsorbed materials when penetrate an organism, they may release the adsorbed materials in cell. These ingested MNPs may be more easily transferred into the circulatory systems. Thus, these contaminants have harmful effects in different part of the organisms.

Figure 4. Factor affecting adsorption behavior of MNPs

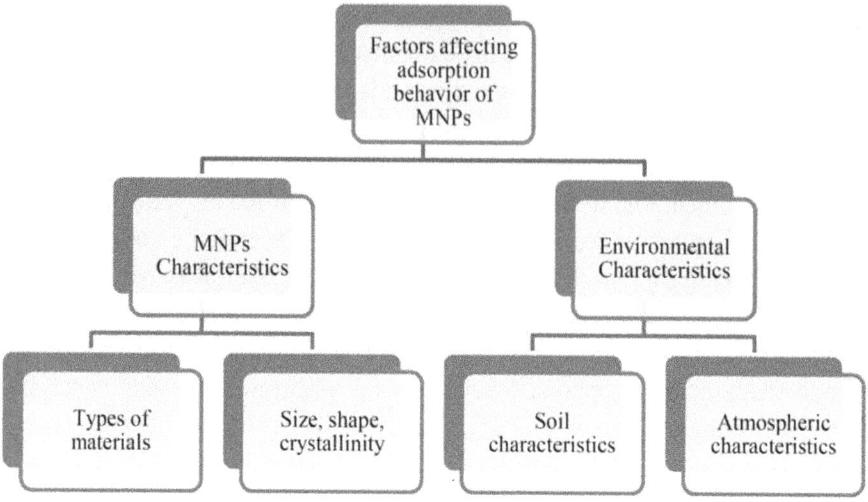

Comprehending the adsorption behavior and interplay between MNPs is crucial for assessing the impact on the environment, transportation, ecological consequences, and enabling risk assessment and regulatory frameworks. It makes it possible for researchers, decision-makers, and environmentalists to more successfully address the problems brought on by MNPs

5. CONCLUSION

Micro/nano plastics (MNPs) have potential threat towards the environment as well as all living beings. Presence of MNPs in soil affects the soil quality, fertility, and other properties. Various contaminants present in soil can be easily adsorbed on the surface of these MNPs and enter into living organism cell. Releasing of these contaminants in cell has toxicological effects in living organisms. Pyro-GCMS, IR, Raman, SEM, AFM, EDX are effectively used techniques for detection and determination of these MNPs in soil. Various research groups have been working for reducing the concentration of these MNPs in soil system. Controlled photo-degradation is one the best solution for mitigation of the MNPs from soil. Photo-degradation in different ways has some positive effect in the soil. Reducing plastics materials in daily life is another best option to reduce the concentration of MNPs in terrestrial and aquatic system. Public awareness towards plastic use is very important to achieve this goal.

Table 1. Different types of detection techniques used for different types of micro/nano-plastics in different samples

S.No.	Study country	Study area	Type of polymer	Detection technique	Reference
1	China	River and bay	PE, PET, PS	GC-MS	Fraser et al 2020
2	Nigeria	Beaches	PET, PS, PP	GC-MS	Benson and Fred-Ahmadu 2020
3	Germany	Seawater	LDPE	GC-MS	O'Donovan et al 2018
4	South Korea	Soil	PVC, PU	FTIR	Choi et al 2021
5	Northwestern Pacific	Sea water	PE, PP, NYLON	Raman spectroscopic	Pan et al 2019
6	Caspian Sea	Sediment	PS, PE	Raman spectroscopic	Mehdinia et al 2020
7	Malaysia	fish meals	PE, PP	Raman spectroscopic	Karbalaei et al 2020
8	Germany	Soil sample	PE, PP, PS	Py-GC/MS	Steinmetz et al 2020.
9	France	Municipal Solid Waste Composts	MP	Py-GC/MS	Watteau et al 2018
10	Taiwan	Environmental sample	Nylon, PE	UV-Visible	Fan et al 2022
11	India	Beach Sand	PE, PVC, PS	SEM	Tiwari et al 2019
12	China	Drinking water plant	PE, PVC	AFM	Li et a. 2024

REFERENCES

Agboola, O. D., & Benson, N. U. (2021). Physisorption and chemisorption mechanisms influencing micro (nano) plastics-organic chemical contaminants interactions: A review. *Frontiers in Environmental Science*, 9, 678574. DOI: 10.3389/fenvs.2021.678574

Andrade, J., Fernández-González, V., López-Mahía, P., & Muniategui, S. (2019). A low-cost system to simulate environmental microplastic weathering. *Marine Pollution Bulletin*, 149, 110663. DOI: 10.1016/j.marpolbul.2019.110663

Andrade, M. C., Winemiller, K. O., Barbosa, P. S., Fortunati, A., Chelazzi, D., Cincinelli, A., & Giarrizzo, T. (2019). First account of plastic pollution impacting freshwater fishes in the Amazon: Ingestion of plastic debris by piranhas and other serrasalmids with diverse feeding habits. *Environmental Pollution*, 244, 766–773. DOI: 10.1016/j.envpol.2018.10.088

Benson, N. U., Fred-Ahmadu, O. H., Ekett, S. I., Basil, M. O., Adebowale, A. D., Adewale, A. G., & Ayejuyo, O. O. (2020). Occurrence, depth distribution and risk assessment of PAHs and PCBs in sediment cores of Lagos lagoon, Nigeria. *Regional Studies in Marine Science*, 37, 101335. DOI: 10.1016/j.rsma.2020.101335

Campanale, C., Savino, I., Massarelli, C. and Uricchio, V.F., 2023. Fourier transform infrared spectroscopy to assess the degree of alteration of artificially aged and environmentally

Choi, Y. R., Kim, Y. N., Yoon, J. H., Dickinson, N., & Kim, K. H. (2021). Plastic contamination of forest, urban, and agricultural soils: A case study of Yeoju City in the Republic of Korea. *Journal of Soils and Sediments*, 21(5), 1962–1973. DOI: 10.1007/s11368-020-02759-0

Corradini, F., Bartholomeus, H., Lwanga, E. H., Gertsen, H., & Geissen, V. (2019). Predicting soil microplastic concentration using vis-NIR spectroscopy. *The Science of the Total Environment*, 650, 922–932. DOI: 10.1016/j.scitotenv.2018.09.101

Enfrin, M., Hachemi, C., Hodgson, P. D., Jegatheesan, V., Vrouwenvelder, J., Callahan, D. L., Lee, J., & Dumée, L. F. (2021). Nano/micro plastics–Challenges on quantification and remediation: A review. *Journal of Water Process Engineering*, 42, 102128. DOI: 10.1016/j.jwpe.2021.102128

Fan, C., Huang, Y. Z., Lin, J. N., & Li, J. (2022). Microplastic quantification of nylon and polyethylene terephthalate by chromic acid wet oxidation and ultraviolet spectrometry. *Environmental Technology & Innovation*, 28, 102683. DOI: 10.1016/j.eti.2022.102683

Feng, S., Lu, H., Tian, P., Xue, Y., Lu, J., Tang, M., & Feng, W. (2020). Analysis of microplastics in a remote region of the Tibetan Plateau: Implications for natural environmental response to human activities. *The Science of the Total Environment*, 739, 140087. DOI: 10.1016/j.scitotenv.2020.140087

Fraser, M. A., Chen, L., Ashar, M., Huang, W., Zeng, J., Zhang, C., & Zhang, D. (2020). Occurrence and distribution of microplastics and polychlorinated biphenyls in sediments from the Qiantang River and Hangzhou Bay, China. *Ecotoxicology and Environmental Safety*, 196, 110536. DOI: 10.1016/j.ecoenv.2020.110536

Fuller, S., & Gautam, A. (2016). A procedure for measuring microplastics using pressurized fluid extraction. *Environmental Science & Technology*, 50(11), 5774–5780. DOI: 10.1021/acs.est.6b00816

Garaba, S. P., & Dierssen, H. M. (2018). An airborne remote sensing case study of synthetic hydrocarbon detection using short wave infrared absorption features identified from marine-harvested macro-and microplastics. *Remote Sensing of Environment*, 205, 224–235. DOI: 10.1016/j.rse.2017.11.023

Giri, S., Dimkpa, C. O., Ratnasekera, D., & Mukherjee, A. (2024). Impact of micro and nano plastics on phototrophic organisms in freshwater and terrestrial ecosystems: A review of exposure, internalization, toxicity mechanisms, and eco-corona-dependent mitigation. *Environmental and Experimental Botany*, 219, 105666. DOI: 10.1016/j.envexpbot.2024.105666

Gupta, N., Parsai, T., & Kulkarni, H. V. (2024). A review on the fate of micro and nano plastics (MNPs) and their implication in regulating nutrient cycling in constructed wetland systems. *Journal of Environmental Management*, 350, 119559. DOI: 10.1016/j.jenvman.2023.119559

Helcoski, R., Yonkos, L. T., Sanchez, A., & Baldwin, A. H. (2020). Wetland soil microplastics are negatively related to vegetation cover and stem density. *Environmental Pollution*, 256, 113391. DOI: 10.1016/j.envpol.2019.113391

Jang, S., Kim, J. H., & Kim, J. (2021). Detection of Microplastics in Water and Ice. *Remote Sensing (Basel)*, 13(17), 3532. DOI: 10.3390/rs13173532

Jia, W., Karapetrova, A., Zhang, M., Xu, L., Li, K., Huang, M., Wang, J., & Huang, Y. (2022). Automated identification and quantification of invisible microplastics in agricultural soils. *The Science of the Total Environment*, 844, 156853. DOI: 10.1016/j.scitotenv.2022.156853

Jung, M. R., Horgen, F. D., Orski, S. V., Rodriguez, V., Beers, K. L., Balazs, G. H., Jones, T. T., Work, T. M., Brignac, K. C., Royer, S. J., & Hyrenbach, K. D. (2018). Validation of ATR FT-IR to identify polymers of plastic marine debris, including those ingested by marine organisms. *Marine Pollution Bulletin*, 127, 704–716. DOI: 10.1016/j.marpolbul.2017.12.061

Karbalaei, S., Golieskardi, A., Watt, D. U., Boiret, M., Hanachi, P., Walker, T. R., & Karami, A. (2020). Analysis and inorganic composition of microplastics in commercial Malaysian fish meals. *Marine Pollution Bulletin*, 150, 110687. DOI: 10.1016/j.marpolbul.2019.110687

Kumar, V., Singh, E., Singh, S., Pandey, A., & Bhargava, P. C. (2023). Micro-and nano-plastics (MNPs) as emerging pollutant in ground water: Environmental impact, potential risks, limitations and way forward towards sustainable management. *Chemical Engineering Journal*, 459, 141568. DOI: 10.1016/j.cej.2023.141568

Li, J., Song, Y., & Cai, Y. (2020). Focus topics on microplastics in soil: Analytical methods, occurrence, transport, and ecological risks. *Environmental Pollution*, 257, 113570. DOI: 10.1016/j.envpol.2019.113570

Li, S., Wang, T., Guo, J., Dong, Y., Wang, Z., Gong, L., & Li, X. (2021). Polystyrene microplastics disturb the redox homeostasis, carbohydrate metabolism and phytohormone regulatory network in barley. *Journal of Hazardous Materials*, 415, 125614. DOI: 10.1016/j.jhazmat.2021.125614

Li, T., Cui, L., Xu, Z., Liu, H., Cui, X., & Fantke, P. (2023). Micro-and nanoplastics in soil: Linking sources to damage on soil ecosystem services in life cycle assessment. *The Science of the Total Environment*, 904, 166925. DOI: 10.1016/j.scitotenv.2023.166925

Li, Y., Zhang, C., Tian, Z., Cai, X., & Guan, B. (2024). Identification and quantification of nanoplastics (20–1000 nm) in a drinking water treatment plant using AFM-IR and Pyr-GC/MS. *Journal of Hazardous Materials*, 463, 132933. DOI: 10.1016/j.jhazmat.2023.132933

Liao, H., Gao, D., Kong, C., Junaid, M., Li, Y., Chen, X., Zheng, Q., Chen, G., & Wang, J. (2023). Trophic transfer of nanoplastics and di (2-ethylhexyl) phthalate in a freshwater food chain (Chlorella pyrenoidosa-Daphnia magna-Micropterus salmoides) induced disturbance of lipid metabolism in fish. *Journal of Hazardous Materials*, 459, 132294. DOI: 10.1016/j.jhazmat.2023.132294

Liu, K., Pang, X., Chen, H., & Jiang, L. (2024). Visual detection of microplastics using Raman spectroscopic imaging. *Analyst*, 149(1), 161–168. DOI: 10.1039/D3AN01270K

Liu, Z., Wang, G., Sheng, C., Zheng, Y., Tang, D., Zhang, Y., Hou, X., Yao, M., Zong, Q. & Zhou, Z. (2024). Intracellular Protein Adsorption Behavior and Biological Effects of Polystyrene Nanoplastics in THP-1 Cells. *Environmental Science & Technology*.

Malinowska, K., Bukowska, B., Piwoński, I., Foksiński, M., Kisielewska, A., Zarakowska, E., Gackowski, D., & Sicińska, P. (2022). Polystyrene nanoparticles: The mechanism of their genotoxicity in human peripheral blood mononuclear cells. *Nanotoxicology*, 16(6-8), 791–811. DOI: 10.1080/17435390.2022.2149360

Mariano, S., Tacconi, S., Fidaleo, M., Rossi, M., & Dini, L. (2021). Micro and nanoplastics identification: Classic methods and innovative detection techniques. *Frontiers in Toxicology*, 3, 636640. DOI: 10.3389/ftox.2021.636640

Mason, S. A., Welch, V. G., & Neratko, J. (2018). Synthetic polymer contamination in bottled water. *Frontiers in Chemistry*, 6, 389699. DOI: 10.3389/fchem.2018.00407

Matthews, S., Mai, L., Jeong, C. B., Lee, J. S., Zeng, E. Y., & Xu, E. G. (2021). Key mechanisms of micro-and nanoplastic (MNP) toxicity across taxonomic groups. *Comparative Biochemistry and Physiology. Toxicology & Pharmacology : CBP*, 247, 109056. DOI: 10.1016/j.cbpc.2021.109056

Mattonai, M., Watanabe, A., & Ribechini, E. (2020). Characterization of volatile and non-volatile fractions of spices using evolved gas analysis and multi-shot analytical pyrolysis. *Microchemical Journal*, 159, 105321. DOI: 10.1016/j.microc.2020.105321

Mecozzi, M., Pietroletti, M., & Monakhova, Y. B. (2016). FTIR spectroscopy supported by statistical techniques for the structural characterization of plastic debris in the marine environment: Application to monitoring studies. *Marine Pollution Bulletin*, 106(1-2), 155–161. DOI: 10.1016/j.marpolbul.2016.03.012

Mehdinia, A., Dehbandi, R., Hamzehpour, A., & Rahnama, R. (2020). Identification of microplastics in the sediments of southern coasts of the Caspian Sea, north of Iran. *Environmental Pollution*, 258, 113738. DOI: 10.1016/j.envpol.2019.113738

Naji, A., Nuri, M., Amiri, P., & Niyogi, S. (2019). Small microplastic particles (S-MPPs) in sediments of mangrove ecosystem on the northern coast of the Persian Gulf. *Marine Pollution Bulletin*, 146, 305–311. DOI: 10.1016/j.marpolbul.2019.06.033

O'Donovan, S., Mestre, N. C., Abel, S., Fonseca, T. G., Carteny, C. C., Cormier, B., Keiter, S. H., & Bebianno, M. J. (2018). Ecotoxicological effects of chemical contaminants adsorbed to microplastics in the clam Scrobicularia plana. *Frontiers in Marine Science*, 5, 143. DOI: 10.3389/fmars.2018.00143

Pan, Z., Guo, H., Chen, H., Wang, S., Sun, X., Zou, Q., Zhang, Y., Lin, H., Cai, S., & Huang, J. (2019). Microplastics in the Northwestern Pacific: Abundance, distribution, and characteristics. *The Science of the Total Environment*, 650, 1913–1922. DOI: 10.1016/j.scitotenv.2018.09.244

Pérez-Reverón, R., Álvarez-Méndez, S. J., González-Sálamo, J., Socas-Hernández, C., Díaz-Peña, F. J., Hernández-Sánchez, C., & Hernández-Borges, J. (2023). Nanoplastics in the soil environment: Analytical methods, occurrence, fate and ecological implications. *Environmental Pollution*, 317, 120788. DOI: 10.1016/j.envpol.2022.120788

Phan, S., Padilla-Gamiño, J. L., & Luscombe, C. K. (2022). The effect of weathering environments on microplastic chemical identification with Raman and IR spectroscopy: Part I. polyethylene and polypropylene. *Polymer Testing*, 116, 107752. DOI: 10.1016/j.polymertesting.2022.107752

Picó, Y., & Barceló, D. (2022). Micro (Nano) plastic analysis: A green and sustainable perspective. *Journal of Hazardous Materials Advances*, 6, 100058. DOI: 10.1016/j.hazadv.2022.100058

Pieniazek, L. S., McKinney, M., & Carr, J. 2023, October. Quantification of Microplastics in Soil Using UV-VIS-NIR Spectroscopy with Ultra High Resolution. In *ASA, CSSA, SSSA International Annual Meeting*. ASA-CSSA-SSSA.

Primpke, S., Godejohann, M., & Gerdts, G. (2020). Rapid identification and quantification of microplastics in the environment by quantum cascade laser-based hyperspectral infrared chemical imaging. *Environmental Science & Technology*, 54(24), 15893–15903. DOI: 10.1021/acs.est.0c05722

Qi, G., Zhao, L., Liu, J., Tian, C., & Zhang, S. (2024). Single particle detection of micro/nano plastics based on recyclable SERS sensor with two-dimensional AuNPs thin films. *Materials Today. Communications*, 38, 108293. DOI: 10.1016/j.mtcomm.2024.108293

Rathod, N. B., Xavier, K. M., Özogul, F., & Phadke, G. G. (2023). Impacts of nano/micro-plastics on safety and quality of aquatic food products. *Advances in Food and Nutrition Research*, 103, 1–40. DOI: 10.1016/bs.afnr.2022.07.001

Renner, G., Schmidt, T. C., & Schram, J. (2018). Analytical methodologies for monitoring micro (nano) plastics: Which are fit for purpose? *Current Opinion in Environmental Science & Health*, 1, 55–61. DOI: 10.1016/j.coesh.2017.11.001

Ruan, X., Xie, L., Liu, J., Ge, Q., Liu, Y., Li, K., You, W., Huang, T., & Zhang, L. (2024). Rapid detection of nanoplastics down to 20 nm in water by surface-enhanced raman spectroscopy. *Journal of Hazardous Materials*, 462, 132702. DOI: 10.1016/j.jhazmat.2023.132702

Russo, M., Oliva, M., Hussain, M. I., & Muscolo, A. (2023). The hidden impacts of micro/nanoplastics on soil, crop and human health. *Journal of Agriculture and Food Research*, 14, 100870. DOI: 10.1016/j.jafr.2023.100870

Shim, W. J., Song, Y. K., Hong, S. H., & Jang, M. (2016). Identification and quantification of microplastics using Nile Red staining. *Marine Pollution Bulletin*, 113(1-2), 469–476. DOI: 10.1016/j.marpolbul.2016.10.049

Silva, A. L. P., Silva, S. A., Duarte, A., Barceló, D., & Rocha-Santos, T. (2022). Analytical methodologies used for screening micro (nano) plastics in (eco) toxicity tests. *Green Analytical Chemistry*, 3, 100037. DOI: 10.1016/j.greeac.2022.100037

Sohail, M., Urooj, Z., Noreen, S., Baig, M. M. F. A., Zhang, X., & Li, B. (2023). Micro-and nanoplastics: Contamination routes of food products and critical interpretation of detection strategies. *The Science of the Total Environment*, 891, 164596. DOI: 10.1016/j.scitotenv.2023.164596

Steinmetz, Z., Kintzi, A., Muñoz, K., & Schaumann, G. E. (2020). A simple method for the selective quantification of polyethylene, polypropylene, and polystyrene plastic debris in soil by pyrolysis-gas chromatography/mass spectrometry. *Journal of Analytical and Applied Pyrolysis*, 147, 104803. DOI: 10.1016/j.jaap.2020.104803

Tang, P., Forster, R., McCumskay, R., Rogerson, M. & Waller, C. (2019). Handheld FT-IR spectroscopy for the triage of micro-and meso-sized plastics in the marine environment incorporating an accelerated weathering study and an aging estimation.

Tian, X., Beén, F., & Bäuerlein, P. S. (2022). Quantum cascade laser imaging (LDIR) and machine learning for the identification of environmentally exposed microplastics and polymers. *Environmental Research*, 212, 113569. DOI: 10.1016/j.envres.2022.113569

Tiwari, M., Rathod, T. D., Ajmal, P. Y., Bhangare, R. C., & Sahu, S. K. (2019). Distribution and characterization of microplastics in beach sand from three different Indian coastal environments. *Marine Pollution Bulletin*, 140, 262–273. DOI: 10.1016/j.marpolbul.2019.01.055

Usman, S., Muhammad, Y., & Chiroman, A. (2016). Roles of soil biota and biodiversity in soil environment–A concise communication. *Eurasian Journal of Soil Science*, 5(4), 255–265. DOI: 10.18393/ejss.2016.4.255-265

Valiyaveettil Salimkumar, A., Kurisingal Cleetus, M. C., Ehigie, J. O., Onogbosele, C. O., Nisha, P., Kumar, B. S., Prabhakaran, M. P., & Rejish Kumar, V. J. (2024). *Adsorption Behavior and Interaction of Micro-Nanoplastics in Soils and Aquatic Environment. Management of Micro and Nano-plastics in Soil and Biosolids: Fate*. Occurrence, Monitoring, and Remedies.

Wang, J., Li, J., Liu, S., Li, H., Chen, X., Peng, C., Zhang, P., & Liu, X. (2021). Distinct microplastic distributions in soils of different land-use types: A case study of Chinese farmlands. *Environmental Pollution*, 269, 116199. DOI: 10.1016/j.envpol.2020.116199

Watteau, F., Dignac, M. F., Bouchard, A., Revallier, A., & Houot, S. (2018). Microplastic detection in soil amended with municipal solid waste composts as revealed by transmission electronic microscopy and pyrolysis/GC/MS. *Frontiers in Sustainable Food Systems*, 2, 81. DOI: 10.3389/fsufs.2018.00081

Xie, L., Gong, K., Liu, Y., & Zhang, L. (2022). Strategies and challenges of identifying nanoplastics in environment by surface-enhanced Raman spectroscopy. *Environmental Science & Technology*, 57(1), 25–43. DOI: 10.1021/acs.est.2c07416

Yoon, J. H., Kim, B. H., & Kim, K. H. (2024). Distribution of microplastics in soil by types of land use in metropolitan area of Seoul. *Applied Biological Chemistry*, 67(1), 15. DOI: 10.1186/s13765-024-00869-8

Chapter 5
Computational Approaches for Identification of Micro/Nano-Plastic Pollution

Kartavya Mathur
https://orcid.org/0000-0003-3563-5666
Gautam Buddha University, India

Eti Sharma
Gautam Buddha University, India

Nisha Gaur
https://orcid.org/0000-0002-8699-6659
Gautam Buddha University, India

Shashank Mittal
O.P. Jindal Global University, India

Shubham Kumar
University of Minnesota, USA

ABSTRACT

The dissemination of miniaturized plastics, both micro- and nano-plastics, athwart diverse ecosystems is an argument of global apprehension. The accretion of these plastics is due to their chemical steadiness. In arrears to their trivial size, frequently

DOI: 10.4018/979-8-3693-3447-8.ch005

identification of miniaturized plastics is very problematic. The foremost approaches for identification of micro- and-nano plastics rely upon their visual inspection through microscopy and chemical analysis. The advent of high-throughput computing has eased the detection of miniaturized plastic pollution. Machine learning and computer vision methods are being readily applied for analyzing microscopy images to identify and classify microplastics. Molecular simulation methods are also being applied for studying the interaction between environment and microplastics. Additionally, remote sensing methods have also been used to collect and analyze suspected locations of microplastic pollution.

1. INTRODUCTION

The last decade has seen an unparalleled uprising in micro- and nano- plastics (MNPs) research enlightening the possibility and scale of MNP pollution athwart the globe (Allen et al., 2022). Around 51 trillion miniaturized plastics particles have been reported to be disposed in aquatic systems like oceans which is affecting environment as well as its living creatures equally (Tirkey & Upadhyay, 2021). Micro- and nano- plastics are those polymeric particles that have a size dithering between 1 μm to 5mm or less than 100 nm, correspondingly. Additionally, they appear in different shapes like microbeads, fibers, fragments, film foam pellets besides filaments (Galloway, 2015; Kim et al., 2023; Oliveira et al., 2022; Tirkey & Upadhyay, 2021). The reduced size of nanoparticles reflects an increased surface-to-volume ratio culminating into their hazardous nature due to their capacity to absorb organic pollutants as well as endure bioaccumulation (da Costa et al., 2016). These particles can either be manufactured resolutely as morsels in cosmetics and personal care products, or may be molded due to crumbling of larger plastics through photodegradation, weathering or pyrolysis in the environment (Kim et al., 2023; Oliveira et al., 2022). The weathering and degradation results into alteration of physiochemical properties like color, surface morphology, crystalline nature along with density, thus impacting the environment. Most of these plastics are from widely used classes like polyethylene, polystyrene, polyvinyl chloride, polyethylene terephthalate, polyamide among various others. Other biodegradable plastics also exists like polyhydroxyalkanoates, polylactic acid etc. (Singh & Kumar, 2024). The presence of miniaturized plastics has been detected in nearly all the ecological landscapes (viz. water bodies, soil, and air), marine organism, salt, beer, as well as in packaged drinking water worldwide (Yee et al., 2021). Evidences have also suggested MNPs may get accrued inside human tissues (viz. lungs, gastrointestinal tract and blood vessels), triggering immune system, cytotoxicity, oxidative processes, mitochondrial dysfunction besides ER-stress and inducing diverse medical impediments like organ-damage, inflammation besides fibrosis and gut microbiota dysbiosis (Bastyans et al., 2022; Tirkey & Upadhyay, 2021).

A global interest in MNPs along with their speculated health hazard makes detection as well as identification of these miniaturized plastics a necessity in environmental samples. Broadly, the detection of plastic particles in samples take three important steps: sample preparation, analysis of samples for MNP detection followed by quality control (Figure 1) (Tirkey & Upadhyay, 2021). The samples for analysis can be acquired from different sources like water bodies, soil, industrial effluents etc. The sample is required to be prepared, before it can be subjected to analysis. Samples can either be prepared through physical methods or by using chemical methods. Physical methods tend to remove unwanted particles like plant matter, sand gravel, organic matter to isolate plastic particles. This is done by sieving, filtering, visual inspection or density separation. Sieving uses meshes of different sizes to separate microplastics. The particles that get accumulated on mesh are collected, while others are discarded. This method is best for liquid samples. The filtration method uses funnel, filter membrane (pore size ranging from 0.7 micron to 11 micron), or vacuum pump to separate solid particles from liquid. Visual sorting uses naked eyes or microscopic methods like optical microscope to remove unwanted non-plastic particles like shell fragments, seaweed, wood, oil residues etc. This method is quite laborious as well as time-consuming due to small size of MNPs, often leading to misidentification. Finally, density gradient separation has also been suggested for separating non-plastics from MNPs as their denseness (0.8-1.6 gcm^{-3}) is lesser in comparison to debris (~2.7 gcm^{-3}). Chemical methods on other hand tend to digest unwanted micro or macroscopic debris using acids (viz. HCl, HNO_3), alkalis (viz. NaOH or KOH) and peroxide. In some cases where microplastics are embedded in biological tissues, enzymatic degradation is often preferred. Once the miniaturized plastics are isolated from the unwanted debris, different analytical methods (viz. SEM, TEM, Raman Spectroscopy, FT-IR, Near-Infrared, NMR etc.) are applied for identification of plastic particles as well as comprehending their chemical composition. Lastly, to elude contaminations from different sources, the process of isolation is suggested to be repeated in laminar flow in a sterile condition, all the equipment needs to disinfected with ethanol, glassware should be rinsed thoroughly with Milli-Q water, control petriplates must be kept and samples must be covered or locked with aluminum sheets to avoid air contamination (Tirkey & Upadhyay, 2021).

Figure 1. Steps involved in detection and identification of miniaturized plastics

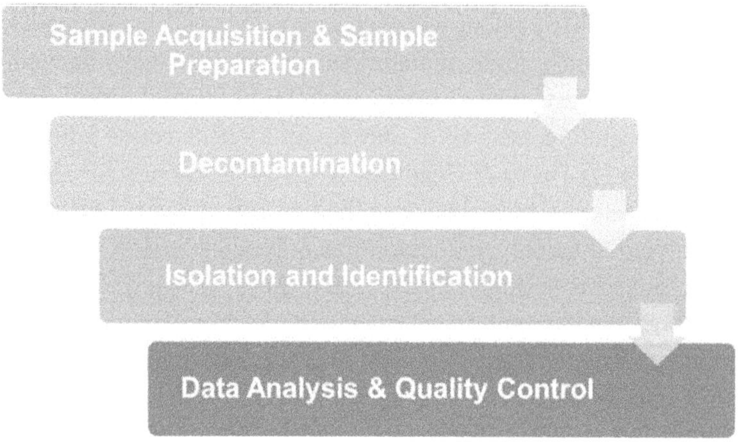

It is widely known that the present analytical methods utilized for MNP detection and characterization are time-consuming, expensive, and laborious. When coupled with unstandardized collection and processing with large-sample size, accurate detection of MNP become unfeasible (Vighi et al., 2021). A paradigm-shift towards automation is seemed to be the key for improving the efficacy of present analytical methods. Comprehending patterns of microplastics is necessary to facilitate identification of their nature as well as localization. Computational analysis of spectroscopic data and computer-assisted visual inspection are two majorly used methods for extracting key features, as well as compare between samples for presence of microscale plastics. These methods have been improved by incorporation of predictive modelling methods like machine learning and deep learning through instance-based and model-based arguments (Abdurahman et al., 2020; Phan & Luscombe, 2023; Waqas et al., 2023). Recently, Lin and colleagues had explored the ML methods and their applications for analysis of spectroscopic datasets (Lin et al., 2022). Lin et al. also underlines necessity for development of new algorithms and databases for facilitating microplastic identification and analysis from environmental samples. Lately, computer-vision has emerged as an assuring tool for uncovering as well as characterization of MNPs (Hufnagl et al., 2022; Kim et al., 2023; Primpke, Godejohann, et al., 2020; Vidal & Pasquini, 2021; Zhou et al., 2022). Freshly, molecular dynamics simulations have also turn out to be a accepted field in twigging the collaboration amongst biomolecules and MNPs for perusing their consequence on health and environment (Chen et al., 2023; Rubin & Zucker, 2022). This chapter would reconnoiter diverse computer-assisted procedures laboring for finding, elucidation as well as characterization of miniaturized plastics.

2. APPROACHES FOR IDENTIFICATION AND CHARACTERIZATION OF PLASTICS UNITS IN MICRO-NANO-SCALE

Uncovering as well as prognostication of MNPs is exceedingly multifaceted. Broadly, the methods utilized for MNP detection can be classified into two groups: (1) on the basis of visual inspection; and (2) on the basis of chemical composition. The methods of visual inspection rely upon microscopy and imaging. Different visual inspection and analytical methods have traditionally been utilized to identify the size, quantity, as well as types of plastic particles in the environment. Some of these widely used techniques include: microscopy, Fourier transform infrared spectroscopy (FTIR), Raman spectroscopy besides thermal analysis like thermogravimetric analysis-gas chromatography-mass spectroscopy (TGA-GC-MS) are utilized for detection as well as analysis of miniaturized plastics (Kiran et al., 2022; Löder & Gerdts, 2015) (Table 1). While microscopy-based techniques like Transmission Electron Microscopy (TEM), Field Emission Scanning Electron Microscopy (FE-SEM), Dynamic Light Scattering (DLS) and Multi-angle light scattering (MALS) are able detect particles in nanoscale (~10 nm) by providing finite images or patterns, FTIR and Raman Spectroscopy are able to identify the functional groups in plastic polymers. TGA along with thermal desorption-gas chromatography can provide insight into the structural aspects of plastics. Sieve pattern of plastics along with their magnitude generate a principal cut-off for elements present in the samples. Most of the aforementioned techniques are sensitive to impurities as well as selection of appropriate pre-treatment processes viz. wet peroxide oxidation and enzymatic digestion for detecting MNPs. Additionally, few studies have also indicated a simpler test – 'hot-point test' wherein a needle is heated and placed in the environmental sample. If plastic is present in the sample, it gets melted and is masked on the needle. Although this method is inexpensive, it does not provide anything on the nature and type of plastic (Kiran et al., 2022).

Although microscopy is a great method for visualizing the presence of MNPs, it lacks characterization and identification of MNPs. Often, studies have coupled microscopy with different analytical methods for improving the characterization of MNPs. Raman spectroscopy as well as FT-IR have been widely applied with microscopy for identifying micro-scaled plastics along with their quality. However, analytical methods are often costly due to expensive instrumentation besides a longer analysis time. Recently, staining has also been presented as viable mechanism for sleuthing incidence of MNPs. A dye, Nile-red explicitly muddles to polymeric materials empowering swift recognition and evaluation of MNP due to its subsuming, discernment and luminous traits. This dye is adsorbed into plastic exteriors permitting their fluorescent properties on irradiation with blue light. This method is far effectual in comparison to other techniques, and necessitates minimalistic time for investigation in disparity to environmental samples (Mariano et al., 2021).

Table 1. Methods used for identification and detection of MNPs

S. No.	Method	Advantages	Disadvantages	Limit of Detection (Particle Size)	References
A	**Microscope**				
1	Stereo Microscopy	Swift and easy Identifies shape, size and color	Cannot confirm nature of plastics and composition of polymers Cannot analyze small or transparent particles	Larger size (> 100 nm)	(Mariano et al., 2021)
2	Transmission Electron Microscopy	High Resolution (< 0.1 mm) Elemental analysis of plastic particles	High time consumption Sample preparation required for larger sized particles	1nm to >100 nm	(Mariano et al., 2021)
3	Scanning Electron Microscopy	Provide clear and high-resolution imaging of particles Elemental analysis possible Small particles can also be detected using STEM	Very expensive Require a good amount of time and efforts for analysis Do not provide information on type of polymer		(Mariano et al., 2021)
4	Atomic Microscopy	No radiation-assisted damage to samples Preserves sample surface Provide 3D details of surface structures of polymers Provide best resolution (0.3nm)	Possibility of contamination Tip of microscope can sometimes damage the polymer		(Mariano et al., 2021)
5	Fluorescence Microscopy	Detects transparent particles with ease Immediate visualization of particles	At times fluorescence lasers can be harmful and toxic for samples Chemicals can interfere with fluorescence		(Mariano et al., 2021)
B	**Spectroscopic Techniques**				
1	Raman Spectroscopy	Non-destructive analysis Minimal false and negative and false positives Can analyze samples in all the states of matter – gas, solution, film, surface, solids or crystals	Costly instrumentation Heavy time-consumption Interferes with pigments	< 1 um to 1 um	(Mariano et al., 2021)

continued on following page

Table 1. Continued

S. No.	Method	Advantages	Disadvantages	Limit of Detection (Particle Size)	References
2	Fourier Transform Infrared (FT-IR) Spectroscopy	Confirms composition of MNPs No false negative or positive detection Non-destructive analysis	Expensive Require very long time for detection Requires a specific wavelength irradiation for detection.	< 20 um	(Mariano et al., 2021)

3. COMPUTATIONAL METHODS FOR IDENTIFICATION AND DETECTION OF MNPS

As mentioned earlier, detection and characterization of MNPs can be done using microscopy and different analytical methods. Both of these methods often are used in combination to improve the accuracy of MNP detection. Few studies have reported the use of either FT-IR or Raman Spectroscope simultaneously for chemical identification of microscopic particles and qualitative confirmation of its type (Muthulakshmi et al., 2023). The high cost, long analysis time, unstandardized collection, processing and analysis of large samples, besides labor-intensive approaches render the accurate detection of MNPs infeasible. For actual superintending to occur, it is crucial to comprehend the arrangements of microplastic transference for augmenting the site as well as operation of advanced examining tools. High-throughput and mechanized observing has become necessary for effacing and large-scale investigation and comprehending microplastic accretion configurations. Studies have presented three different methods for elucidating microplastic behaviors. Mathematical modelling provides a unique approach for modelling microplastic behavior for gaining insight into their distribution, sources, sink besides pathways. Recently, predictive modelling techniques have been utilized for solving a diverse-problems concerning large amounts of information unconventionally, thus saving time. Image-based predictive modelling has been pragmatic to material science findings that encompass a large number of images to identify the configuration as well as property associations in materials. These methods are still evolving for enhancement until their full latent is apprehended (Figure 2). Contemporary predictive learning as well as computer-vision tools emphasize on mechanizing the enumeration besides categorizing of miniaturized plastics. Cataloguing is a dispensation phase which can deliver evidences around the magnitude and contour of miniature plastics originating in the environment. Dependent on microplastic physical properties such as shape, density, and size, microplastics can exhibit dissimilar behaviours in the marine environment and can endure unalike transport progressions such as tempestuous transport and settling velocity (Phan & Luscombe, 2023). Studies have also highlighted the importance of remote sensing for monitoring environmental distribution of MNPs (Waqas et al., 2023). The development of new imagery systems like hyperspectral imaging, flow control imaging, confocal laser microscopy, electron microscopy etc. have enabled monitoring and detection of MNPs (W. Li et al., 2020). Additionally, computational analysis have also been suggested for comprehending interaction between MNP and environmental variables (Abdurahman et al., 2020; Chen et al., 2023; H. Luo et al., 2022; Rubin & Zucker, 2022; X. Wang et al., 2022). Recently, the efforts Jung et al. (2023) developed a robust database of analytical methods like Raman spectroscopy, FTIR, gas chromatography, thermogravimetric analysis etc. for MNP detection (Jung et al., 2023). Similarly, Nyadjro et al. (2023) presented a database for marine microplastic datasets (https://www.ncei.noaa.gov/) (Nyadjro et al., 2023). In this chapter, we shall explore, different computational vision approaches aligned for detecting MNPs.

Figure 2. Different methods identification MNPs

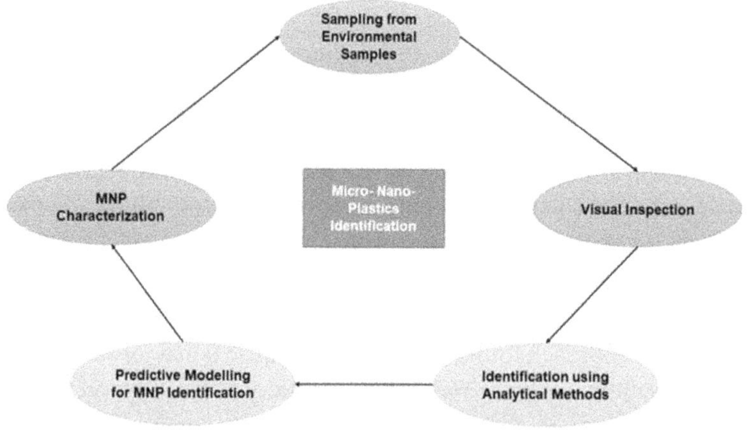

4. RELEVANCE OF MACHINE LEARNING AND ARTIFICIAL INTELLIGENCE IN IDENTIFICATION OF MNPS

The advent of predictive modelling has enabled dynamic processing of vast and complex data both linear and non-linear in nature from a variety of sources. The machine learning and artificial intelligence tends to extract meaningful patterns from experimental data obtained from different resources to comprehend potential relationships between relevant factors for solving classification and regression problems. The high speed, automation, and precision in results have prompted wide usage of predictive modelling for advancing scientific research in the areas of water treatment as well as catalysis. Machine learning has been suggested to be a reliable tool for automation especially for identification and detection of microplastics.

We have seen, different analytical methods for detecting microplastics from environment. While thermal-based approaches are much swifter, prone to lower contamination and could provide information on concentration as well as chemical composition of MNPs, spectroscopic methods are better in providing information related to their size, which is crucial for hazard assessment. ML algorithms trained on spectroscopy data or sensor images of microplastics have potential to attain enhanced-accurateness for identification and classification on the basis of dissimilar MNP characteristics (viz. peaks, shape, RGB value of images etc.). Hyperspectral imaging has also been fruitful for gaining information via multi-channels for envisaging equivalent locations. Hence, it is speculated, amalgamation of hyperspectral imaging with predictive modelling shall provide an unprecedented growth to MNP identification. Predictive modelling can further be used to explore potential implications of MNP toxicity (Su et al., 2023). Different predictive modelling approaches like random forest, support vector networks, regression models, k-nearest neighbor besides deep learning techniques have been applied for extracting necessary features pertaining to traits and qualities of MNPs. Although, a wide-range of studies have been performed, there remain only select few, that have shown promising results like Kedzierski et al. (2019), Hufnagl et al. (2022), Lei et al. (2022), Luo et al. (2023) (Weber et al., 2023). Kedzierski and colleagues analyzed FTIR-attenuated total reflection (ATR) spectrum using k-nearest neighbor algorithm on a set training set of 969 spectra from marine MNPs. This method was suitable for identifying common plastic particles like polyethylene, polypropylene, polyvinyl chloride among a few others. On a validation set of around 4000 spectra, it achieved 90.5% accurate results. Manual intervention and analysis increased the accuracy of this method to 97% (Kedzierski et al., 2019). Similarly, the study by Hufnagl and colleagues using random forest classifier and Monte Carlo cross validation on a dataset obtained from focal plane array (FPA) micro-FTIR spectrum for 20 dissimilar MNPs achieved sensitivity ranging between 92.5% to 100% and specificity ranging between 99.84% to 100% (Hufnagl et al., 2022). Similarly, a study by Back et al. (2022) developed a support vector-based classifier to differentiate between 13 classes of polymers. They sampled water from sea using FTIR. They applied log-loss minimization function to estimate the performance of the model (Back et al., 2022). The use of Raman spectroscopy in conjunction with predictive modelling algorithms have been very minimal. Lei and colleagues had shown previously achieved ~95% accuracy in identifying microplastics analyzed using Raman spectroscopy. They applied different algorithms like k-nearest neighbors, multiple layer perceptron, random forest classifiers for artificially manufactured plastic particles (Lei et al., 2022). In contrast to this, Luo and colleagues collected samples from rain water and surface water, and mixed commercially available microplastics into it. With a spectral database of 3675, coupled sparse autoencoder supplemented with softmax framework attained a test accuracy of 99.1% (Y. Luo et al., 2023). Another study by Weber et al. (2023) also investigated potential of micro-Raman spectra in detecting microplastics from 47 dissimilar environmental and waste water samples (~64000 spectras). They employed deep learning methods with one solitary model and one model per class using Rectified Linear Unit Function (ReLU) hidden layer activation function beside sigmoid function-based output layer. On comparison with human-guided identification, they were able to analyze polyethylene, polypropylene, polystyrene, polyvinyl chloride etc. in spectra with 99.4% sensitivity and 97.1% specificity in considerably reduced time (Weber et al., 2023).

With these examples, we can categorize implementation of predictive modelling in MNP identification and detection into two groups: model/archetypal-based or instance/illustration-based predictive modelling methodologies. The archetypal-based method comprehends a statistical prototype using spectroscopical reference data. This method is then tested on a entirely new spectrum. This method has been widely accepted to group samples on polymer type or matrix component. In contrast to model-based algorithms, instances-based methods use similarity estimates for classifying an unidentified spectrum. Instance-based method uses Pearson correlation coefficient to compute similarity measure by using spectral libraries. Both of these methods have their own set of pros and cons. While instance-based methods promote adaptation of reference data by altering the reference spectrum library, the model-based learning is dependent on chemometric expertise, and therefore, is difficult to alter. Model-based approaches are far better in performance, and speed in comparison to instance-based learning (Hufnagl et al., 2022). Usually, the instance-based algorithms are memory-driven instead of explicit generalization like k-nearest neighbor, kernel machines etc. (Keogh, 2010) whereas model-based algorithms are more laborious but with better generalizability like regression algorithms, autoencoders etc. We shall explore these illustrations of machine learning in detail in subsequent sections.

5. COMPUTATION-ASSISTED SPECTROSCOPIC ANALYSIS

From the argument above, we can firmly state the significance of spectroscopic techniques in detection of miniaturized plastics based on a diverse set of vibration bands. The examination of miniaturized plastic in the microscale necessitates dispensation of samples to separate and distillate the microplastic units on sieves. The existence of non-plastic particles on filters influences successive solitary point scales of the spectra of each unit, making it inefficient. Diverse findings have recommended a focal plane array (FPA)-based micro-FTIR imaging to simplify production of chemical images by concurrently detailing multitude of spectra within a minute time-frame. In spite of this, at times, manual comparison of received spectra with a reference-spectra may take a lengthier time, not insignificant for examining wherein large sample size is used. Therefore, regardless of which analytical techniques are used for MNP assessment, there is a need for automation to process spectral information (Hufnagl et al., 2022).

A diverse set of free as well as proprietary software have been developed for estimating microplastic presence in environmental samples like siMPle, PlasticNet, OMNIC, OPUS, SMACC etc. Systemic Identification of Microplastics in Environment (siMPle) software stemmed at Aalborg University is a freeware program for identifying microplastics using micro-FT-IR data. It performs the comparison of provided data with an already established reference library, returning the material name along with probability score reflecting the accuracy of prediction (https://simple-plastics.eu/). This tool was developed by combining the MPhunter software along with an automated analysis by Primpke et al. (2017) (Liu et al., 2019; Primpke et al., 2017). Primpke et al. (2020) performed a cross-validation of siMPle on different FT-IR systems (viz. Burker Hyperion 3000, Agilent Cary 620/670, PerkinElmer Spotlight and Thermo Fischer Scientific Nicolet iN10) for a harmonized analysis of MNPs on a dataset and reference set (Primpke, Cross, et al., 2020). MPhunter was another program established by the same group which performed cataloguing of materials using an algorithm that performs manual assessment based on thresholding to recognize particles (Liu et al., 2019). OMNIC Paradigm Software is another widely software for identification of MNPs. This software developed by Thermo Fischer enables detection of microscopic-sized plastic particles based on FT-IR and Raman spectroscopy using a reference library CMDR polymer kit. The method analyses images of microplastics attached on filters (https://www.thermofisher.com). Another tool widely used for automated microplastic analysis is OPUS by Burker systems. It performs quantification, dispensation as well as assessment of infrared, near-infrared, besides Raman spectroscopy data. The chemical imaging from micro-FTIR have been shown to provide MNP identification when coupled with FPA. Recently, Primpke (2017) developed an automated pipeline by combining the FTIR results obtained from OPUS© Software with image analysis using Python and Simple ITK image processing module by exploring 1.8 million single spectra. The images obtained as false colors were assessed using different algorithms like standard search or full-width at half maximum. This pipeline was able to reduce computation time significantly (Primpke et al., 2017).

Recently, Hufnagl et al. (2022) had utilized a combination of micro-FTIR and machine learning for detecting MNPs in environmental samples. The method described a model-based algorithm for identifying microplastics. They used water samples and applied different preparation techniques. The samples were subjected to FPA-based micro-FTIR analysis using Burker Hyperion 3000 followed by FTIR imaging. They used a spiked samples with 21 different polymers as a training set for Random Forest Classifier. The data was obtained from 22 different classes. They used a large set of hyperspectral images (HSI) representing different weathering patterns and applied Monte Carlo Cross Validation to optimizing the model performance. For data-analysis, they used Microplastics Finder based on Epina ImageLab Engine with slight modifications for streamlining the MNP detection. This new pipeline Bayreuth Microplastics Finder (BMF) performed poorly in classifying polymers into right classes (Hufnagl et al., 2022). Previously, in another study, Renner et al. (2019) had acquired microplastic samples from different polymers, and beach samples. They performed micro-FTIR and ATR-FTIR experiments using IRTracer-100 FTIR spectrophotometer. The similarity between reference and test samples were identified using hit quality index (HQI). They applied different algorithms like Euclidean distance, Pearson correlation and micro-IDENT and estimated correctly-identified rate, incorrectly identified rate, and not-identified rate. They were able to correctly identify ~98% microplastics using their automation protocol (Renner et al., 2019). In another study, Renner et al. (2020) provided a python-based framework for development of an automated and intelligent microplastics using FTIR based analysis. They presented a non-FPA based approach using infrared microplastic identification algorithm (micro-IDENT) to measure spots on filters which were highly reliable to provide a spectrum. This approach considerably reduced computation time from ~50hr to jut 7 hours (Renner et al., 2020). Another method recently developed by Escalona-Segura et al. (2022) focused on identification of microplastics in bird tests. They searched for bird nests, crumbled them and used stacked sieves of different mesh sizes to separate MNPs to prepare the samples. FTIR analysis was performed using Nicolet iS5 (Thermo) FTIR spectrophotometer and explored using OMNIC package and its reference collections. Kurskal-Wallis statistical test was performed to determine differences between microplastic size present in the nest using Statistica® 6.0. The study identified 2535.3 ± 2175.9 items per kilogram of microplastics (Escalona-Segura et al., 2022). Many other studies have also recommended amalgamation of FTIR with other microscopy practices in MNP recognition for reducing detection time and efforts (Chen et al., 2020).

6. IMAGE-BASED DETECTION AND REMOTE SENSING OF MNPS

Imagery analysis has turn out to be a favorable practice for trailing plastic pollution from large-scale to a miniscule scale, over a large geography using remote sensing. Remote sensing have been applied for recognizing, enumerating as well as sleuthing plastic litter in earthly as well as aquatic milieu. This has been possible due to development of high-throughput sensors like high-resolution remote sensing information in Visible (VIS), Near-Infrared (NIR), Short-Wave Infrared (SWIR), Thermal Infrared (TIR), Synthetic aperture radar (SAR) besides Light Detection and Ranging (LiDAR). This have been enhanced by availability of new-age imaging equipment such as hyperspectral imaging, spatial imaging, flow imaging etc. Some of these have been reviewed in Waqas et al. (2023). For illustration, studies by Gnann et al. (2022), Mukonza and Chiang (2022), and Maximenko et al. (2019) have used in-situ data and findings from spectral indices for training a diverse set of supervised and unsupervised image classification models. Studies have also used satellite imaging and high-resolution aerial imaging for training microplastic classifiers (Waqas et al., 2023). Previously, Davassuren and colleagues have suggested the use of Sentinel-1A and COSMO-SkyMed SAR imaging in North Pacific and Atlantic oceans. They performed contextual imaging to identify polyethylene and polyethylene terephthalate microplastics in ocean samples (Davaasuren et al., 2018). Recently, Vidal and Pasquini (2021) had also suggested a hyperspectral imaging based Near-Infrared sensing for detecting microplastics from beach sand. They used a multivariate supervised soft independent modelling of class analogy (SIMCA) by scanning 75 cm^2 area for 156X156 micrometer pixels. Their model studied the effect of size, color and weathering on microplastic content, and achieved a sensitivity and specificity ~99% for microplastic classes like polyethylene, polypropylene, polyamide etc. (Vidal & Pasquini, 2021). Likewise, Primpke and contemporaries (2020) also accomplished quantum-cascade laser-based microscopy to perceive MNPs in water pipelines. They engaged a python-based architecture along with Pearson association for investigating 8 million images in a 144mm^2 area with 4.2 microns pixels. They were able to attain good recall for sleuthing MNP existence (Primpke, Godejohann, et al., 2020). One more study by Zhou and contemporaries secondhanded fluorescence generation tomography procedures for sleuthing synthetic polymers (Zhou et al., 2022). Also, in alternative study, Kim et al. utilized flow imaging microscopy for investigating the occurrence of MNPs alongside natural organic matters (Kim et al., 2023). In one more noteworthy report by Giradino et al. (2022), samples from Po River water as well as Borgio Verezzi cave residues were utilized for microscale plastic comprehension. They accomplished image analysis of sieved samples with as well as without Nile Red staining. The section of awareness were acknowledged by implementing Canny-based thresholding to regulate geomorphology besides group of MNPs. This established architecture reduced labor for investigation substantially (Giardino et al., 2023). Lately, Hong et al. (2024) presented an inversive archetypal tactic grounded on multiple regression for sleuthing MNPs. They obtained samples from Bohai Sea and employed feature selection algorithms succeeding prognoses process, band grouping technique, as well as remote sensing index with SPA outperforming others, for assessing spatial as well as temporal distribution of microplastics. This model achieved a determination coefficient of 0.75, and root mean square error of 0.38 items per mm^3 (Hong et al., 2024). Lorenzo-Navarro (2020) also presented a new method, System for Microplastic Automatic Counting and Classification (SMACC) for enumerating as well as categorizing MNP units. The samples were obtained from coastal areas. The images of samples obtained in transparent and RGB mode were acquired using Epson Perfection V800 with VueScan image scanner. Image segmentation method of Sauvola threshold was used for identifying region of interest. Feature selection was performed using SURF detectors using RANSAC algorithm. Local binary patterns were chosen to extract essential patterns. Finally, classification model was developed using random forest, decision trees, K-nearest neighbors and support vector networks along with VGG-16 deep learning framework. The framework tested on 12 beach samples and 2507 microplastic particles. Manual computation of particles was also done to be used as ground truth. The study was able to achieve an accuracy of 96% (Lorenzo-Navarro et al., 2020). All of these studies indicate a potential of computational methods in improving MNP detection by reducing computation time. Although, still human intervention is key to improving the accuracy, larger data, with high quality may provide valuable results. Few more revolutionary studies are provided in Table 2.

Table 2. Selected studies for computational investigation of micro-nano-plastics

S. No.	Location	Data Type	Method	Type of Plastics	Algorithm	Performance	Reference
1	China	Imaging using Stereo Microscope with a digital camera	Polarization Holographic Imaging	PC, PET, PVC, PP, PS, PMMA	YOLO v5 model with lightweight CNN combined with Strong-SORT network	Not Provided	(Y. Li et al., 2024)

continued on following page

Table 2. Continued

S. No.	Location	Data Type	Method	Type of Plastics	Algorithm	Performance	Reference
2	Great Lakes – Saint Lawerence Waterway at Lake Ontario in Kingston	Microscopy	Nanodigital In-line Holographic Microscopy (Nano-DIHM)	PE, PP, PS, PET, PVC, PUR	Neural Network	Classified 2% waterborne particles as nanoplastics and 1% as microplastics	(Z. Wang et al., 2024)
3	Playa del Poris in Tenerife Island, Spain	Imaging	Digital Camera	Not Provided	U-NET neural network in Stage-I VGG16 neural network in Stage-2	Accuracy = 98.11% for classification of microplastics	(Lorenzo-Navarro et al., 2021)
4		Optical Microscope and Holographic microscope	Holographic Plastic Identification	PS, PE, PVC, PP, PET	Support vector network	Accuracy ~99%	(Bianco et al., 2020)
5	Koh Yo in Songkhla province, Thailand	Spectroscopy and Imaging	GF/C filter	Fauna, Alkyd, CPB, LDPE, PC, PP, PVC, polyester	Faster-RCNN-FPN with a ResNet-50-FPN backbone	Precision = 85.5 to 87.8% mAP ~ 33.9A% on internal test and 35.7% on external test	(Thammasanya et al., 2024)
6	Dublin Ireland	Optical photothermal infrared (O-PTIR)	Aluminium oxide filters	Nylon plasti from tea bags	SVM	Accuracy = 91.33%	(Yang et al., 2024)
7		Spetrosopy	u-FTIR	PE PP	Transforer neural network	Marine Accuracy = 98.7% Marine F1 Score = 98.7% Standard Accuray = 92.3% Standard F1 = 92.0%	(Barker et al., 2022)
8		Microscopy	Scanning Electron Microscopy using electron micrographs	Polyester and polyacrylic	U-Net, MultiResUNet for semantic segmentation and VGG16 neural network for shape	Jaccard index = 0.75 Accuracy = 98.33%	(Shi et al., 2022)
9	Ofanto Rier South Italy	Imaging	Smartphone aera / digital Microscope		KNN	Auray = 922%	(Massarelli et al., 2021)

7. MOLECULAR DYNAMICS SIMULATION

It is evident that miniaturized plastics with or without being absorbed can pose health hazard to various species. Bioaccumulation of these miniscule units in maritime lifeforms have indicated a unswerving direction for interaction with humans via food-web. Some studies have also indicated existence of pollutants clung to the superficial of microplastics can result into coactive, opposed, as well as addictive consequences underneath real-life scenario. Freshly, the utility of molecular simulation (MS) procedures viz. molecular dynamics (MD) and Monte Carlo (MC) has been enlightened for assessing the interaction between MNPs and environmental constituents. These methods can help in comprehending potential health effects of miniaturized plastics on human. MS is a powerful tool for investigating phenomena at miniscule-level. It has been efficacious in foretelling macroscopical thermodynamical as well as dynamical observables for assorted systems. The role of MS in unfolding system properties for conditions that are experimentally unachievable is increasing rapidly. MS is also now being used in conjunction with investigational outcomes for comprehending the mechanistic spectacle of interest. MC and MD both are able to describe microscopic particles through pairwise interaction potentials. Molecular dynamics simulation uses an arithmetic simulation technique for estimating the thermodynamical as well as conveyance assets of multiparticle systems. The chronological progression of a set of intermingling units is supplemented by the addition of traditional equations of motion. The sequential averages of the trajectories along with their oscillations are interrelated with the macroscopical assets of the deliberate system. On the other hand, Monte Carlo (MC) is grounded on unearthing preferable state with most minuscule energy at random. This is achieved by alternating the system by one move for each MC step, followed by computation of free energy change of the system. If energy is at lower extreme, the step is accepted, otherwise rejected. Equally MD and MC are being pragmatic in comprehending the adsorption of impurities to miniaturized plastics besides investigating the mechanism of attraction between plastics and biomolecules like DNA, RNA, phospholipids as well as proteins. This enables development of a comprehensive understanding of probable effect of miniscule polymers on human health as well as on ecosystem (Oliveira et al., 2022).

A wide range of force fields have been utilized in literature for studying molecular interactions using MD and MC like Optimized Potentials for Liquid Systems, All-Atoms (OPLS-AA), Assisted Model Building and Energy Refinement (AMBER), Chemistry at Harvard Macromolecular Mechanics (CHARMM), General AMBER Force Field (GAFF), Transferable Potentials for Phase Equilibrium (TraPPE), COMPASS forcefield, GROningen Molecular Simulation (GROMOS), Large-scale Atomic/Molecular Massively Parallel Simulator (LAMMPS), Gabedit program among various others. These forcefields estimate interaction potentials using different parameters like particle size, bonds, dihedral angles etc. Usually a water model with triple point charge like TIP3P or SPC/E have been suggested to provide a natural setting of the biomolecule. A MD of nearly 100 to 500 ps or more time-duration is set for exploring the energetics of the system, its stability, and adverse effects (Oliveira et al., 2022).

Humic acids (HA) a major constituent of soils, formed from decomposition of organic matters reflect upon the soil health. Miniaturized plastics have been understood to interact with humic acids in soil. Upon release into soil, microplastics tend to corrode and undergo aging through abrasive forces, UV-radiation, chemical oxidation, and biodegradation. This leads them to travel inside the soil, towards the water reservoirs, thus polluting it. Studies have indicated at higher ionic strength, hydrophobicity of microplastics is altered, and it can get retained in unsaturated media composed of humic acid (Abdurahman et al., 2020; X. Wang et al., 2022). Structural changes in humic acid upon binding with microplastics were also validated using spectroscopic techniques (H. Luo et al., 2022). Another study based on molecular dynamics simulations suggested increased diffusion rate of microplastics post interactions with humic acid resulting alternation of hydrogen bonding (Chen et al., 2023).

8. CONCLUSION

In this chapter, we explored and reviewed a diverse set of analytical and computational methods such as imaging modalities, spectroscopic techniques, molecular dynamic simulations that have been applied for sleuthing the presence of miniaturized plastics from environmental samples. We observed the utility of predictive modelling techniques alongside computer vision methodologies for improved detection of microplastics in different environments. FTIR and Raman spectroscopy are key techniques utilized along with imaging sensors for efficient identification of MNPs. However, despite the advancements in computational mechanisms, the lack of large datasets in public domain, high variability in the peaks of spectrum, and improper standardization of protocols in preparation of environmental samples along with quality control, poses a threat to achieving an effective classification of microplastics.

REFERENCES

Abdurahman, A., Cui, K., Wu, J., Li, S., Gao, R., Dai, J., Liang, W., & Zeng, F. (2020). Adsorption of dissolved organic matter (DOM) on polystyrene microplastics in aquatic environments: Kinetic, isotherm and site energy distribution analysis. *Ecotoxicology and Environmental Safety*, 198, 110658. DOI: 10.1016/j.ecoenv.2020.110658

Allen, S., Allen, D., Karbalaei, S., Maselli, V., & Walker, T. R. (2022). Micro(nano)plastics sources, fate, and effects: What we know after ten years of research. *Journal of Hazardous Materials Advances*, 6, 100057. DOI: 10.1016/j.hazadv.2022.100057

Back, H. de M., Vargas Junior, E. C., Alarcon, O. E., & Pottmaier, D. (2022). Training and evaluating machine learning algorithms for ocean microplastics classification through vibrational spectroscopy. *Chemosphere*, 287, 131903. DOI: 10.1016/j.chemosphere.2021.131903

Barker, M., Willans, M., Pham, D.-S., Krishna, A., & Hackett, M. (2022). Explainable Detection of Microplastics Using Transformer Neural Networks. In Aziz, H., Corrêa, D., & French, T. (Eds.), *AI 2022: Advances in Artificial Intelligence* (pp. 102–115). Springer International Publishing. DOI: 10.1007/978-3-031-22695-3_8

Bastyans, S., Jackson, S., & Fejer, G. (2022). Micro and nano-plastics, a threat to human health? *Emerging Topics in Life Sciences*, 6(4), 411–422. DOI: 10.1042/ETLS20220024

Bianco, V., Memmolo, P., Carcagnì, P., Merola, F., Paturzo, M., Distante, C., & Ferraro, P. (2020). Microplastic Identification via Holographic Imaging and Machine Learning. *Advanced Intelligent Systems*, 2(2), 1900153. DOI: 10.1002/aisy.201900153

Chen, Y., Tang, H., Cheng, Y., Huang, T., & Xing, B. (2023). Interaction between microplastics and humic acid and its effect on their properties as revealed by molecular dynamics simulations. *Journal of Hazardous Materials*, 455, 131636. DOI: 10.1016/j.jhazmat.2023.131636

Chen, Y., Wen, D., Pei, J., Fei, Y., Ouyang, D., Zhang, H., & Luo, Y. (2020). Identification and quantification of microplastics using Fourier-transform infrared spectroscopy: Current status and future prospects. *Current Opinion in Environmental Science & Health*, 18, 14–19. DOI: 10.1016/j.coesh.2020.05.004

da Costa, J. P., Santos, P. S. M., Duarte, A. C., & Rocha-Santos, T. (2016). (Nano)plastics in the environment – Sources, fates and effects. *The Science of the Total Environment*, 566–567, 15–26. DOI: 10.1016/j.scitotenv.2016.05.041

Davaasuren, N., Marino, A., Boardman, C., Alparone, M., Nunziata, F., Ackermann, N., & Hajnsek, I. (2018). Detecting Microplastics Pollution in World Oceans Using Sar Remote Sensing. *IGARSS 2018 - 2018 IEEE International Geoscience and Remote Sensing Symposium*, 938–941. DOI: 10.1109/IGARSS.2018.8517281

Escalona-Segura, G., Borges-Ramírez, M. M., Estrella-Canul, V., & Rendón-von Osten, J. (2022). A methodology for the sampling and identification of microplastics in bird nests. *Green Analytical Chemistry*, 3, 100045. DOI: 10.1016/j.greeac.2022.100045

Galloway, T. S. (2015). Micro- and Nano-plastics and Human Health. In Bergmann, M., Gutow, L., & Klages, M. (Eds.), *Marine Anthropogenic Litter* (pp. 343–366). Springer International Publishing. DOI: 10.1007/978-3-319-16510-3_13

Giardino, M., Balestra, V., Janner, D., & Bellopede, R. (2023). Automated method for routine microplastic detection and quantification. *The Science of the Total Environment*, 859, 160036. DOI: 10.1016/j.scitotenv.2022.160036

Hong, P., Xiao, J., Liu, H., Niu, Z., Ma, Y., Wang, Q., Zhang, D., & Ma, Y. (2024). An inversion model of microplastics abundance based on satellite remote sensing: A case study in the Bohai Sea. *The Science of the Total Environment*, 909, 168537. DOI: 10.1016/j.scitotenv.2023.168537

Hufnagl, B., Stibi, M., Martirosyan, H., Wilczek, U., Möller, J. N., Löder, M. G. J., Laforsch, C., & Lohninger, H. (2022). Computer-Assisted Analysis of Microplastics in Environmental Samples Based on µFTIR Imaging in Combination with Machine Learning. *Environmental Science & Technology Letters*, 9(1), 90–95. DOI: 10.1021/acs.estlett.1c00851

Jung, S., Raghavendra, A. J., & Patri, A. K. (2023). Comprehensive analysis of common polymers using hyphenated TGA-FTIR-GC/MS and Raman spectroscopy towards a database for micro- and nanoplastics identification, characterization, and quantitation. *NanoImpact*, 30, 100467. DOI: 10.1016/j.impact.2023.100467

Kedzierski, M., Falcou-Préfol, M., Kerros, M. E., Henry, M., Pedrotti, M. L., & Bruzaud, S. (2019). A machine learning algorithm for high throughput identification of FTIR spectra: Application on microplastics collected in the Mediterranean Sea. *Chemosphere*, 234, 242–251. DOI: 10.1016/j.chemosphere.2019.05.113

Keogh, E. (2010). Instance-Based Learning. In Sammut, C., & Webb, G. I. (Eds.), *Encyclopedia of Machine Learning* (pp. 549–550). Springer US. DOI: 10.1007/978-0-387-30164-8_409

Kim, S., Hyeon, Y., & Park, C. (2023). Microplastics' Shape and Morphology Analysis in the Presence of Natural Organic Matter Using Flow Imaging Microscopy. *Molecules (Basel, Switzerland)*, 28(19), 6913. DOI: 10.3390/molecules28196913

Kiran, B. R., Kopperi, H., & Venkata Mohan, S. (2022). Micro/nano-plastics occurrence, identification, risk analysis and mitigation: Challenges and perspectives. *Reviews in Environmental Science and Biotechnology*, 21(1), 169–203. DOI: 10.1007/s11157-021-09609-6

Lei, B., Bissonnette, J. R., Hogan, Ú. E., Bec, A. E., Feng, X., & Smith, R. D. L. (2022). Customizable Machine-Learning Models for Rapid Microplastic Identification Using Raman Microscopy. *Analytical Chemistry*, 94(49), 17011–17019. DOI: 10.1021/acs.analchem.2c02451

Li, W., Luo, Y., & Pan, X. (2020). Identification and Characterization Methods for Microplastics Basing on Spatial Imaging in Micro-/Nanoscales. In He, D., & Luo, Y. (Eds.), *Microplastics in Terrestrial Environments: Emerging Contaminants and Major Challenges* (pp. 25–37). Springer International Publishing. DOI: 10.1007/698_2020_446

Li, Y., Zhu, Y., Huang, J., Ho, Y.-W., Fang, J. K.-H., & Lam, E. Y. (2024). High-throughput microplastic assessment using polarization holographic imaging. *Scientific Reports*, 14(1), 2355. DOI: 10.1038/s41598-024-52762-5

Lin, J., Liu, H., & Zhang, J. (2022). Recent advances in the application of machine learning methods to improve identification of the microplastics in environment. *Chemosphere*, 307, 136092. DOI: 10.1016/j.chemosphere.2022.136092

Liu, F., Olesen, K. B., Borregaard, A. R., & Vollertsen, J. (2019). Microplastics in urban and highway stormwater retention ponds. *The Science of the Total Environment*, 671, 992–1000. DOI: 10.1016/j.scitotenv.2019.03.416

Löder, M. G. J., & Gerdts, G. (2015). Methodology Used for the Detection and Identification of Microplastics—A Critical Appraisal. In Bergmann, M., Gutow, L., & Klages, M. (Eds.), *Marine Anthropogenic Litter* (pp. 201–227). Springer International Publishing. DOI: 10.1007/978-3-319-16510-3_8

Lorenzo-Navarro, J., Castrillón-Santana, M., Sánchez-Nielsen, E., Zarco, B., Herrera, A., Martínez, I., & Gómez, M. (2021). Deep learning approach for automatic microplastics counting and classification. *The Science of the Total Environment*, 765, 142728. DOI: 10.1016/j.scitotenv.2020.142728

Lorenzo-Navarro, J., Castrillón-Santana, M., Santesarti, E., De Marsico, M., Martínez, I., Raymond, E., Gómez, M., & Herrera, A. (2020). SMACC: A System for Microplastics Automatic Counting and Classification. *IEEE Access, 8*, 25249–25261. DOI: 10.1109/ACCESS.2020.2970498

Luo, H., Liu, C., He, D., Sun, J., Zhang, A., Li, J., & Pan, X. (2022). Interactions between polypropylene microplastics (PP-MPs) and humic acid influenced by aging of MPs. *Water Research*, 222, 118921. DOI: 10.1016/j.watres.2022.118921

Luo, Y., Su, W., Xu, X., Xu, D., Wang, Z., Wu, H., Chen, B., & Wu, J. (2023). Raman Spectroscopy and Machine Learning for Microplastics Identification and Classification in Water Environments. *IEEE Journal of Selected Topics in Quantum Electronics*, 29(4), 1–8. Advance online publication. DOI: 10.1109/JSTQE.2022.3222065

Mariano, S., Tacconi, S., Fidaleo, M., Rossi, M., & Dini, L. (2021). Micro and Nanoplastics Identification: Classic Methods and Innovative Detection Techniques. *Frontiers in Toxicology*, 3, 636640. DOI: 10.3389/ftox.2021.636640

Massarelli, C., Campanale, C., & Uricchio, V. F. (2021). A Handy Open-Source Application Based on Computer Vision and Machine Learning Algorithms to Count and Classify Microplastics. *Water (Basel)*, 13(15), 15. Advance online publication. DOI: 10.3390/w13152104

Muthulakshmi, L., Mohan, S., & Tatarchuk, T. (2023). Microplastics in water: Types, detection, and removal strategies. *Environmental Science and Pollution Research International*, 30(36), 84933–84948. DOI: 10.1007/s11356-023-28460-6

Nyadjro, E. S., Webster, J. A. B., Boyer, T. P., Cebrian, J., Collazo, L., Kaltenberger, G., Larsen, K., Lau, Y. H., Mickle, P., Toft, T., & Wang, Z. (2023). The NOAA NCEI marine microplastics database. *Scientific Data*, 10(1), 726. DOI: 10.1038/s41597-023-02632-y

Oliveira, Y. M., Vernin, N. S., Maia Bila, D., Marques, M., & Tavares, F. W. (2022). Pollution caused by nanoplastics: Adverse effects and mechanisms of interaction via molecular simulation. *PeerJ*, 10, e13618. DOI: 10.7717/peerj.13618

Phan, S., & Luscombe, C. K. (2023). Recent trends in marine microplastic modeling and machine learning tools: Potential for long-term microplastic monitoring. *Journal of Applied Physics*, 133(2), 020701. DOI: 10.1063/5.0126358

Primpke, S., Cross, R. K., Mintenig, S. M., Simon, M., Vianello, A., Gerdts, G., & Vollertsen, J. (2020). Toward the Systematic Identification of Microplastics in the Environment: Evaluation of a New Independent Software Tool (siMPle) for Spectroscopic Analysis. *Applied Spectroscopy*, 74(9), 1127–1138. DOI: 10.1177/0003702820917760

Primpke, S., Godejohann, M., & Gerdts, G. (2020). Rapid Identification and Quantification of Microplastics in the Environment by Quantum Cascade Laser-Based Hyperspectral Infrared Chemical Imaging. *Environmental Science & Technology*, 54(24), 15893–15903. DOI: 10.1021/acs.est.0c05722

Primpke, S., Lorenz, C., Rascher-Friesenhausen, R., & Gerdts, G. (2017). An automated approach for microplastics analysis using focal plane array (FPA) FTIR microscopy and image analysis. *Analytical Methods*, 9(9), 1499–1511. DOI: 10.1039/C6AY02476A

Renner, G., Sauerbier, P., Schmidt, T. C., & Schram, J. (2019). Robust Automatic Identification of Microplastics in Environmental Samples Using FTIR Microscopy. *Analytical Chemistry*, 91(15), 9656–9664. DOI: 10.1021/acs.analchem.9b01095

Renner, G., Schmidt, T. C., & Schram, J. (2020). Automated rapid & intelligent microplastics mapping by FTIR microscopy: A Python–based workflow. *MethodsX*, 7, 100742. DOI: 10.1016/j.mex.2019.11.015

Rubin, A. E., & Zucker, I. (2022). Interactions of microplastics and organic compounds in aquatic environments: A case study of augmented joint toxicity. *Chemosphere*, 289, 133212. DOI: 10.1016/j.chemosphere.2021.133212

Shi, B., Patel, M., Yu, D., Yan, J., Li, Z., Petriw, D., Pruyn, T., Smyth, K., Passeport, E., Miller, R. J. D., & Howe, J. Y. (2022). Automatic quantification and classification of microplastics in scanning electron micrographs via deep learning. *The Science of the Total Environment*, 825, 153903. DOI: 10.1016/j.scitotenv.2022.153903

Singh, B., & Kumar, A. (2024). Advances in microplastics detection: A comprehensive review of methodologies and their effectiveness. *Trends in Analytical Chemistry*, 170, 117440. DOI: 10.1016/j.trac.2023.117440

Su, J., Zhang, F., Yu, C., Zhang, Y., Wang, J., Wang, C., Wang, H., & Jiang, H. (2023). Machine learning: Next promising trend for microplastics study. *Journal of Environmental Management*, 344, 118756. DOI: 10.1016/j.jenvman.2023.118756

Thammasanya, T., Patiam, S., Rodcharoen, E., & Chotikarn, P. (2024). A new approach to classifying polymer type of microplastics based on Faster-RCNN-FPN and spectroscopic imagery under ultraviolet light. *Scientific Reports*, 14(1), 3529. DOI: 10.1038/s41598-024-53251-5

Tirkey, A., & Upadhyay, L. S. B. (2021). Microplastics: An overview on separation, identification and characterization of microplastics. *Marine Pollution Bulletin*, 170, 112604. DOI: 10.1016/j.marpolbul.2021.112604

Vidal, C., & Pasquini, C. (2021). A comprehensive and fast microplastics identification based on near-infrared hyperspectral imaging (HSI-NIR) and chemometrics. *Environmental Pollution*, 285, 117251. DOI: 10.1016/j.envpol.2021.117251

Vighi, M., Bayo, J., Fernández-Piñas, F., Gago, J., Gómez, M., Hernández-Borges, J., Herrera, A., Landaburu, J., Muniategui-Lorenzo, S., Muñoz, A.-R., Rico, A., Romera-Castillo, C., Viñas, L., & Rosal, R. (2021). Micro and Nano-Plastics in the Environment: Research Priorities for the Near Future. In de Voogt, P. (Ed.), *Reviews of Environmental Contamination and Toxicology* (Vol. 257, pp. 163–218). Springer International Publishing. DOI: 10.1007/398_2021_69

Wang, X., Diao, Y., Dan, Y., Liu, F., Wang, H., Sang, W., & Zhang, Y. (2022). Effects of solution chemistry and humic acid on transport and deposition of aged microplastics in unsaturated porous media. *Chemosphere*, 309, 136658. DOI: 10.1016/j.chemosphere.2022.136658

Wang, Z., Pal, D., Pilechi, A., & Ariya, P. A. (2024). Nanoplastics in Water: Artificial Intelligence-Assisted 4D Physicochemical Characterization and Rapid In Situ Detection. *Environmental Science & Technology*, 58(20), 8919–8931. DOI: 10.1021/acs.est.3c10408

Waqas, M., Wong, M. S., Stocchino, A., Abbas, S., Hafeez, S., & Zhu, R. (2023). Marine plastic pollution detection and identification by using remote sensing-meta analysis. *Marine Pollution Bulletin*, 197, 115746. DOI: 10.1016/j.marpolbul.2023.115746

Weber, F., Zinnen, A., & Kerpen, J. (2023). Development of a machine learning-based method for the analysis of microplastics in environmental samples using µ-Raman spectroscopy. *Microplastics and Nanoplastics*, 3(1), 9. DOI: 10.1186/s43591-023-00057-3

Yang, C., Xie, J., Gowen, A., & Xu, J.-L. (2024). Machine learning driven methodology for enhanced nylon microplastic detection and characterization. *Scientific Reports*, 14(1), 3464. DOI: 10.1038/s41598-024-54003-1

Yee, M., Hii, L.-W., Looi, C., Lim, W. M., Wong, S. F., Kok, Y.-Y., Tan, B.-K., Wong, C., & Leong, C.-O. (2021). Impact of Microplastics and Nanoplastics on Human Health. *Nanomaterials (Basel, Switzerland)*, 11(2), 496. DOI: 10.3390/nano11020496

Zhou, F., Wang, X., Wang, G., & Zuo, Y. (2022). A Rapid Method for Detecting Microplastics Based on Fluorescence Lifetime Imaging Technology (FLIM). *Toxics*, 10(3), 118. DOI: 10.3390/toxics10030118

Chapter 6
Advanced Oxidation Processes (AOPs) for the Degradation of Micro and Nano Plastic

Aqsa Rukhsar
Lahore Garrison University, Pakistan

Muhammad Shahzeb Khan
Sulaiman Bin Abdullah Aba Al-Khail Centre for Interdisciplinary Research in Basic Sciences (SA-CIRBS), Faculty of Basic and Applied Sciences, International Islamic University Islamabad, Islamabad, Pakistan

Zeenat Fatima Iqbal
Department of Chemistry, University of Engineering and Technology, Lahore, Pakistan

ABSTRACT

Micro-nano plastics, or MNPs, are a growing concern due to their widespread presence in the environment. To tackle this issue, advanced oxidation processes (AOPs) offer promising solutions. The focus on employing advanced oxidation processes (AOPs) to remove microplastic nanoparticles (MNPs) from water is increasing among scientists. This study compiles advancements in various AOPs such as photocatalysis, UV photolysis, ozone oxidation, electrocatalysis, Fenton oxidation, plasma oxidation, and persulfate oxidation for MNPs removal. It covers oxidation mechanisms, reaction pathways, removal efficiencies, and influencing factors. However, most AOPs achieve only modest mineralization rates, necessitating further optimization for improved performance. Exploring different AOPs is

DOI: 10.4018/979-8-3693-3447-8.ch006

crucial for complete MNPs breakdown in water, highlighting the future importance of AOPs in MNP elimination

1. INTRODUCTION

Different polymers are used to make plastic and are used extensively in daily life and industrial production because they are easy to process and mold, offer solvent resistance, and provide insulation (Jeong et al., 2023). Usually, Plastics play a ubiquitous role in various applications, with notable examples including polyethylene (PE), polypropylene (PP), and polyvinyl chloride (PVC) (Rose et al., 2023). These plastics typically have a considerable hydrophobicity and higher molecular weight, which makes it challenging for microbial and chemical processes to break them down (J. Chen et al., 2022). These plastic products therefore continue to exist in soil and water, precipitating numerous ecological health concerns and becoming a central focus of environmental research efforts. Sometimes, microplastics break down into smaller particles called nanoplastics (NPs) (Monira et al., 2023). While there is no exact range of sizes for nanoplastics, most research defines them as particles having a diameter of below 0.1 μm or 1 μm (Z. Chen et al., 2021). MNPs have various shapes and colors, such as pellets, films, fibers, foams, and fragments (Murphy, Ewins, Carbonnier, & Quinn, 2016; Nakanishi, Yamaguchi, Hirata, Nakashima, & Fujiwara, 2021). They have different sources and also have different impacts on the human body as shown in Figure 1.

Figure 1. Sources of Micro-Nano plastic and their effects on human health

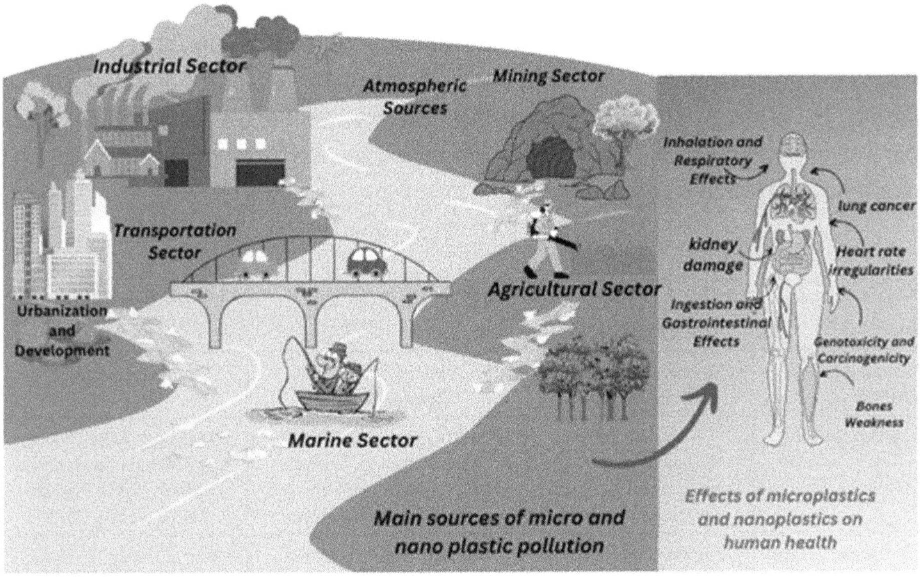

Once released into the environment, micro/nanoplastics (MNPs) pose risks to human health and can have detrimental effects on aquatic life and ecosystems (Xinjie Wang et al., 2021). It is also observed that heavy metals, antibiotics, and organic pollutants can be absorbed by MNPs from water, potentially reducing their negative effects on public health and the environment (Cole & Galloway, 2015; J. Liu et al., 2018; Wegner, Besseling, Foekema, Kamermans, & Koelmans, 2012) . For the treatment of micro/nano plastics, the biological method is the fundamental method (Ahmed et al., 2021; Henoumont, Devreux, & Laurent, 2023; Lv et al., 2019). In addition to biological methods, some physical and chemical methods are also used to eliminate micro and nano plastic from water, physical methods include such as sieving, mechanical filtration, and adsorption, and chemical are electrocoagulation, traditional coagulation, advanced oxidation processes (AOPs), and sedimentation (Heo, Lee, & Lee, 2022; Long et al., 2023; Tadsuwan & Babel, 2022; Zhang, Chen, & Li, 2020) in this text current

technologies with their drabacks and why we need for advanced ones is mention in table 1. AOPs have become more well-known in the management of micro/nanoplastics (MNPs) in recent years. In MNP treatment studies, AOPs have been recognized for their effectiveness, usability, and environmental friendliness. As a result, AOPs are becoming more well-acknowledged for their part in removing MNPs from water (Kim et al., 2022). For the purpose of removing micro/nanoplastics (MNPs) from water, this chapter pays attention to the lack of comprehensive analyses of relevant technologies and AOPs. The review covers a number of AOP treatment methods, such as direct photolysis, photocatalysis, Fenton oxidation, electrocatalysis, sulfate oxidation, ozone oxidation, and plasma oxidation, to fill this gap. This chapter explores the processes and MNPs breakdown mechanisms by AOPs and looks at the factors that affect MNP removal by AOPs. The final section of the article explores the various methods and approaches employed by AOPs to remove micro- and nanoplastics from the environment.

Table 1. Challenges, current technologies with their drawbacks and advanced ones for MNPs

Challenges Of MNPs	Current Technologies to Remove MNPs	Advanced Methods to Remove MNPs	References
Health Risks	Traditional Sedimentation: Not efficient on suspended particles in water	Hybrid Approaches: combines photocatalysts or oxidants with UV radiation to create a synergistic enhancement	(Mohamed Noor, Wong, Ngadi, Mohammed Inuwa, & Opotu, 2022; Nosike, Zhang, & Wu, 2021)
Regulatory Gaps	Screening and Filtration (Often, mechanical techniques are unable to collect the smallest particles)	Constant Monitoring and Adjustment: Uses real-time tracking to make adjustments to operations in real-time for optimum efficiency.	(Wong, Vuong, & Chow, 2021)
Ecosystem Disruption	Inconsistent Regulations: different enforcement and standards in different areas.	Bio-inspired Methods: Use natural enzymes to degrade MNPs	(Janakiraman et al., 2024; Kleit, 1992; Malik et al., 2023)
Complex Removal	UV Irradiation: when mixed with oxidants such as H_2O_2, increases the rates of breakdown	Photocatalytic Oxidation: Use ZnO and TiO_2 photocatalysts for the degradation of MNPs	(Babaei et al., 2017; Garrido et al., 2019)
Environmental Persistence	Chemical Decomposition (it may breakdown MNPs but also generates harmful hazardous products	Nanotechnology: Creates nanomaterials to break the MNPs	(Anusha et al., 2024; Dube & Okuthe, 2023)
Bioaccumulation	Adsorption onto Activated Carbon: it has a confined capacity to remove MNPs	AOPs (generates a significant amount of reactive oxygen species (ROS) to break down MNPs into less hazardous molecules.	(Kouchakipour, Hosseinzadeh, Qaretapeh, & Dashtian, 2024; Singh, Singh, Vasishth, Kumar, & Pattnaik, 2024)

2. USE OF AOPS IN DEGRADATION OF MNPS

The polymeric structure of MNPs makes them stable in processes such as flotation, coagulation, sedimentation, and screening. But under certain chemical reactions, they might break down (Yang Li et al., 2022). AOPs have the ability to generate potent reactive oxygen species (ROSs), including $SO^{4·-}$ ($E_0 = +2.5, +3.1$ V, against NHE) and hydroxyl radical (·OH, $E_0 = +1.8, +2.7$ V), among others. With their strong redox potential, these ROSs can break down MNPs into small molecule intermediates and even further mineralize into CO_2 and H_2O. AOPs have emerged as a key technology for MNP degradation, utilizing various indicators such as mass, particle size, shape, composition, functional group changes (vinyl and carbonyl indices), and mineralization efficiency to gauge degradation rates (Li, Li, Xiong, Yao, & Lai, 2019) Here are some applications of AOPs for MNP degradation.

2.1 Wet Oxidation

The wet oxidation method breaks down micro/nanoplastics (MNPs) by using air (WAO) or oxygen (WOO) as an oxidant at high temperatures and pressures. Free radical chain reactions, which are harmless and do not require dangerous chemical reagents, are usually used in this process. However, because of its high temperature (150–325°C) and pressure (10–200 bar), it needs substantial oxidation levels to be obtained in a moderate amount of time. The WOO method has demonstrated encouraging outcomes despite its limitations, lowering MP concentration by 86.3 ± 1.7% at working conditions of 50–60 bar and 200°C. The combined effects of temperature, O_2, and water have an impact on this degradation process. Chain cracking is caused by the oxidative breakdown of hydroperoxides, which is the result of a reaction between the polymer and O_2 in the solution that produces peroxide radicals. MPs' thermal degradation is similarly influenced by temperature. The potential for the formation of free radicals increases with temperature (Solís-Balbín, Sol, Laca, Laca, & Díaz, 2023).

2.2 Photocatalytic Oxidation

Usually, photocatalysis speeds up reactions by transforming light energy into compounds that act on semiconductor catalysts, without consuming the catalyst in the reaction (El Baraka et al., 2020). When light is absorbed with more energy than the intrinsic band gap of the photocatalyst, the photocatalyst is activated in the conduction band (Eq. (1), in valence band produces e^- and h^+. These substances can then combine with OH and H_2O to form radicals like O_2 and.OH.$^-$Equations (2) and (3) (Llorente-García, Hernández-López, Zaldívar-Cadena, Siligardi, &

Cedillo-González, 2020; Yuwendi, Ibadurrohman, Setiadi, & Slamet, 2022). In the photocatalytic process, equation (4)–(7) shows that the highly productive generation of H_2O_2 is an essential step towards the subsequent creation of •OH radicals in water (Vital-Grappin et al., 2021). Furthermore, by assembling R• radicals from the capture of hydrogen atoms on the polymer's side chains, these reactive oxygen species (ROSs) have the ability to damage MNPs. This process breaks down polymer chains and forms intermediate groups, which in turn causes the mineralization of H_2O and CO_2Eq. (8) (Bacha, Nabi, & Zhang, 2021; Bhatnagar & Asija, 2016; u Zhao, Li, Chen, Shi, & Zhu, 2007). MNPs molecules are essential for indirect photolysis, facilitating the production of ROSs. However, even in the absence of MNPs, ROSs can be generated directly and efficiently during the photocatalytic process. The principal mechanism behind MNP degradation during photocatalytic oxidation is the action of ROSs produced by photocatalysts.

$$(\text{Photocatalyst}) + hv \rightarrow e^- + h^+ \quad (1)$$

$$H_2O + h^+ \rightarrow \cdot OH + H^+$$
(2)

$$O_2 + e^- \rightarrow O_2^{\cdot}$$
(3)

$$O_2\cdot^- + H_2O \rightarrow HO_2\cdot^- + OH^-$$
(4)

$$2HO_2\cdot^- \rightarrow H_2O_2 + O_2$$
(5)

$$H_2O_2 + O_2\cdot^- \rightarrow \cdot OH + O_2 + OH^-$$
(6)

$$hv + H_2O_2 \rightarrow 2\cdot OH$$
(7)

$$MNPs + ROS \rightarrow CO_2 + H_2O \quad (8)$$

This photocatalytic process produces an abundance of active free radicals as compared to the less potent direct photolysis method, Through a variety of processes, these radicals can engage with the MNPs surface, including charge transfer and adsorption, facilitating MNP deconstruction and substantially accelerating the degradation process.

2.3 Bismuth-Based Oxides

Bismuth-based oxides have significant attention for environmental remediation as promising solar light-activated photocatalysts. For degrading HDPE (High-Density Polyethylene) MPs (200-250 μm), mannitol was incorporated into the synthesis of ultra-thin, hydroxyl-rich BiOCl nanosheets. These nanosheets exhibited an impressive 5.38% mass loss when exposed to solar radiation, a value that surpasses the mass loss of typical BiOCl nanosheets by 24 times (Jiang et al., 2021).

Furthermore, Du et al. modified BiOBr by incorporating potassium hexafluorophosphate (KPF6). They found that the modified BiOBr nanosheets can produce high H_2O_2, and promote further degradation of MNPs. Interestingly, several micromotors based on photocatalysis have been employed for MNP degradation (Du, Xie, & Wang, 2021). The contact efficiency with contaminants is improved by these micromotors, compensating for the poor aggregation and dispersion commonly observed with conventional photocatalysts. These micromotors not only move in aqueous solutions, but also demonstrate photocatalytic activity under UV or solar light exposure (Villa, Děkanovský, Plutnar, Kosina, & Pumera, 2020). The photocatalytic reaction of O_2 and H_2O_2 bubbles, occurring on the micromotors' surface, directly propels the micromotors forward (Beladi-Mousavi, Hermanova, Ying, Plutnar, & Pumera, 2021). Similar to pure BiOI, the BiOI–Fe_3O_4 micromotor degraded 56% of Polystyrene MPs in 24 hours, approximately 1.47 times faster (Khairudin, Bakar, & Osman, 2022). Thus, photocatalyst-based micromotors are anticipated to improve MNP degrading efficiency by overcoming the intrinsic diffusion-related constraints.

2.4 TiO_2

Various organic contaminants have been effectively degraded through photocatalysis using TiO_2(a widely used photocatalyst) (Skocaj, Filipic, Petkovic, & Novak, 2011). It is clear that TiO_2's structure and shape have a large impact on its photocatalytic activity. For the degradation of PS NPs (314.8 nm), Domínguez-Jaimes et al. (2021) reported three distinct structures of TiO_2. According to his findings, the TiO_2/M structure underwent a substantially greater mineralization rate (1.7%) when exposed to UV irradiation. Charge separation on TiO_2/M could be facilitated by surface defects occurring, thereby enhancing the rate of PS NPS photocatalytic degradation and reducing the rate of photo-generated carrier recombination (Domínguez-Jaimes, Cedillo-González, Luévano-Hipólito, Acuña-Bedoya, & Hernández-López, 2021). Its general mechanism for the degradation of plastic is shown in Figure 2.

Figure 2. A general mechanism for the degradation of plastic through TiO_2 base photocatalysts (Nabi, Ahmad, & Zhang, 2021)

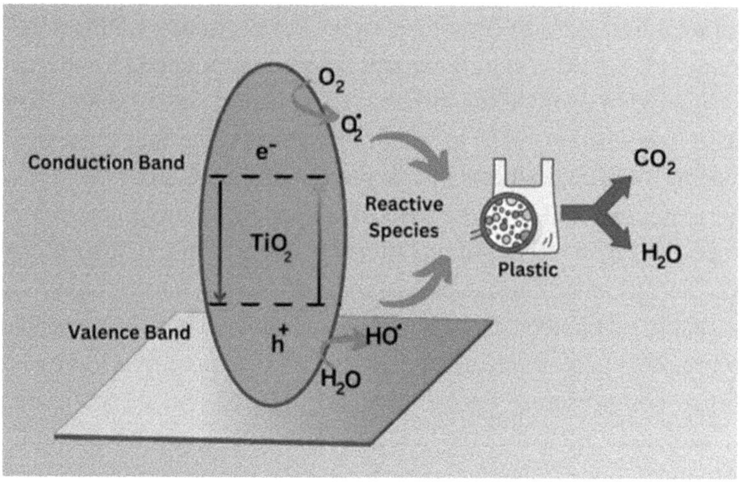

It has been found that doping with non-metals and metals, such as reduced graphene oxide (RGO) or graphene oxide (GO), and silver or nitrogen, enhances TiO_2's photocatalytic degradation efficiency for MNPs (Uogintė, Pleskytė, Skapas, Stanionytė, & Lujanienė, 2023; L. Wang, Kaeppler, Fischer, & Simmchen, 2019). For example, nanocomposites of Ag/TiO_2 can totally degrade PS MPs (sized between 100 and 250 μm) when exposed to UV light for 120 minutes, resulting in total mass loss (Maulana & Ibadurrohman, 2021). Furthermore, Fadli et al. (2021) noted that in contrast to Ag/TiO_2-RGO (76%) and Ag/TiO_2 (68%), PE MPs mass loss was relatively low when degraded with unmodified TiO_2 (56%). This improvement is attributed to the enhanced electron-hole separation and charge transfer facilitated by RGO and Ag. However, TiO_2 photocatalytic activity is primarily limited to the ultraviolet spectrum. Therefore, metal-free modifications such as N–TiO_2 may further enhance its performance (Fadli, Ibadurrohman, & Slamet, 2021). In extension of this study Ariza-Tarazona et al., successfully degraded HDPE MPs (700–1000 μm) using N–TiO_2. While 1.85% of the mass was lost in the air during an 18-hour period under solar light irradiation, a 6.4% mass loss was recorded. It has been demonstrated in earlier research that the photocatalytic degradation of MNPs requires ·OH radicals produced by water adsorbed on the semiconductor surface. (Ariza-Tarazona, Villarreal-Chiu, Barbieri, Siligardi, & Cedillo-González, 2019). Despite extensive research on TiO_2 photocatalysis, its relevance remains significant in addressing the ongoing issue of fresh MNP contamination in the real world.

2.5 Fenton Oxidation

Photo-/heat-assisted Fenton and Fenton processes use the environmentally benign oxidant H_2O_2 (E_0 = +1.77 V versus NHE) in situ to produce highly reactive ·OH radicals when they are supported by a Fe salt catalyst. These radicals can effectively attack MNPs, leading to their further breakdown into smaller molecules, including CO_2 and H_2O (Feng et al., 2011; J.-c. Wang, Wang, Huang, & Wang, 2017). Equations (9)–(11) delineate the general Fenton oxidation reaction mechanism responsible for the degradation of MNPs (J.-c. Wang et al., 2017).

$$Fe^{2+} + H_2O_2 \rightarrow Fe^{3+} + OH^- + \cdot OH \tag{9}$$

$$RH + \cdot OH \rightarrow H_2O + R\cdot \tag{10}$$

$$R\cdot + H_2O_2 \rightarrow ROH + \cdot OH \tag{11}$$

Fenton oxidation has been observed to induce chemical structural changes in MPs, leading to particle size reductions, and even mineralization (Ortiz et al., 2022). Wang et al. discovered that hydrophilic oxygen-containing groups were added to MP surfaces by Fenton oxidation, which significantly reduced the contact angle of water of both PC and PS MPs (J.-c. Wang et al., 2017). Moreover, PS and PE MPs' diameters reduced by 57% and 63% sequentially, to less than 20 μm following Fenton oxidation (P. Liu et al., 2019). According to another study, under rather harsh working circumstances when the temperature is 80°C and pH is 3, the mineralization rate of PS NPs (140 nm) reached 70% after 7.5 hours of Fenton oxidation (Ortiz et al., 2022).

Other oxidation techniques, such as photocatalysis, and photolysis can be conducted with the Fenton process. For instance, after being exposed to sunlight for seven days, PVC (73 μm) and PP (155 μm) MPs in a photo-assisted Fenton system were destroyed. SnOx was used to coat the ZnO nanorods, and Fe^0 nanoparticles were added as catalysts. PVC was shrunk to 28 μm in size, while PP was lowered to 52 μm (Piazza et al., 2022). In the Fenton process, Fe^0 releases Fe^{2+} upon oxidation in the presence of H_2O_2. To continuously supply Fe^{2+}, the produced Fe^{3+} reacts with Fe^0. Furthermore, as illustrated in Figure 3, Fe^{3+} can recycle both ions by absorbing the excited e- in the ZnO conduction band and regenerating Fe^{2+} (Raji, Mirbagheri, Ye, & Dutta, 2021). In conclusion, Fenton processes accelerate MNP surface aging, and their mass loss rate remains moderate, accompanied by sludge containing residual iron. Coupling with other technologies enhances degradation but increases energy consumption and reactor complexity. Further research is needed to optimize energy use and improve efficacy.

Figure 3. Mechanism diagram for the photo-assisted Fenton degradation of MNPs using ZnO/SnOx/Fe0 catalyst (Wang, Dai, Li, & Yin, 2023)

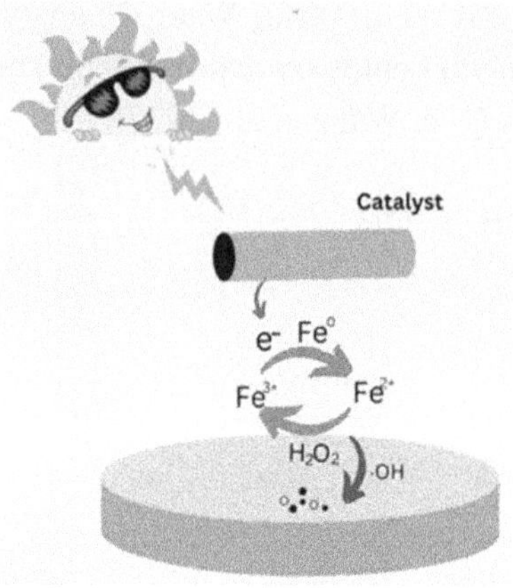

2.6 Direct Photolysis

When MNPs are exposed to sunlight, they undergo a phenomenon known as photo-aging, which alters their physicochemical characteristics and microstructure. This includes changes in their surface charge distribution, particle size, and functional group composition (W. Liu, Pan, Liu, Jiang, & Zhang, 2023; Ouyang et al., 2023; Ouyang, Zhang, et al., 2022). After 150 days of exposure to sunlight, 71.4% of PS MPs (150 μm) decreased in size to less than 80 μm(Xiaojie Wang et al., 2023). Figure 4 Environmentally persistent free radicals (EPFRs) were responsible for producing reactive oxygen species (ROS) on photo-aged PS MPs. The aged PS MPs were found to stimulate triplet states PS (3PS*) in the presence of sunlight, which transferred energy to dissolved O_2 to produce singlet oxygen ($^1O_2$). In the meantime, radiation will cause the H_2O_2 produced by the O_2·- to further transform into ·OH. According to Zhu et al., Many ROS have the potential to damage most MPs, leading to an increase in negative charges and oxidative functional groups on PS MPs' surfaces (Zhu et al., 2020).

Figure 4. The production of reactive oxygen species when polystyrene microplastics are suspended in sunlight (Zhu et al., 2020)

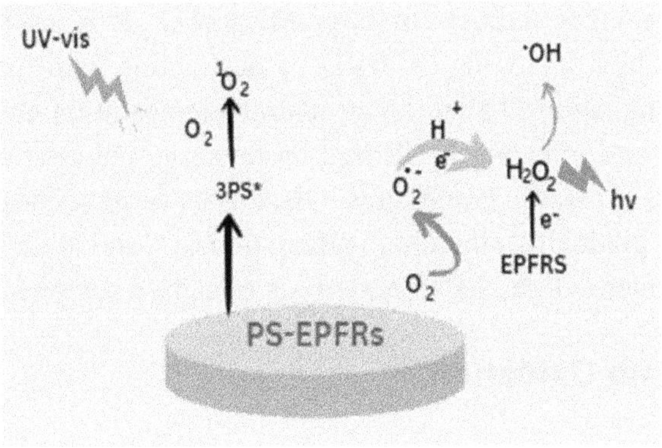

The photolysis of MNPs may be sped up by exposure to UV light (Z. Lin et al., 2022; Shi et al., 2021; Tian et al., 2019). Based on the study, when compared to PE, PVC, and PET MPs, The O/C percentage increased most quickly in PS MPs. As intensities increased from 0 to 1080 mJ cm^{-2}, this ratio climbed from 4% to 62% (J. Lin, Yan, Fu, Chen, & Ou, 2020). This susceptibility is attributed to the PS MPs benzene ring groups, which are easily oxidized by radicals and are sensitive to UV light with a wavelength of 254 nm. Tian et al. utilized the ^{14}C radioisotope tracer approach to evaluate the mineralization and breakdown of PS NPs under 254 nm UV light. Control tests carried out in both environments revealed that the mineralization rate of ^{14}C-labeled PS NPs in water was higher than that in air, mostly due to quicker interactions with water's light-induced ·OH (Tian et al., 2019). MNPs usually undergo a slow direct photolysis process because of their chemical stability and the limited photolysis capacity of UV light. Artificial light stabilizers are added to various plastic manufacturing processes, further hindering this rate (Duan et al., 2021; Luo et al., 2020). UV radiation causes H_2O_2 to change into ·OH. As a result, it is found that applying UV and H_2O_2 together has a synergistic impact that promotes MNP degradation. When exposed to UV/H_2O_2, PS NPs (100 nm) experienced both chemical and physical changes, which increased surface brittleness and roughness because of ·OH oxidation. Following UV/H_2O_2 treatment, PS NPs' surface carbonyl index was nearly 60% more than it was following UV oxidation. Additionally, research was done on PS NP oxidation by UV/Cl_2. It was noted that the pristine PS NPs' carbonyl index increased from 0.05 to 0.2 (X. Liu et al., 2021). Furthermore, MPs radical cations can be stabilized by Lewis base sites on mineral surfaces, especially kaolinite surfaces, which stops them from recombining with hydrated electrons. This stabilization facilitates aerobic ·OH production, which in turn speeds up MP breakdown (Ding et al., 2022).

Under near-UV radiation, MNPs undergo light-induced activation, releasing hydrogen and forming R· radicals. In the presence of O_2, R· reacts to produce ROO·, which generates hydroperoxide (ROOH) by abstracting hydrogen from polymer molecules. UV irradiation also induces structural fragmentation, creating mesopores, and micropores, and reducing MNP size by breaking polymer chains (He et al., 2023; X. Wang et al., 2020). Eventually, ·OH attacks various low molecular weight chain fragments, gradually mineralizing them into CO_2 and H_2O. However, even in the presence of intense light, MNP photolysis remains a somewhat slow process.

2.7 Persulphate Oxidation

When water ionizes with persulfate, it generates the ions of persulfuric acid ($S_2O_8^{2-}$, $E_0 = +2.01$ V vs NHE). These ions have oxidation potential and functional groups such as peroxy (-O-O-), however, their oxidation ability is limited and their reaction rate is slow (Luo et al., 2021). However, under specific conditions like UV irradiation, high temperature, or transition metal ion catalysis (M), persulfate can be activated to form stronger oxidizing active radicals, such as ·OH and $SO_4^{\bullet-}$, as shown by Eq. (12)- (15). Consequently, persulfate-accelerated oxidation shows great promise for degrading MNPs in water (Gao, Wang, Ji, & Li, 2022; Ouyang, Li, et al., 2022).

$$S_2O_8^{2-} + \text{heat} \rightarrow 2\,SO_4^{\bullet-} \qquad (12)$$

$$SO_4^{\bullet-} + H_2O \rightarrow \bullet OH + H^+ + SO_4^{2-} \qquad (13)$$

$$S_2O_8^{2-} + h\nu \rightarrow 2\,SO_4^{\bullet-} \qquad (14)$$

$$S_2O_8^{2-} + M \rightarrow M^+ + SO_4^{\bullet-} + SO_4^{2-} \qquad (15)$$

According to Liu et al., PE and PS MPs with initial diameters of 50.4 μm and 45.5 μm, were treated with 100 mM $K_2S_2O_8$ solutions at 70°C an activation temperature. After 30 days, 80.1% and 97.4% of the MPs, respectively, were below 20 μm. PE exhibited a more significant reduction in average size compared to PS. Additionally, the fragmentation times differed between PE (5-30 days) and PS (0-20 days), likely due to variations in their tensile strength and molecular makeup. PS often has lower tensile strength than PE because the carbon atoms connected to the phenyl group in

PS are less resistant to chemical oxidation than the secondary carbon atoms in the PE main chain (P. Liu et al., 2019). Mn@NCNTs (Manganese carbide nanoparticles) and hydrothermal conditions were used by Kang et al. to activate peroxymonosulfate, which then broke down MPs (PE (0.45 mm)) in water. This method resulted in a 41% mass loss rate for PE MPs, significantly higher than rates achieved with MnO_2 (25%) or NCNTs (18%) alone. Persulfate activation yields $SO_4^{\bullet -}$ alongside •OH, offering superior redox potential and stability, with less susceptibility to pH changes. Current research emphasizes persulfate activation via catalysts, heat, and UV light (Kang et al., 2019). While persulfate accelerated oxidation is a burgeoning area of research, most studies currently delve into mechanism and process analysis. There's a predominant focus on understanding the production and conversion of various reactive oxygen species (ROS), with limited attention given to their role in MNP degradation. To broaden the applications of persulfate advanced oxidation, future research should aim to deepen both theoretical understanding and technological advancements in MNP degradation techniques (Kang et al., 2019).

2.8 Electrochemical Oxidation

Micro/nanoparticles (MNPs) can be effectively degraded in water through eco-friendly electrochemical advanced oxidation processes (EAOPs), generating highly active oxidizing agents such as H_2O_2 and ·OH (Paździor, Bilińska, & Ledakowicz, 2019). PVC was subjected to electro-Fenton-like degradation in a study by Miao et al. (2020) reported a TiO_2/graphite cathode, which within six hours yielded a mass loss of 56% and a dechlorination efficiency of 75%. PVC was oxidized and fragmented by ·OH, whereas direct reduction at the cathode was the main force for dechlorination. Adsorbed as an intermediate product on the TiO_2/graphite cathode surface, H_2O_2 enabled the in situ production of ·OH without the need for desorption (Miao et al., 2020). According to Kendrebeogo et al. (2021), PS MPs (25 μm) suspended in a solution may undergo a mass loss of up to 89 ± 8% in just 6 hours if boron-doped diamond (BDD) is used as the anode electrode. It was proposed that anodic oxidation at the BDD electrode was the mechanism underlying PS MP deterioration (Kiendrebeogo, Estahbanati, Mostafazadeh, Drogui, & Tyagi, 2021). According to certain theories, anodic oxidation at the BDD electrode is the process underlying PS MP deterioration. Ph indicates phenolic group. Eventually, this decomposition could lead to mineralization into CO_2 and H_2O.

$$BDD + H_2O \rightarrow BDD(\cdot OH) + H^+ + e \quad (16)$$

$$-(CH_2\text{-}CHPh\text{–}CH_2)\text{-} + BDD(\cdot OH) \rightarrow \text{–}(CH_2\text{–}CO\cdot Ph\text{–}CH_2)\text{–} + H^+ + e^- + BDD \quad (17)$$

$-(CH_2–CO·Ph–CH_2)– \rightarrow CH_2–CO\text{-}Ph + –CH_2·$
(18)

$–(CH_2–CO·Ph–CH_2) –\rightarrow CH_2–CO–CH_2 + Ph·$
(19)

$CH_2–CO\text{-}Ph; –CH_2·; CH_2–CO–CH_2; Ph·; + BDD(·OH) \rightarrow CO_2 + H_2O$ (20)

EAOP effectively degrades MNPs in water, offering easy operation and low pollution. Using BDD as an anode enhances oxidative capabilities, but direct oxidation might not fully decompose MNPs, requiring high energy. Introducing a cathode to produce H_2O_2 improves efficiency and reduces costs, making EO-H_2O_2 a promising technology for MNPs removal (Cai, Niu, Shi, & Zhao, 2019).

2.9 Ozone Oxidation

Because of its potent oxidizing abilities, ozone (O_3) can break down organic pollutants in water by reacting with a variety of chemical bonds, such as double bonds that comprise atoms of phosphorus, nitrogen, and oxygen, organic molecules, and C–C bonds (Hao et al., 2023; Z. Liu, Demeestere, & Van Hulle, 2021; Yu et al., 2024). Exposure to ozone (O_3) altered the physicochemical properties of MNPs, strengthening the –OH and CO bands on the MNP surface. With a constant H_2O_2 concentration of 1 M and O_3 gas flow rates of 1, 3, and 5 L min^{-1}, researchers examined the effects of O_3/H_2O_2 treatment on the chemical structure of PE MPs (0.85 mm). Injecting H_2O_2 can cause O_3 to decompose into ·OH more quickly. The carbonyl index with the highest value (1.33), at the starting pH of 12 and the flow rate of 3 L min^{-1}, was observed (Amelia et al., 2022). Ozone (O_3) oxidation enhances the hydrophilicity of PS NPs (100 nm) by adding oxygen-containing groups, offering a potential solution for NP removal. Achieving a mineralization rate of 42.7% in 240 minutes, with an average O_3 dose of 4.1 mg L^{-1}, demonstrates the efficacy of this approach (Yu Li et al., 2022).

Advanced wastewater treatment facility studies have shown that O_3 oxidation speeds up MP embrittlement and surface degradation, generating smaller particles and increasing the quantity of 3-5 μm MPs by 16.0% following treatment (Z. Wang, Lin, & Chen, 2020). Moreover, the removal effectiveness rose by 22.2% when O_3-oxidized MPs were later filtered via granular activated carbon (GAC). Therefore, the precise mechanism by which GAC boosts the removal efficacy of O_3-oxidized MPs remains unknown. Interestingly, PET fibers disappeared entirely in the O_3 oxidation effluent, while PET particle content increased. PET's increased crystallinity, which makes it more brittle and prone to breaking into smaller particles, maybe the cause of this O3 oxidation. When employing existing detection techniques, it is important to take into account the possibility of analysis and detection mistakes caused by the reduction in particle size, as the real increase in MPs that are smaller may exceed the recorded values (Wu, Zhang, & Tang, 2022). Furthermore, O_3 treatment has been found to directly affect the surfaces of PE films (75–500 μm) (Zafar, Park, & Kim, 2021). For 180 minutes, at a 7 mg/min O_3, flow rate, a more than 20% increase in carbonyl groups was observed. It was revealed that O_3 and ROSs eroded the polymer chain of PE MPs, generating peroxyl radicals that subsequently produced hydrogen peroxide, ketones, and alcohols. The specific process is illustrated by Eqs. (18)–(21) (Staehelin & Hoigne, 1982; Yamauchi, Yamaoka, Ikemoto, & Matsui, 1991). When ≅OH radicals remove hydrogen atoms from the polymer chain (Eq. (21) and create functional groups that include oxygen, alkyl radicals (R·) are created.

$$RH + O_3 \rightarrow ROO\cdot + \cdot OH \quad (21)$$

$$2ROO\cdot + RH \rightarrow ROOH + R\cdot$$
(22)

$$2ROO\cdot \rightarrow RC{—}O + RC\text{-}OH + O_2$$
(23)

$$RH + \cdot OH \rightarrow H_2O + R\cdot \quad (24)$$

Ozone, a potent oxidant, is widely used in Advanced Oxidation Processes (AOPs) for microplastic (MPs) removal. Combining ozone with conventional wastewater treatment processes, according to Hideayaturahman and Lee, achieved significant MPs removal, up to 99.2% (Hidayaturrahman & Lee, 2019). Chen et al., demonstrated >90% degradation of polymers within 60 minutes at 35–40°C using ozone alone. Ozone treatment altered MPs' adsorption behaviors, enhancing sorption capacity. However, complete MPs degradation often requires post or combined processes (R. Chen, Qi, Zhang, & Yi, 2018).

3. CHALLENGES AND CONSIDERATIONS IN ADVANCED OXIDATION PROCESSES (AOPS) FOR MICROPLASTICS REMOVAL

Comparing the effectiveness of different AOPs for removing MNPs from water is challenging due to varying reaction conditions such as oxidant concentration, energy intensity, pH, and temperature. While direct photolysis of MNPs is slow, combining

UV with oxidants like H_2O_2, O_3, or Cl_2 can accelerate the process. Photocatalytic degradation shows promise but is complex and costly due to catalyst synthesis and recycling challenges. O_3 and Fenton oxidation have limited efficiency for MNPs removal and require additional energy or reactants, increasing costs. Electrochemical oxidation offers high efficiency but consumes significant energy. Coupled technologies like photo-Fenton or electro-peroxidation can enhance mineralization rates. Economic feasibility and energy consumption are crucial considerations for large-scale adoption. In order to attain high removal efficiency with minimal energy consumption, optimization is necessary because treatment cost is still a barrier.

4. FUTURE PERSPECTIVE

Further research is needed to remove MNPs from water it should cover several areas including (i) Interaction between substances (existing in polluted water practically) and MNPs, also during MNPs degradation process, harmful substances absorbed on MNPs surface their transformation and migration behavior

(ii) utilizing a variety of techniques to break down MNPs, maximizing the effectiveness of oxidants and catalysts to provide a cost-effective and efficient treatment, and controlling reaction conditions to improve treatment efficacy

(iii) highlighting the toxicological impacts of MNPs, especially the byproducts produced in between AOP treatments Selecting the right technologies for the efficient treatment of MNPs will be made easier with a thorough understanding of the essential characteristics of MNPs at various phases of their life cycle. To minimize the detrimental effect of MNPs sources, it is needed to make environment-friendly and degradable plastic

CONCLUSION

In order to control MNPs pollution, AOPs play a significant role. MNPs can also be removed by coagulation and filtration but complete eradication from the environment is not possible. So rather than being limited to the removal of MNPs, AOPs are an essential technology for their destruction. For the degradation of MNPs, photocatalysis is the most viewed study. Researcher hopes to remove MNPs and accelerate their degradation because MNPs degradation mostly based on the process of direct photolysis of the sun in the natural environment. As a result of photocatalysis, MNPs can break down more quickly, resulting in smaller particles and higher mass losses. lower energy is efficient for photolysis, and for practical application decomposition process becomes very slow. Additionally, many other technologies such as electrocatalysis, oxidant-based (e.g., O_3 and H_2O_2), and Fenton oxidation, have demonstrated significant degrading effects, with significant mass loss, and even a significant rate of mineralization. Electrolysis can be combined with different technologies and it shows higher degradation efficiency. The main thing preventing its widespread use, meanwhile, is still its excessive energy consumption. Therefore, given the current state of technological advancement, it is not possible to totally eradicate MNPs with AOPs alone. To effectively remove MNPs from water, a variety of combination techniques should be employed.

REFERENCES

Ahmed, M. B., Rahman, M. S., Alom, J., Hasan, M. S., Johir, M., Mondal, M. I. H., Lee, D.-Y., Park, J., Zhou, J. L., & Yoon, M.-H. (2021). Microplastic particles in the aquatic environment: A systematic review. *The Science of the Total Environment*, 775, 145793. DOI: 10.1016/j.scitotenv.2021.145793

Amelia, D., Karamah, E. F., Mahardika, M., Syafri, E., Rangappa, S. M., Siengchin, S., & Asrofi, M. (2022). Effect of advanced oxidation process for chemical structure changes of polyethylene microplastics. *Materials Today: Proceedings*, 52, 2501–2504. DOI: 10.1016/j.matpr.2021.10.438

Anusha, J., Citarasu, T., Uma, G., Vimal, S., Kamaraj, C., Kumar, V., & Sankar, M. M. (2024). Recent advances in nanotechnology-based modifications of micro/nano PET plastics for green energy applications. *Chemosphere*, 352, 141417. DOI: 10.1016/j.chemosphere.2024.141417

Ariza-Tarazona, M. C., Villarreal-Chiu, J. F., Barbieri, V., Siligardi, C., & Cedillo-González, E. I. (2019). New strategy for microplastic degradation: Green photocatalysis using a protein-based porous N-TiO2 semiconductor. *Ceramics International*, 45(7), 9618–9624. DOI: 10.1016/j.ceramint.2018.10.208

Babaei, A. A., Kakavandi, B., Rafiee, M., Kalantarhormizi, F., Purkaram, I., Ahmadi, E., & Esmaeili, S. (2017). Comparative treatment of textile wastewater by adsorption, Fenton, UV-Fenton and US-Fenton using magnetic nanoparticles-functionalized carbon (MNPs@ C). *Journal of Industrial and Engineering Chemistry*, 56, 163–174. DOI: 10.1016/j.jiec.2017.07.009

Bacha, A.-U.-R., Nabi, I., & Zhang, L. (2021). Mechanisms and the engineering approaches for the degradation of microplastics. *ACS ES&T Engineering*, 1(11), 1481–1501. DOI: 10.1021/acsestengg.1c00216

Beladi-Mousavi, S. M., Hermanova, S., Ying, Y., Plutnar, J., & Pumera, M. (2021). A maze in plastic wastes: Autonomous motile photocatalytic microrobots against microplastics. *ACS Applied Materials & Interfaces*, 13(21), 25102–25110. DOI: 10.1021/acsami.1c04559

Bhatnagar, N., & Asija, N. (2016). *Durability of high-performance ballistic composites Lightweight ballistic composites*. Elsevier.

Cai, J., Niu, T., Shi, P., & Zhao, G. (2019). Boron-doped diamond for hydroxyl radical and sulfate radical anion electrogeneration, transformation, and voltage-free sustainable oxidation. *Small*, 15(48), 1900153. DOI: 10.1002/smll.201900153

Chen, J., Wu, J., Sherrell, P. C., Chen, J., Wang, H., Zhang, W., & Yang, J. (2022). How to build a microplastics-free environment: Strategies for microplastics degradation and plastics recycling. *Advanced Science (Weinheim, Baden-Wurttemberg, Germany)*, 9(6), 2103764. DOI: 10.1002/advs.202103764

Chen, R., Qi, M., Zhang, G., & Yi, C. (2018). *Comparative experiments on polymer degradation technique of produced water of polymer flooding oilfield.* Paper presented at the IOP Conference Series: Earth and Environmental Science. DOI: 10.1088/1755-1315/113/1/012208

Chen, Z., Huang, Z., Liu, J., Wu, E., Zheng, Q., & Cui, L. (2021). Phase transition of Mg/Al-flocs to Mg/Al-layered double hydroxides during flocculation and polystyrene nanoplastics removal. *Journal of Hazardous Materials*, 406, 124697. DOI: 10.1016/j.jhazmat.2020.124697

Cole, M., & Galloway, T. S. (2015). Ingestion of nanoplastics and microplastics by Pacific oyster larvae. *Environmental Science & Technology*, 49(24), 14625–14632. DOI: 10.1021/acs.est.5b04099

Ding, L., Yu, X., Guo, X., Zhang, Y., Ouyang, Z., Liu, P., Zhang, C., Wang, T., Jia, H., & Zhu, L. (2022). The photodegradation processes and mechanisms of polyvinyl chloride and polyethylene terephthalate microplastic in aquatic environments: Important role of clay minerals. *Water Research*, 208, 117879. DOI: 10.1016/j.watres.2021.117879

Domínguez-Jaimes, L. P., Cedillo-González, E. I., Luévano-Hipólito, E., Acuña-Bedoya, J. D., & Hernández-López, J. M. (2021). Degradation of primary nanoplastics by photocatalysis using different anodized TiO2 structures. *Journal of Hazardous Materials*, 413, 125452. DOI: 10.1016/j.jhazmat.2021.125452

Du, H., Xie, Y., & Wang, J. (2021). Microplastic degradation methods and corresponding degradation mechanism: Research status and future perspectives. *Journal of Hazardous Materials*, 418, 126377. DOI: 10.1016/j.jhazmat.2021.126377

Duan, J., Bolan, N., Li, Y., Ding, S., Atugoda, T., Vithanage, M., Sarkar, B., Tsang, D. C. W., & Kirkham, M. (2021). Weathering of microplastics and interaction with other coexisting constituents in terrestrial and aquatic environments. *Water Research*, 196, 117011. DOI: 10.1016/j.watres.2021.117011

Dube, E., & Okuthe, G. E. (2023). Plastics and Micro/Nano-Plastics (Mnps) in the environment: Occurrence, impact, and toxicity. *International Journal of Environmental Research and Public Health*, 20(17), 6667. DOI: 10.3390/ijerph20176667

El Baraka, N., Laknifli, A., Saffaj, N., Addich, M., Taleb, A. A., Mamouni, R., . . . Baih, M. A. (2020). *Study of coupling photocatalysis and membrane separation using tubular ceramic membrane made from natural Moroccan clay and phosphate.* Paper presented at the E3S Web of Conferences. DOI: 10.1051/e3sconf/202015001007

Fadli, M. H., Ibadurrohman, M., & Slamet, S. (2021). *Microplastic pollutant degradation in water using modified TiO2 photocatalyst under UV-irradiation.* Paper presented at the IOP Conference Series: Materials Science and Engineering. DOI: 10.1088/1757-899X/1011/1/012055

Feng, H.-M., Zheng, J.-C., Lei, N.-Y., Yu, L., Kong, K. H.-K., Yu, H.-Q., Lau, T.-C., & Lam, M. H. (2011). Photoassisted Fenton degradation of polystyrene. *Environmental Science & Technology*, 45(2), 744–750. DOI: 10.1021/es102182g

Gao, Y., Wang, Q., Ji, G., & Li, A. (2022). Degradation of antibiotic pollutants by persulfate activated with various carbon materials. *Chemical Engineering Journal*, 429, 132387. DOI: 10.1016/j.cej.2021.132387

Garrido, I., Pastor-Belda, M., Campillo, N., Viñas, P., Yañez, M. J., Vela, N., Navarro, S., & Fenoll, J. (2019). Photooxidation of insecticide residues by ZnO and TiO2 coated magnetic nanoparticles under natural sunlight. *Journal of Photochemistry and Photobiology A Chemistry*, 372, 245–253. DOI: 10.1016/j.jphotochem.2018.12.027

Hao, T., Miao, M., Wang, T., Xiao, Y., Yu, B., Zhang, M., Ning, X., & Li, Y. (2023). Physicochemical changes in microplastics and formation of DBPs under ozonation. *Chemosphere*, 327, 138488. DOI: 10.1016/j.chemosphere.2023.138488

He, J., Han, L., Wang, F., Ma, C., Cai, Y., Ma, W., Xu, E. G., Xing, B., & Yang, Z. (2023). Photocatalytic strategy to mitigate microplastic pollution in aquatic environments: Promising catalysts, efficiencies, mechanisms, and ecological risks. *Critical Reviews in Environmental Science and Technology*, 53(4), 504–526. DOI: 10.1080/10643389.2022.2072658

Henoumont, C., Devreux, M., & Laurent, S. (2023). Mn-based MRI contrast agents: An overview. *Molecules (Basel, Switzerland)*, 28(21), 7275. DOI: 10.3390/molecules28217275

Heo, Y., Lee, E.-H., & Lee, S.-W. (2022). Adsorptive removal of micron-sized polystyrene particles using magnetic iron oxide nanoparticles. *Chemosphere*, 307, 135672. DOI: 10.1016/j.chemosphere.2022.135672

Hidayaturrahman, H., & Lee, T.-G. (2019). A study on characteristics of microplastic in wastewater of South Korea: Identification, quantification, and fate of microplastics during treatment process. *Marine Pollution Bulletin*, 146, 696–702. DOI: 10.1016/j.marpolbul.2019.06.071

Janakiraman, V., Manjunathan, J., SampathKumar, B., Thenmozhi, M., Ramasamy, P., Kannan, K., Ahmad, I., Asdaq, S. M. B., & Sivaperumal, P. (2024). Applications of fungal based nanoparticles in cancer therapy-A review. *Process Biochemistry (Barking, London, England)*, 140, 10–18. DOI: 10.1016/j.procbio.2024.02.002

Jeong, Y., Gong, G., Lee, H.-J., Seong, J., Hong, S. W., & Lee, C. (2023). Transformation of microplastics by oxidative water and wastewater treatment processes: A critical review. *Journal of Hazardous Materials*, 443, 130313. DOI: 10.1016/j.jhazmat.2022.130313

Jiang, R., Lu, G., Yan, Z., Liu, J., Wu, D., & Wang, Y. (2021). Microplastic degradation by hydroxy-rich bismuth oxychloride. *Journal of Hazardous Materials*, 405, 124247. DOI: 10.1016/j.jhazmat.2020.124247

Kang, J., Zhou, L., Duan, X., Sun, H., Ao, Z., & Wang, S. (2019). Degradation of cosmetic microplastics via functionalized carbon nanosprings. *Matter*, 1(3), 745–758. DOI: 10.1016/j.matt.2019.06.004

Khairudin, K., Bakar, N. F. A., & Osman, M. S. (2022). Magnetically recyclable flake-like BiOI-Fe3O4 microswimmers for fast and efficient degradation of microplastics. *Journal of Environmental Chemical Engineering*, 10(5), 108275. DOI: 10.1016/j.jece.2022.108275

Kiendrebeogo, M., Estahbanati, M. K., Mostafazadeh, A. K., Drogui, P., & Tyagi, R. D. (2021). Treatment of microplastics in water by anodic oxidation: A case study for polystyrene. *Environmental Pollution*, 269, 116168. DOI: 10.1016/j.envpol.2020.116168

Kim, S., Sin, A., Nam, H., Park, Y., Lee, H., & Han, C. (2022). Advanced oxidation processes for microplastics degradation: A recent trend. *Chemical Engineering Journal Advances*, 9, 100213. DOI: 10.1016/j.ceja.2021.100213

Kleit, A. N. (1992). Enforcing time-inconsistent regulation. *Economic Inquiry*, 30(4), 639–648. DOI: 10.1111/j.1465-7295.1992.tb01286.x

Kouchakipour, S., Hosseinzadeh, M., Qaretapeh, M. Z., & Dashtian, K. (2024). Sustainable large-scale Fe3O4/carbon for enhanced polystyrene nanoplastics removal through magnetic adsorption coagulation. *Journal of Water Process Engineering*, 58, 104919. DOI: 10.1016/j.jwpe.2024.104919

Li, J., Li, Y., Xiong, Z., Yao, G., & Lai, B. (2019). The electrochemical advanced oxidation processes coupling of oxidants for organic pollutants degradation: A mini-review. *Chinese Chemical Letters*, 30(12), 2139–2146. DOI: 10.1016/j.cclet.2019.04.057

Li, Y., Li, J., Ding, J., Song, Z., Yang, B., Zhang, C., & Guan, B. (2022). Degradation of nano-sized polystyrene plastics by ozonation or chlorination in drinking water disinfection processes. *Chemical Engineering Journal*, 427, 131690. DOI: 10.1016/j.cej.2021.131690

Li, Y., Liu, Y., Liu, S., Zhang, L., Shao, H., Wang, X., & Zhang, W. (2022). Photoaging of baby bottle-derived polyethersulfone and polyphenylsulfone microplastics and the resulting bisphenol S release. *Environmental Science & Technology*, 56(5), 3033–3044. DOI: 10.1021/acs.est.1c05812

Lin, J., Yan, D., Fu, J., Chen, Y., & Ou, H. (2020). Ultraviolet-C and vacuum ultraviolet inducing surface degradation of microplastics. *Water Research*, 186, 116360. DOI: 10.1016/j.watres.2020.116360

Lin, Z., Jin, T., Zou, T., Xu, L., Xi, B., Xu, D., & Peng, J. (2022). Current progress on plastic/microplastic degradation: Fact influences and mechanism. *Environmental Pollution*, 304, 119159. DOI: 10.1016/j.envpol.2022.119159

Liu, J., Ma, Y., Zhu, D., Xia, T., Qi, Y., Yao, Y., Guo, X., Ji, R., & Chen, W. (2018). Polystyrene nanoplastics-enhanced contaminant transport: Role of irreversible adsorption in glassy polymeric domain. *Environmental Science & Technology*, 52(5), 2677–2685. DOI: 10.1021/acs.est.7b05211

Liu, P., Qian, L., Wang, H., Zhan, X., Lu, K., Gu, C., & Gao, S. (2019). New insights into the aging behavior of microplastics accelerated by advanced oxidation processes. *Environmental Science & Technology*, 53(7), 3579–3588. DOI: 10.1021/acs.est.9b00493

Liu, W., Pan, T., Liu, H., Jiang, M., & Zhang, T. (2023). Adsorption behavior of imidacloprid pesticide on polar microplastics under environmental conditions: Critical role of photo-aging. *Frontiers of Environmental Science & Engineering*, 17(4), 41. DOI: 10.1007/s11783-023-1641-0

Liu, X., Sun, P., Qu, G., Jing, J., Zhang, T., Shi, H., & Zhao, Y. (2021). Insight into the characteristics and sorption behaviors of aged polystyrene microplastics through three type of accelerated oxidation processes. *Journal of Hazardous Materials*, 407, 124836. DOI: 10.1016/j.jhazmat.2020.124836

Liu, Z., Demeestere, K., & Van Hulle, S. (2021). Comparison and performance assessment of ozone-based AOPs in view of trace organic contaminants abatement in water and wastewater: A review. *Journal of Environmental Chemical Engineering*, 9(4), 105599. DOI: 10.1016/j.jece.2021.105599

Llorente-García, B. E., Hernández-López, J. M., Zaldívar-Cadena, A. A., Siligardi, C., & Cedillo-González, E. I. (2020). First insights into photocatalytic degradation of HDPE and LDPE microplastics by a mesoporous N–TiO2 coating: Effect of size and shape of microplastics. *Coatings*, 10(7), 658. DOI: 10.3390/coatings10070658

Long, Y., Zhou, Z., Wen, X., Wang, J., Xiao, R., Wang, W., & Deng, C. (2023). Microplastics removal and characteristics of a typical multi-combination and multi-stage constructed wetlands wastewater treatment plant in Changsha, China. *Chemosphere*, 312, 137199. DOI: 10.1016/j.chemosphere.2022.137199

Luo, H., Zeng, Y., Zhao, Y., Xiang, Y., Li, Y., & Pan, X. (2021). Effects of advanced oxidation processes on leachates and properties of microplastics. *Journal of Hazardous Materials*, 413, 125342. DOI: 10.1016/j.jhazmat.2021.125342

Luo, H., Zhao, Y., Li, Y., Xiang, Y., He, D., & Pan, X. (2020). Aging of microplastics affects their surface properties, thermal decomposition, additives leaching and interactions in simulated fluids. *The Science of the Total Environment*, 714, 136862. DOI: 10.1016/j.scitotenv.2020.136862

Lv, X., Dong, Q., Zuo, Z., Liu, Y., Huang, X., & Wu, W.-M. (2019). Microplastics in a municipal wastewater treatment plant: Fate, dynamic distribution, removal efficiencies, and control strategies. *Journal of Cleaner Production*, 225, 579–586. DOI: 10.1016/j.jclepro.2019.03.321

Malik, S., Bora, J., Nag, S., Sinha, S., Mondal, S., Rustagi, S., & Minkina, T. (2023). Fungal-based remediation in the treatment of anthropogenic activities and pharmaceutical-pollutant-contaminated wastewater. *Water (Basel)*, 15(12), 2262. DOI: 10.3390/w15122262

Maulana, D. A., & Ibadurrohman, M. (2021). *Synthesis of nano-composite Ag/TiO2 for polyethylene microplastic degradation applications.* Paper presented at the IOP Conference Series: Materials Science and Engineering. DOI: 10.1088/1757-899X/1011/1/012054

Miao, F., Liu, Y., Gao, M., Yu, X., Xiao, P., Wang, M., Wang, S., & Wang, X. (2020). Degradation of polyvinyl chloride microplastics via an electro-Fenton-like system with a TiO2/graphite cathode. *Journal of Hazardous Materials*, 399, 123023. DOI: 10.1016/j.jhazmat.2020.123023

Mohamed Noor, M., Wong, S., Ngadi, N., Mohammed Inuwa, I., & Opotu, L. (2022). Assessing the effectiveness of magnetic nanoparticles coagulation/flocculation in water treatment: A systematic literature review. *International Journal of Environmental Science and Technology*, 19(7), 6935–6956. DOI: 10.1007/s13762-021-03369-0

Monira, S., Roychand, R., Hai, F. I., Bhuiyan, M., Dhar, B. R., & Pramanik, B. K. (2023). Nano and microplastics occurrence in wastewater treatment plants: A comprehensive understanding of microplastics fragmentation and their removal. *Chemosphere*, 334, 139011. DOI: 10.1016/j.chemosphere.2023.139011

Murphy, F., Ewins, C., Carbonnier, F., & Quinn, B. (2016). Wastewater treatment works (WwTW) as a source of microplastics in the aquatic environment. *Environmental Science & Technology*, 50(11), 5800–5808. DOI: 10.1021/acs.est.5b05416

Nabi, I., Ahmad, F., & Zhang, L. (2021). Application of titanium dioxide for the photocatalytic degradation of macro-and micro-plastics: A review. *Journal of Environmental Chemical Engineering*, 9(5), 105964. DOI: 10.1016/j.jece.2021.105964

Nakanishi, Y., Yamaguchi, H., Hirata, Y., Nakashima, Y., & Fujiwara, Y. (2021). Micro-abrasive glass surface for producing microplastics for biological tests. *Wear*, 477, 203816. DOI: 10.1016/j.wear.2021.203816

Nosike, E. I., Zhang, Y., & Wu, A. (2021). Magnetic hybrid nanoparticles for environmental remediation *Magnetic Nanoparticle-BasedHybrid Materials*, 591–615.

Ortiz, D., Munoz, M., Nieto-Sandoval, J., Romera-Castillo, C., de Pedro, Z. M., & Casas, J. A. (2022). Insights into the degradation of microplastics by Fenton oxidation: From surface modification to mineralization. *Chemosphere*, 309, 136809. DOI: 10.1016/j.chemosphere.2022.136809

Ouyang, Z., Li, S., Xue, J., Liao, J., Xiao, C., Zhang, H., & Guo, X. (2023). Dissolved organic matter derived from biodegradable microplastic promotes photo-aging of coexisting microplastics and alters microbial metabolism. *Journal of Hazardous Materials*, 445, 130564. DOI: 10.1016/j.jhazmat.2022.130564

Ouyang, Z., Li, S., Zhao, M., Wangmu, Q., Ding, R., Xiao, C., & Guo, X. (2022). The aging behavior of polyvinyl chloride microplastics promoted by UV-activated persulfate process. *Journal of Hazardous Materials*, 424, 127461. DOI: 10.1016/j.jhazmat.2021.127461

Ouyang, Z., Zhang, Z., Jing, Y., Bai, L., Zhao, M., Hao, X., Li, X., & Guo, X. (2022). The photo-aging of polyvinyl chloride microplastics under different UV irradiations. *Gondwana Research*, 108, 72–80. DOI: 10.1016/j.gr.2021.07.010

Paździor, K., Bilińska, L., & Ledakowicz, S. (2019). A review of the existing and emerging technologies in the combination of AOPs and biological processes in industrial textile wastewater treatment. *Chemical Engineering Journal*, 376, 120597. DOI: 10.1016/j.cej.2018.12.057

Piazza, V., Uheida, A., Gambardella, C., Garaventa, F., Faimali, M., & Dutta, J. (2022). Ecosafety screening of photo-fenton process for the degradation of microplastics in water. *Frontiers in Marine Science*, 8, 791431. DOI: 10.3389/fmars.2021.791431

Raji, M., Mirbagheri, S. A., Ye, F., & Dutta, J. (2021). Nano zero-valent iron on activated carbon cloth support as Fenton-like catalyst for efficient color and COD removal from melanoidin wastewater. *Chemosphere*, 263, 127945. DOI: 10.1016/j.chemosphere.2020.127945

Rose, P. K., Jain, M., Kataria, N., Sahoo, P. K., Garg, V. K., & Yadav, A. (2023). Microplastics in multimedia environment: A systematic review on its fate, transport, quantification, health risk, and remedial measures. *Groundwater for Sustainable Development*, 20, 100889. DOI: 10.1016/j.gsd.2022.100889

Shi, Y., Liu, P., Wu, X., Shi, H., Huang, H., Wang, H., & Gao, S. (2021). Insight into chain scission and release profiles from photodegradation of polycarbonate microplastics. *Water Research*, 195, 116980. DOI: 10.1016/j.watres.2021.116980

Singh, A., Singh, J., Vasishth, A., Kumar, A., & Pattnaik, S. S. (2024). *Emerging Materials in Advanced Oxidation Processes for Micropollutant Treatment Process. Advanced Oxidation Processes for Micropollutant Remediation*. CRC Press.

Skocaj, M., Filipic, M., Petkovic, J., & Novak, S. (2011). Titanium dioxide in our everyday life; is it safe? *Radiology and Oncology*, 45(4), 227–247. DOI: 10.2478/v10019-011-0037-0

Solís-Balbín, C., Sol, D., Laca, A., Laca, A., & Díaz, M. (2023). Destruction and entrainment of microplastics in ozonation and wet oxidation processes. *Journal of Water Process Engineering*, 51, 103456. DOI: 10.1016/j.jwpe.2022.103456

Staehelin, J., & Hoigne, J. (1982). Decomposition of ozone in water: Rate of initiation by hydroxide ions and hydrogen peroxide. *Environmental Science & Technology*, 16(10), 676–681. DOI: 10.1021/es00104a009

Tadsuwan, K., & Babel, S. (2022). Unraveling microplastics removal in wastewater treatment plant: A comparative study of two wastewater treatment plants in Thailand. *Chemosphere*, 307, 135733. DOI: 10.1016/j.chemosphere.2022.135733

Tian, L., Chen, Q., Jiang, W., Wang, L., Xie, H., Kalogerakis, N., Ma, Y., & Ji, R. (2019). A carbon-14 radiotracer-based study on the phototransformation of polystyrene nanoplastics in water versus in air. *Environmental Science. Nano*, 6(9), 2907–2917. DOI: 10.1039/C9EN00662A

Uogintė, I., Pleskytė, S., Skapas, M., Stanionytė, S., & Lujanienė, G. (2023). Degradation and optimization of microplastic in aqueous solutions with graphene oxide-based nanomaterials. *International Journal of Environmental Science and Technology*, 20(9), 9693–9706. DOI: 10.1007/s13762-022-04657-z

Villa, K., Děkanovský, L., Plutnar, J., Kosina, J., & Pumera, M. (2020). Swarming of perovskite-like Bi2WO6 microrobots destroy textile fibers under visible light. *Advanced Functional Materials*, 30(51), 2007073. DOI: 10.1002/adfm.202007073

Vital-Grappin, A. D., Ariza-Tarazona, M. C., Luna-Hernández, V. M., Villarreal-Chiu, J. F., Hernández-López, J. M., Siligardi, C., & Cedillo-González, E. I. (2021). The role of the reactive species involved in the photocatalytic degradation of hdpe microplastics using c, n-tio2 powders. *Polymers*, 13(7), 999. DOI: 10.3390/polym13070999

Wang, J., Wang, H., Huang, L., & Wang, C. (2017). Surface treatment with Fenton for separation of acrylonitrile-butadiene-styrene and polyvinylchloride waste plastics by flotation. *Waste Management (New York, N.Y.)*, 67, 20–26. DOI: 10.1016/j.wasman.2017.05.009

Wang, L., Kaeppler, A., Fischer, D., & Simmchen, J. (2019). Photocatalytic TiO2 micromotors for removal of microplastics and suspended matter. *ACS Applied Materials & Interfaces*, 11(36), 32937–32944. DOI: 10.1021/acsami.9b06128

Wang, X., Bolan, N., Tsang, D. C., Sarkar, B., Bradney, L., & Li, Y. (2021). A review of microplastics aggregation in aquatic environment: Influence factors, analytical methods, and environmental implications. *Journal of Hazardous Materials*, 402, 123496. DOI: 10.1016/j.jhazmat.2020.123496

Wang, X., Dai, Y., Li, Y., & Yin, L. (2023). Application of advanced oxidation processes for the removal of micro/nanoplastics from water: A review. *Chemosphere*, 140636.

Wang, X., Zheng, H., Zhao, J., Luo, X., Wang, Z., & Xing, B. (2020). Photodegradation elevated the toxicity of polystyrene microplastics to grouper (Epinephelus moara) through disrupting hepatic lipid homeostasis. *Environmental Science & Technology*, 54(10), 6202–6212. DOI: 10.1021/acs.est.9b07016

Wang, Z., Lin, T., & Chen, W. (2020). Occurrence and removal of microplastics in an advanced drinking water treatment plant (ADWTP). *The Science of the Total Environment*, 700, 134520. DOI: 10.1016/j.scitotenv.2019.134520

Wegner, A., Besseling, E., Foekema, E. M., Kamermans, P., & Koelmans, A. A. (2012). Effects of nanopolystyrene on the feeding behavior of the blue mussel (Mytilus edulis L.). *Environmental Toxicology and Chemistry*, 31(11), 2490–2497. DOI: 10.1002/etc.1984

Wong, E. L., Vuong, K. Q., & Chow, E. (2021). Nanozymes for environmental pollutant monitoring and remediation. *Sensors (Basel)*, 21(2), 408. DOI: 10.3390/s21020408

Wu, J., Zhang, Y., & Tang, Y. (2022). Fragmentation of microplastics in the drinking water treatment process-A case study in Yangtze River region, China. *The Science of the Total Environment*, 806, 150545. DOI: 10.1016/j.scitotenv.2021.150545

Yamauchi, J., Yamaoka, A., Ikemoto, K., & Matsui, T. (1991). Reaction mechanism for ozone oxidation of polyethylene as studied by ESR and IR spectroscopies. *Bulletin of the Chemical Society of Japan*, 64(4), 1173–1177. DOI: 10.1246/bcsj.64.1173

Yu, S.-Y., Xie, Z.-H., Wu, X., Zheng, Y.-Z., Shi, Y., Xiong, Z.-K., & Pan, Z.-C. (2024). Review of advanced oxidation processes for treating hospital sewage to achieve decontamination and disinfection. *Chinese Chemical Letters*, 35(1), 108714. DOI: 10.1016/j.cclet.2023.108714

Yuwendi, Y., Ibadurrohman, M., Setiadi, S., & Slamet, S. (2022). Photocatalytic degradation of polyethylene microplastics and disinfection of E. coli in water over Fe-and Ag-modified TiO2 nanotubes. *Bulletin of Chemical Reaction Engineering & Catalysis*, 17(2), 263–277. DOI: 10.9767/bcrec.17.2.13400.263-277

Zafar, R., Park, S. Y., & Kim, C. G. (2021). Surface modification of polyethylene microplastic particles during the aqueous-phase ozonation process. *Environmental Engineering Research*, 26(5), 200412. DOI: 10.4491/eer.2020.412

Zhang, X., Chen, J., & Li, J. (2020). The removal of microplastics in the wastewater treatment process and their potential impact on anaerobic digestion due to pollutants association. *Chemosphere*, 251, 126360. DOI: 10.1016/j.chemosphere.2020.126360

Zhao, X., Li, Z., Chen, Y., Shi, L., & Zhu, Y. (2007). Solid-phase photocatalytic degradation of polyethylene plastic under UV and solar light irradiation. *Journal of Molecular Catalysis A Chemical*, 268(1-2), 101–106. DOI: 10.1016/j.molcata.2006.12.012

Zhu, K., Jia, H., Sun, Y., Dai, Y., Zhang, C., Guo, X., Wang, T., & Zhu, L. (2020). Long-term phototransformation of microplastics under simulated sunlight irradiation in aquatic environments: Roles of reactive oxygen species. *Water Research*, 173, 115564. DOI: 10.1016/j.watres.2020.115564

Chapter 7
Phenotypic and Genotypic Alterations in Plants in Response to Micro/Nano-Plastics

Ridhi Pandey
Gautam Buddha University, India

Aaradhya Pandey
Gautam Buddha University, India

Nishtha Kaushik
Gautam Buddha University, India

Nisha Gaur
https://orcid.org/0000-0002-8699-6659
Gautam Buddha University, India

Eti Sharma
Gautam Buddha University, India

Andrea Naziri
https://orcid.org/0000-0002-6090-308X
University of Cyprus, Cyprus

ABSTRACT

The ubiquitous presence of MPs/NPs has led to multifarious changes which have become a serious concern due to their persistence and existence in the environment. This chapter investigates the complex interactions between the plants and micro/nano-plastics. The primary focus of the chapter is the impact on morphology as well as the molecular framework of the plant as the consequence attributed by MPs/NPs. The origins, fate, absorption, translocation, and physiological impacts of MPs and NPs in plants are highlighted in this review. Furthermore, the idea shifts towards the genotypic landscape wherein the plant gene expression patterns are studied due to the stress level caused by plastics. With the benefit of uncapping technology, many signaling pathways associated with the coping mechanism can be comprehended

DOI: 10.4018/979-8-3693-3447-8.ch007

and the migration of MPs/NPs to the plant tissues and their presence in the seeds are also clarified. Therefore, the phenotypic and genotypic findings are concluded, along with a discussion of the wider implications for ecosystem health.

1. INTRODUCTION

Plastics are flexible materials that can be moulded into solid objects of different sizes and configurations. They are composed of different organic polymers, either synthetic or semi-synthetic. They have a widespread utilization owing to its flexible features, high stability, and simplicity of manufacturing. As a result, worldwide plastics are produced annually at a rate that increased to 359 million tons in 2018 (Azeem et al., 2021). Thus, its uncontrolled movement from terrestrial ecosystems to aquatic habitats has created havoc, even though rules for recycling and managing plastic waste are improving, inappropriate disposal practices are still a global concern. (Geyer et al., 2017). The presence of this resistant polymer and its further degradation by mechanical abrasion, photochemical oxidation, and biological deterioration, (Wang et al., 2021) into tiny plastic particles have become new pollutants. These pollutants are categorized based on the diameter of the plastic pieces or particles, Microplastics are classified as having a diameter of less than 5 mm, whereas Nano-plastics are classified as having a diameter of 1 to 100 or 1000 nm and are produced from larger plastic. In comparison, NPs create a more harmful environment to the plants as they seep deeper into the cells due to their smaller size, they can easily permeate into the plant membrane and destroy its machinery to the core (Ng et al., 2018) (Hahladakis et al., 2018). Microplastics and nano-plastics have a great significance in plant's morphology, reproduction, and genetic parameters. It has shown negative correlation with the toxicity of MPs/NPs.

Most of the recent articles have focused upon MPs/NPs accumulation in aquatic and soil ecosystems and due to insufficient knowledge regarding the phenotypic and genotypic altercations in plants. Therefore, this paper highlights a vivid analysis of how micro- and nano-plastics are absorbed, transported, and accumulated by terrestrial plants. It specifically provides an overview of the presence of these contaminants in the atmosphere and agricultural soil, systematically examines the pathways through which MPs/NPs are absorbed by terrestrial plants, explaining various factors influencing its uptake process, it also, outlines the mechanisms regulating the gene pattern, DNA damage and epigenetic modifications caused due to subsequent accumulation within terrestrial plants; and lastly, outlines potential directions for future research in this area.

2. PHENOTYPIC ALTERATIONS IN PLANTS

The ecosystem of agriculture has been highly polluted by these plastic particles, which have adverse effects on plant development and growth. (Azeem et al., 2021) The MP /NPs uptake by plant roots via absorption and penetrate easily in other plant tissues through a transcriptional pull (Chen et al., 2022). Besides causing oxidative stress, these interactions negatively affect photosynthesis, genetic expression, metabolism, and other growth parameters as described in Figure 1. Certain factors affecting the phytotoxicity of MP/NPS are characteristics of plastic particles (like size, dose of exposure, surface charge, and type), the plants (species, growth stage, and tissues), and the environmental condition (X. Li et al., 2023a) (Chen et al., 2022). They can also modify plant growth, seed germination, and soil properties like its structure, capacity to hold water, and pH of the rhizosphere via different electrical, physical, and chemical properties (Chen et al., 2022) (X. Li et al., 2023a). Furthermore, research was conducted to understand their toxicity, penetration in plants, and their link with electron shutters via a theoretical foundation to better understand their environmental pattern and impact on plants (X. Li et al., 2023a).

2.1. Morphological Changes

The harmful effects of MP/NPs are initiated by their absorption and accumulation via the hydrophobic system. They can be incorporated into plant roots present within the environment which can land on the exterior part of foliage, flowers, and fruits. These accumulations can lead to morphological changes that appear in the form of a decreased plant growth rate, affecting root length, root cell viability, stem, and even the plant's reproductive organs. The MP (30-600 μm) of polyethylene makeup affects the root cell viability and root length, root number, particularly leaf growth rate, and chlorophyll a and b composition of the leaf *Lemna minor* (water duckweed) (Kalčíková et al., 2017). In *Lepidium sativum* (cress seed) at the time of germination plastic particles gather on the surface of a seed, then proceed to track the growing root hairs and radicle (Bosker et al., 2019). In contradiction, there are certain studies conducted by Dovidat.et.al which prove that plastic particles ranging from 50 to 500 nm do not harm the plant development and chlorophyll production of duckweed (*S.polyrhiza*) (Dovidat et al., 2020). Therefore, it is not always necessary that the accumulation of MPs/NPs leads to morphological changes in plants; however sometimes there is no modification in plant morphology. Phenotypic alteration involves the effect of the photosynthetic pigment of plants, transpiration, and nutrient uptake of the plants which are directly related to morphological changes. For example, when MPs and NPs have an impact on the leaf morphology of plants, they lead to various effects such as stomatal blockage and are further discussed in detail.

2.2. Growth Patterns

The growth pattern of plants acts as an important indicator that provides information related to the developing stages of plants and ecological adaptation. This pattern involves information about the variety of procedures such as germination, reproduction stage, and vegetative development of a plant, which are highly affected by various extension factors (i.e. light, temperature, soil, nutrients), and internal factors (genetic factors). It also helps to understand the complex relationship of balancing between internal and external factors. Therefore, it becomes vital to comprehend this pattern for advancement in agricultural processes and efforts done for environmental conservation. MPs/NPs have diverse biological impacts on plant growth all over the plant life cycle, whose impact can be harmful, inconsequential, or favourable. The most significant and starting phase of plant development is seed germination when plastic particles gather on the seed's surface, leading to a decrease in the nutrient and water uptake (Bosker et al., 2019). For example, when *Ceratopteris pteridoides* comes in contact with the PS-NPs (100 nm) immersion and germination of spores suffer a harmful impact, and also involve modification in the gametophyte sex differentiation as it tries to adjust to ecological stress caused by the PS-NPs (Yuan et al., 2019). Various factors like size, shape, surface charge, and concentration of plastic particles, plant species along with their development stage have an impact by the phytotoxicity of NPs/MPs. (Chen et al., 2022) The absorption of plastic particles in the plant depends upon the size of the particles. The particles of plastic contain different types of chemical and physical properties which are successfully fragmented by using physicochemical and biological action (Roy et al., 2023).

A higher concentration of NPs/MPs leads to their maximum interaction with the plants and smaller sizes of particles have greater bioavailability, both factors can induce toxicity at a higher potential. (Chen et al., 2022) For elucidation, the study conducted by Azeem et al. demonstrates that NPs less than 100 nm have harmful effects on the plant's growth parameters whereas MPs >100 nm affect the biochemical indicators of the plant. They have also observed that exposure to NPs harms the germination rate along the root morphology of plants, whereas in comparison MPs have an adverse effect on the germination of plants by 14% and generally they do not have harmful effects on the root morphology. Moreover, a greater exposure period of NPs leads to a reduction in plant health. (Azeem et al., 2022) This can be supported by studies conducted by de Souza Machado. et.al, that higher concentrations of polyethylene microplastic (PE-MP) possess a major effect on the water-holding capacity and some other properties of the soil. This restricts the amount of water that is absorbed by seed cells which causes the seed cells to form protoplasmic colloids which are incapable of changing from a gel phase into a sol state (De Souza Machado et al., 2019). Furthermore, it is also observed when PE-MP (2%,950μm) is in sandy soil, that reduces the soil circularity and permeability and affects the air exchange in soil, which eventually impacts the oxygen supply needed by seed for respiration leading to forming of adverse condition for the seed germination (Wang et al., 2023).

MPs/NPs have a major impact on the interior biological circumstances of seed development. Predominantly they can affect the amount and decrease the rate of inner nutrients. As in the process of seed germination, a "source-sink" association is developed between the endosperm and cotyledons which acts as the storage house of nutrients for the development organs, such as embryonic root and shoot. (X. Li et al., 2023a) According to Bosker. et.al research demonstrates that when *Lepidium sativum* is exposed to the PE-MP of different size ranges 50, 500 and 4800 nm via utilizing consistent bioassay of 72 hrs, they have observed a major impact on the germination rate by reducing it after the exposure to all 3 types of MPs. When seeds are exposed to 4800 nm MPs significant impacts are observed on the germination rate of plants. At the same time, major changes were observed in a plant root within 24 hrs of exposure, which were not seen 48 or 72 hrs after exposure. (Bosker et al., 2019)

In contrast, research by Liang. et.al represent that NP of carbon dots (CDs) plastic of diameter less than 10 nm developed via a waste composed of non-biodegradable plastic, are capable of working as nano-seed germination promoters. But if the quantity of CDs reaches a certain level causes a reduction in soil pH, which leads to erosion on the exterior part of the seed coat of a pea. These "erosion impacts" have beneficial outcomes on seed coat i.e. enlargement, weakening, gas exchange of seed, and water absorption even if MNPs are attached to the surface of the seed coat (Liang et al., 2023). These favourable impacts of CDs on NP are overshadowed by their strong attachment property being obturate on the water and nutrient absorption pathway in the seed.(X. Li et al., 2023a) Therefore, their many factors of MPs/NPs that can lead to phytotoxicity in plants such as size, concentration, and period of exposure which can have adverse effects on the overall plant development that include negative on the germination rate, respiration, water-holding capacity of seed, root morphology and so on. Whereas there are certain studies conducted by (Liang et al., 2023; and C. Sun et al., 2023b) some favourable effects of MPs/NPs on the growth of plants. However, these favourable effects are surpassed by phytotoxicity caused by them.

2.3. Leaf Morphology

Leaf Morphology ongoing research has demonstrated that MP/NPs accumulate in the leaves of a plant via various pathways such as upward migration from the roots toward the leaves, attachment on the leaf surface, direct atmospheric deposit, and penetration by stomatal apertures (Rillig et al., 2019) (De Souza Machado et al., 2019). The particles of MP/NPs are absorbed by a plant's leaf, which can migrate the entire plant via utilizing vascular tissues of plants. MP/NPs are capable of penetrating leaf tissues through stomatal cuticles and can migrate throughout the plant via vascular tissues which make them arrive in different parts of a plant like root, stem, and reproductive organs.(Rillig et al., 2019). These accumulations lead to a primary impact on the photosynthesis of plants by reducing the rate of photosynthesis, and stomatal conductivity. Moreover, also affects the rate of transpiration by making it rapid and also includes certain physiological effects such as enhancing the intercellular concentration of CO_2 .(H. Yu et al., 2021) According to Azeem et al., the accumulation of large particles of the size of MP/NPs on plants' stomata can lead to blockage of stomatal pores, whereas small-size of MP/NPs are removed via the transpiration method. (Azeem et al., 2021) Plants shed water via their stomata (pores present in a leaf) during a process called transpiration. This procedure produces a negative pressure that pulls soluble materials involving MP/NPs from roots to shoots. This kind of strain transfer by the apoplastic pathway involves the transfer of substances via cell wall gaps without penetrating the cell. Nevertheless, the vascular bundle is only able to transmit trace amounts of MP/NP to the stems and leaves (Z. Yu et al., 2024).

According to research conducted by Sun. et.al, MP/NPs may impact gene expression, including protein synthesis linked with photosynthesis (X.-D. Sun et al., 2020). Researchers have also shown that chlorophyll a and chlorophyll b levels decrease in lettuce when exposed to PE micro/nano plastic enhanced (Y. Gao & Collins, 2009). The studies performed by Dong. et.al have demonstrated that PS-MP of size 100-700 nm can travel by carrot stems to various plant parts like flowers, leaves, and fruits; they have a variety of effects on plant photosynthesis (Dong et al., 2021). There are various reasons why MP/NPs affect photosynthesis. Amongst the several insights, one of the major reasons is MP/NP directly impedes a photochemical reaction and charge separation of photosynthesis reaction centers by decreasing the electron transfer rate, increasing the gathering of electrons in the plant's photosynthetic reaction centers, and enhancing the ROS rate (Ullah et al., 2023). Another reason can be when MP/NPs are added it leads to a decrease in the carbon assimilation in photosynthesis, impacting plant photosynthesis. The decrease in assimilation of carbon can restrict the transition of carbon inside the soil and roots of plants, which causes nutrient inadequacy and also weak root development. The dispersion of macro and trace substances in the plant can reduce photosynthetic efficiency and leave development and MP /NP can also act as photoinhibition (Dong et al., 2021). Thereby modification caused by the accumulation of MPs/NPs in leaf morphology affects the growth and development of plants.

2.4. Root System

The MP/NP accumulation on root systems shows various effects on the plant's development like enhancement of biomass, root structure, and physical-biochemical features, highlighting the many facts of the nature of the character of the study (X. Li et al., 2023a). The MP/NPs with small sizes and primarily those with razor-sharp edges and rough surfaces have particularly harmful effects on the roots due to their ease of absorption by the surface of the root (Azeem et al., 2021). The research has demonstrated that micro/nano plastic of PS and PTEE can have mechanical harm on the roots of rice, which enhances the reactive oxygen species (ROS) in roots and leaves, harming the cell membrane and lipid peroxidation (X.-D. Sun et al., 2020). The present investigation done by Li. et.al shows that micrometer-sized plastic particles can penetrate in roots of plants by a crack entrance at the position of lateral root budding, travel with water flow, and nutrient transport in the other parts of the plant (Z. Li et al., 2021). These discoveries are proven by a report that shows both negatively and positively charged nano-plastics are capable of gathering in Arabidopsis thaliana, and positively charged NP can accumulate at a low rate on the root tips in comparison to negatively charged NP, which are usually present at the apoplast and xylem (X.-D. Sun et al., 2020). These accumulations of MP/NPs in plant roots result in the blockage of pores of the cell wall or restrict cell interaction, reducing nutrient as well as water transport and adoption. When plants like onions and lettuce were revealed to MP/NP demonstrated a major decrease in root biomass (below-ground biomass) and above-ground biomass, whereas in ryegrass they restricted an excitation on root biomass (M. Gao et al., 2021) (Cui et al., 2022). Additionally, plants such as tomatoes are revealed to MP/NP, they reduce both root length and fresh mass too, where PS has a great influence on the root length and PE affects the root weight (H. Yu et al., 2021). Thus, the accumulation of MP/NPs on the roots of plants can cause various effects on the overall plant's development such as damaging the cell membrane and cell wall, decreasing the nutrient uptake in plants, increasing the biomass of roots, and so on as mentioned above.

2.5. Nutrient Uptake

As MP/NPs have a small size and act as a stable colloid, which makes them persevere in the environment, therefore they can easily migrate. They have major effects on the water and salt movement in soil and plant nutrient uptake. MP/NP affects the soil moisture level and electrical conductivity, which impacts moisture movement and salt transport. By the process of deprotonation, MP/NPs can neutralize the negatively charged soil and decrease the conductivity of substances. MPs enhances the soil contact angle by filling the soil, it reduces the water infiltration and soil permeability. Due to this, they are hydrophobic and cannot hold soil moisture (Thompson et al., 2021) (Gu et al., 2022) (Diao et al., 2023). They are capable of restricting plant roots by causing hetero-aggregation of opposite-charge particles via blocking ion channels and cell wall gaps. For example, rapeseed roots are not capable of absorbing nutrients, due to the result of the breakdown of root cell integrity. They can reduce micronutrient levels (like iron, manganese, zinc, and copper) which affects the plant's disease resistance (Xu et al., 2022) (T. Li et al., 2023) (C.-Q. Zhou et al., 2021).

Figure 1. Diagrammatical representation of the phenotypic and genotypic alterations in plants due to MPs/NPs

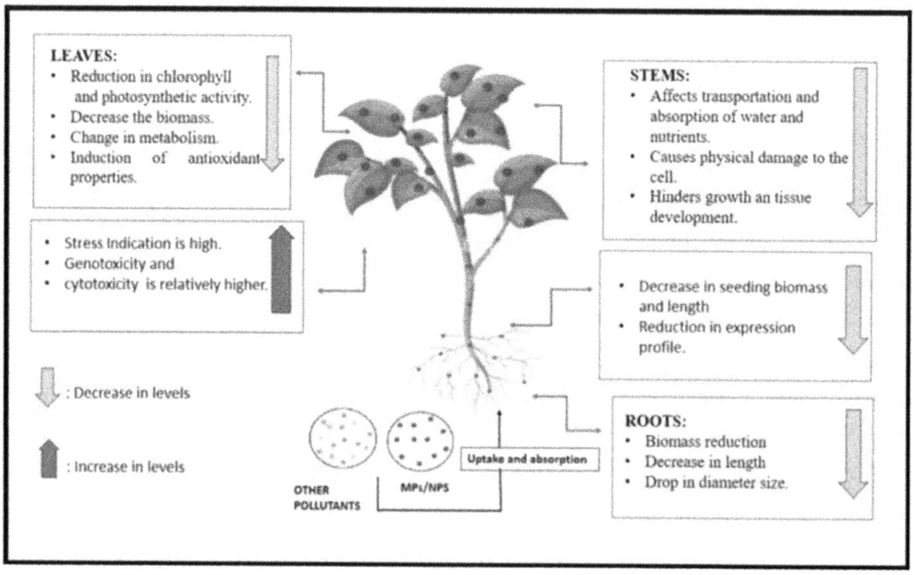

3. GENOTYPIC ALTERATIONS IN PLANTS

Since plants play a vital role in the ecology of an ecosystem, the effects of pollutants on them are particularly significant as shown in Figure 1. Plants absorb NPs through the roots, different observations of transport from the xylem to the upper sections of the plant have been studied (X. Li et al., 2023a). Gene expression has been influenced in a variety of plants by MPs and NPs. Root cells showed a decrease in mitotic rate as a result of a dose-dependent drop in gene expression. Impaired cell viability has been demonstrated by various in vitro investigations, probably as a result of oxidative stress with an increase in reactive oxygen species (ROS) and detoxifying enzymes (Ekner-Grzyb et al., 2022) (Maity et al., 2020) (Yang & Gao, 2022). A decline in mitotic division has been observed by a particular factor alteration due to accumulation of MNPs (Ekner-Grzyb et al., 2022) (Maity et al., 2020). According to the studies conducted by Z. Yu et al., 2024 and Q. Zhou et al., 2020, they have concluded that the alterations in chromosomal regions involves the regression of chromosomal length which results in the dysregulation of the genes involved. A variety of methods including microscopy, cDNA synthesis and spectrophotometry can be employed for the study of these variations in the genes (Yu et al., 2024) (Zhou et al., 2020).

3.1. Identification of Target Genes

Spectroscopy and optical and electron microscopy comprise the methods employed for the detection of the accumulation of microplastics. These strategies facilitate the accurate assessment and quantification of micro and nano plastics (Z. Yu et al., 2024). Weighted Gene Co- expression Network Analysis (WGCNA) is a software implemented to improve the study of the pathways of microplastics (Poma et al., 2023) (Zhou et al., 2020). The approach of estimating gene expression requires the extraction of RNA, development of libraries and the sequencing the reads. To yield the reliable signals and homogenous data, the removal of adaptors and poly N sequences in the resulted reads is employed. The method utilized to validate the transcribed genes and assess the potential of sequencing of total RNA is Reverse Transcription real time PCR of the sample (Maity et al., 2020) (Zhou et al., 2020).

The application of microscopic slides is crucial for the investigation of nuclear modifications and chromosomal abnormalities. A statistical analysis comprising Mitotic Index (MI), Nuclear Abnormality Index (NAI), Micronucleus Index (MN) and Chromosomal Abnormality Index (CAI) is the methodology for the characterization of these abnormalities (Kaur et al., 2022) (Maity et al., 2020). The proportion of the cells acquired experiencing mitosis is Mitotic Index (MI). This is estimated by dividing whole number of cells constantly dividing by the total number of cells. The ration is multiplied with 100 to obtain MI%. The ratio of the cells with chromosomal aberrations and the total number of dividing cells is portrayed by Chromosomal Aberration Index and to access CAI%, it is multiplied with 100. Nuclear Aberration Index indicates the proportion of the cells having nuclear abnormalities to the total number of cells. With the statistical approach, it can be calculated by dividing the cells with abnormalities by total number of cells and can be multiplied with 100 for the NAI%. Micronucleus index or MN (%), are tiny extra nuclei that can occur during cell division and may or may not contain whole or fragmented chromosomes. The fraction of dividing cells with micronuclei is shown by this index. It is computed, similarly to the others, by multiplying the total number of dividing cells detected by the total number of cells having micronuclei, and then dividing the result by 100.

The DPPH radical scavenging activity is used to quantify the antioxidant activity. A stable free radical that resembles intracellular reactive oxygen species is called DPPH. When oxidized, the methanolic solution of DPPH is a deep blue color; however, antioxidants reduce this color to a colourless state. This technique entails the roots homogenization followed by supernatant collection after centrifugation. After the addition of methanol, the absorbance is measured for the determination of antioxidant activity. By measuring the blue Mono Formazan produced by NBT reduction, the superoxide radical was detected indirectly (Ekner-Grzyb et al., 2022), (Maity et al., 2020).

3.2. Epigenetic Modifications and Gene Expression

The study of heritable effects that are not caused by DNA sequence changes is known as epigenetics. Developmental processes are strongly related to epigenetics, which explains how several cells in an organism can arise from the same set of genomic instructions. Epigenetic changes brought on by environmental influences are the subject of environmental epigenetics (Ekner-Grzyb et al., 2022). It has been noted that a hazardous substance's Mito depressive action arises from impeding DNA replication and inhibiting cell cycle regulators, hence preventing cells from entering the G2 phase by damaging cdc2 factor, and the slow decline in the mitotic index over time and in response to varying doses indicates MPS is cytotoxic in *Alium cepa* (Ekner-Grzyb et al., 2022) (Poma et al., 2023) (Maity et al., 2020). It was suggested that epigenetic changes like DNA hypomethylation control a portion of a plant's reaction to environmental stressors. Further research validates that elevated levels of oxidative stress may result in DNA hypomethylation (Awasthi et al., 2018). Oxidative response can also be induced by NPs in plants and cause damage to cells (Ekner-Grzyb et al., 2022; C. Sun et al., 2023a). Excess of reactive oxygen species

(ROS) like superoxide anion radical, hydroxyl radical and hydrogen peroxide is the indication of oxidative stress, attributing to MNP.

Superoxide anion is created in the electron transport chain of the chloroplast and is regarded as the principal ROS. The primary source of oxidative damage to the photosystem in chloroplasts is the hydroxyl radical, which is also the least stable ROS in *Triticum aestivum and Alium cepa* (Adamczyk et al., 2023) (Maity et al., 2020). Plants have developed enzymatic and nonenzymatic antioxidant mechanisms that scavenge reactive oxygen species (ROS). Catalase (CAT), peroxidases (POX), and superoxide dismutase (SOD) are included in the primary antioxidant enzyme and ascorbic acid, glutathione, tocopherol, carotenoids and phenolic compounds are the examples of non- enzymatic antioxidants. SOD helps in the catalyzation of imbalance of oxygen in singlet for the production of H2O2, and is followed by the breakdown of H2O2 to water and oxygen with the help of CAT and APX (S. Li et al., 2021). To restore redox equilibrium, plants must activate their antioxidant defense system because MP stress enhances the production of ROS with the involvement of several enzymatic and non-enzymatic antioxidants (Jia et al., 2023). The result showed that high concentrations of PE MPs exceeded the antioxidant enzyme system's regulatory capacity demonstrated by the treatment of wheat roots with 1% and 5% PE MPs indicating increase in the activity of CAT, POD, and SOD whereas when wheat roots were treated with 8% PE MPs, the activities of CAT and SOD decreased (Jia et al., 2023) (Liu et al., 2021). The process behind the activity of ROS, including the oxidation of C residues, alteration in signaling networks, and lipid peroxidation, has been shown in Table 1.

Table 1. Components involved in the different activities of ROS

Activity of ROS	Mechanism	Components involved	References
Oxidation of C residue	It creates disulfide bridges, which alter the way proteins are shaped and function.	H_2O_2 detecting HPCA1 protein, mitogen activated protein kinase (MAPK), glyceraldehyde-3-phosphate dehydrogenase (GAPDH)	(Mittler, 2017)
Change in signal network	Interaction with other components to change the signal network.	Calcium ion (Ca2+), Reactive Nitrogen Species (RNS), Reactive Sulfur Species (RSS), plant hormones, mitogen activated protein kinase, transcription factor.	(Castro et al., 2021)
Lipid peroxidation	It has an impact on the breakdown of the membrane protein and limiting the ionic transport capacity. The stability of the membrane is lowered as a result.	Malondialdehyde (MDA) and thiobarbituric acid reactive substances (TBARS).	(Giorgetti et al., 2020)

It has been observed in *Alium* cepa and *Oryza sativa* that the addition of nano-plastics, lipid peroxidation, and the production of hydrogen peroxide and ROS scavenging enzymes exerted higher activities (Poma et al., 2023). In addition to producing hydroxyl and superoxide radicals and increasing DPPH scavenging activity and lipid peroxidation, MPS dramatically shortened the root length (Maity et al., 2020) (Yang & Gao, 2022).

Since oxidative response of plants serves as the defense mechanism, in several studies, melatonin was examined to enhance the response. Melatonin, also known as MEL, N-acetyl-5-methoxytryptamine, is an evolutionary conserved molecule helping the regulation of biological processes including responses to stress (Kanwar et al., 2018). Melatonin helps in the enhancement of antioxidant capacity and developing the antioxidative and pro-oxidative enzymes which result in the protection of photosynthetic electron transport systems (Kanwar et al., 2018; Zuo et al., 2017). It has been investigated that NPs treated plants have shown the genetic modifications and changed the oxidative pathways. It was observed that CAT was increased in leaves and APX and SOD were elevated in roots, respectively, when melatonin was treated with NPs affected *Triticum* (S. Li et al., 2021). Since these enzymes play a vital role in decomposition of increased hydrogen peroxide, melatonin depresses the ROS scavenging system against nano-plastics. It can be noted that melatonin activates the gene expression leading to favourable impact on activation of the antioxidant system.

Furthermore, ROS can alter the DNA methylation pattern, which is crucial for the way plants respond to environmental cues since it modifies the way genes are expressed. Hyperaccumulators have been shown to have epigenetic polymorphisms, primarily involving DNA methylation in pokeweed (Jing et al., 2022). Plants use DNA methylation as a defensive mechanism to withstand the stress of heavy metals. This is an extremely stable stress memory mechanism that is passed down to the progeny. This promotes transcription, whereas methylation in the transposon area is linked to the silencing of the transposon. At cytosine, it appears in three distinct sequences. This activity is catalyzed by a variety of methyltransferases and glycosylases in *Alium cepa* (Maity et al., 2020) (Jing et al., 2022). For genome stability in plants, DNA methylation of cytosines at position 5 of the pyrimidine ring (5-Me-C) has been studied (Zhang et al., 2018). An increase in the levels of -CHG contexts has been postulated as a mechanism connected to the protection of damage in DNA in *S. polyrhiza* (Pasaribu et al., 2023). There are different enzymes involved in the DNA methylation mechanisms, these are described below in Table 2.

Table 2. Enzymes involved in different methylation sites

DNA methylation sequences	Subsets of enzymes involved
C-G	Methyltransferase1 (MET1)
C-H-G	Chromomethylase2 or 3 (CMT2/3)
C-H-H	Domains rearranged methylase 2 (DRM2)

Here, 'H' denotes the presence of adenine, cytosine or thymidine (Jing et al., 2022)

The expression of genes which are involved in electron transport, light harvesting and phptosynthesis, have been affected by MNPs. In the study provided by X. Li et al., 2023, it has been demonstrated that during MNP stress, the overexpression of rbcL and psbA, genes involved in photosynthetic pigments, resulting in the watercress inherited toxicity (X. Li et al., 2023a). Moreover, the study also depicts that the chromosomal abnormalities and nuclear distortions are the impact of MNPs (X. Li et al., 2023a). The development of PS NPs (~100nm) in the roots of *Vicia faba* have been observed resulting in the formation of micronuclei and cellular toxins (Jiang et al., 2019; Yang & Gao, 2022). Some nuclear deviations such as micronucleus, binucleated cells and nuclear bud were observed in the cells suggesting chemical genotoxicity, mutagenicity and aneugenic effects in *Alium cepa*. These abnormalities were suggesting the favorable correlation with the different MPs concentrations and its exposure time (Ekner-Grzyb et al., 2022; Maity et al., 2020). Chromosomal aberrations such as lagging ring, disrupted metaphase or anaphase, vagrant, multi-polarity, precious movement, clumped, sticky bridge and disoriented spindle pole are linked to *Alium cepa*. The stickiness, fragmentation and the fusion of chromosomes and chromatin resulting chromosomal bridges and some permanent changes in chromosomal morphology are the results of the above chromosomal alterations (Li et al., 2023b) (Maity et al., 2020).

4. LINK BETWEEN GENETIC AND PHENOTYPIC ALTERATIONS

The environment becomes severely impacted by the amalgamation of genotypic and phenotypic alterations caused in plants due to accumulation of MPs/NPs. The phenotypic changes attributes to the visible traits of an organism that arise from the interplay between its genotype and its surroundings. When plants are exposed to nano-plastics, they experience alterations such as modified growth patterns, decreased photosynthetic efficiency, and heightened vulnerability to pests and diseases. These alterations may have a domino effect on the production and health of plants, potentially affecting entire ecosystems (de Souza Machado et al., 2018) (Xu et al., 2023).

Furthermore, genotypic changes are adjustments to an organism's genetic composition. There is evidence that suggests nano-plastics can produce epigenetic alterations in plants, wherein causing direct genetic modifications from exposure with nano-plastics are not well known. Long-term repercussions to plant physiology and adaptability might result from epigenetic modifications, which can further modify expression patterns without changing the underlying DNA sequence. Extended exposure to nano-plastics may also lead to inherited modifications in gene regulation that would impact plant populations' ability to withstand environmental stresses (C. Sun et al., 2023). Moreover, epigenetic modifications often disrupt regulatory circuits involved in the defensive response, exacerbating the deleterious effects of nano-plastics on plant health (Ferrante et al., 2022) (Kaur et al., 2022)

The effects of altered plant health caused by nano-plastics affect individual organisms as well as entire ecosystems. In order for ecosystems to operate properly, plants must sequester carbon, stabilize the soil, and provide supplies and habitat for other living things any disruptions to plant communities brought on by nano-plastics may have an effect on biodiversity and ecosystem dynamics. For instance, a decrease in primary production brought on by stress from nano-plastics on plants may have an impact on the distribution and abundance of herbivores, predators, and decomposers, changing the structure of ecosystems and trophic relationships (Ferrante et al., 2022) (Maity et al., 2020). Comprehending these intricate relationships is crucial to reducing the ecological damage caused by nano-plastics and preserving the integrity of ecosystems.

5. CASE STUDIES AND EXPERIMENTAL FINDINGS

According to the studies performed by Zhou. et.al (2021) research has demonstrated that when *Oryza sativa* (rice) seeds culture is in the hydroponic condition and are exposed to the NP of PS (polystyrene) size range between 19 $\pm$ 0.16 nm and the concentration of exposure of NP given was 10, 50 and 100mg/L, period of exposure is 16 days, through Laser confocal scanning micrographs they found out that NPs can have an impact on the plant morphology, physiology (such as phenotypic and transcriptomic) (C.-Q. Zhou et al., 2021). As a result, it shows the major impact of MP on the enhanced root length and nodules, while NP increases the carbon metabolic activity and antioxidant activity, and is also involved in the various root-related gene expression. This report was supported by studies conducted by Siyuan. et.al, in which they have observed the harmful impact of the 5 mg/L of quinolinic as an herbicide with PS-NPs of the size range of 50 nm on *Oryza sativa*

L(rice) via conducting a hydroponic experiment for 7 days. The report concluded that QNC (herbicide quinolinic) and PS combination as well as single have an impact on the plant development, length, and biomass. They both decreased the length of root and shoot and root biomass, a major decrease was observed in the shoot length by 11.43% when a combination of PS and QNC was applied. As well as PS primarily enhances the chlorophyll composition in the rice. The different effects were observed on the chlorophyll b content in rice when treated with the PS, QNC, and their combination. It can also be concluded by this report that PS has a major effect on the pigment of photosynthesis by enhancing it (Lu et al., 2023).

Another report by Meng. et.al, where they performed a pot experiment on a *Phaseolus vulgaris L.* (common bean) at a net house under natural conditions by using two plastic-type polylactic acids (PLA) combined with poly-butylene-adipate-co-terephthalate (PBAT, Bio-MPs) and low-density polyethylene at different concentrations and period for exposure is 12 weeks. At 0.5% of LDPE-MPs do not have a major effect on the shoot, root, and fruit biomass but reduce chloroplast content in a leaf, whereas at 1.0% majorly higher specific root nodules and also enhance the leaf area, 2.5% only shows impact on the root length. In Bio-MPs, 1.5% have major limits on the shoot and root biomass, and 2.0% major impact on limited leaf area and fruit biomass (Meng et al., 2021).

Moreover, MPs can also impact soil properties and microbial communication as shown in the studies conducted by Ren et al., on *Brassica rapa* by using plastic-type PS at different sizes 70 nm and 5μm at a concentration of 10 mg/kg where the exposure period was 21 days, showed that PS affects the plant's photosynthesis and development framework whereas MPs have an impact on the soil characteristics and microbe-community constitution of the rhizosphere (Ren et al., 2021). These results can be supported by the studies conducted by de Souza et.al where various types of microplastic on an *Allium fistulosum* (spring onions) where the main polyamide (PA) beads by Cambridge, UK at a diameter ranging between 15-20 μm. By cutting 100% of the polyester wool of “Dolphin Baby” provided as polyester (PES) fibres from a turkey at a diameter range 8 μm, and a length of 5000 μm, the test soil was isolated from the University of Berlin. They exposed the test soil to the various MPs for ~ 2 months duration, then inoculated the seeds of spring onion for 1.5 days. PES enhances the root biomass, whereas PA reduces the ratio of root and leaf dry biomass. PA majorly affects nitrogen-rich soil. They also enhance root length and have an impact on the root symbiosis system, like PA reduces the non-AMF structures but PES has a significant impact on the root's interaction with their surrounding microbial communication (De Souza Machado et al., 2019).

Conversely, research conducted by Chendong et.al, shows various effects of PS when applied at different concentrations (0,1,10, 50 mg L^{-1}) on the development of strawberries. In this experiment, they isolated the strawberries from various locations present in China such as Yuexiu, Jiandehong, Benihpppe, and Akihime. They found out that lower concentrations of PS-NPs act as inducers for root and shoot development, this was proved by increasing dry weight and shoot length and minute impact on the roots of strawberries. But a higher concentration of PS has a major impact on the repressing of both shoots and roots of strawberries (C. Sun et al., 2023).

Table 3. Data about phenotypic alterations for different plant species

Species of plant	Type of a plastic	Size of plastic	Exposure duration	Phenotypic effects	References
Zea mays	PE	3μm	10 -15days	• PS-MPs develop ROS and cause oxidative damage. • MPs lead to a reduction in transpiration, nitrogen content, and develop	Jiang et al., 2019
Lemna minor	PS	30-600μm	7 days	• No major effect in leave and photosynthesis of a plant but maximum reduction seen in root length.	Kalčíková et al., 2017
Plans community	PET	1.28 +-0.03mm	2 months	• Increase in the root and shoot mass by microfibers at a community level. • Plant community structure effect via microfiber.	Lozano and Rillig, 2020
Vicia	PS	5μm,100nm	48hrs	• Causes oxidative as well as genotoxic. • MPs such as PS can lead to ROS and also cause oxidative damage.	Jiang et al., 2019

In a study conducted by Dainelli et al., the effects of PET-MNPs were evaluated using *Spirodela polyrhiza* (L.) Schleid as a model freshwater species. It focuses on potential epigenetic changes caused by particles (Dainelli et al., 2024). Using repeated cycles of homogenization, MNPs of about 200–300 nm in size were created as aqueous dispersions in PET bottles. These MNPs were then utilized to prepare N-medium at two environmentally relevant concentrations: approximately 0.05 g L^{-1} and approximately 0.1 g L^{-1} (Ekvall et al., 2019). Before testing, *S. polyrhiza* strain 9509 was acclimated for two weeks in a climate room under the following carefully regulated conditions such as temperature, light intensity, photoperiod, and relative humidity. 32 samples in total were obtained for the extraction of genomic DNA (Vos et al., 1995). The number of U-loci was significantly reduced by PET particles at both doses, signifying that *S. polyrhiza* was experiencing DNA hypermethylation in the presence of the pollutants under investigation but only one DNA strand. The primary cause of the PET-induced hypermethylation was hemi-methylation of the exterior cytosine of one of the -CHG sites, or 5′-CCGG sites, where H = A, T, or C (Muyle et al., 2022). The particle-imposed change in oxidative state in *S. polyrhiza* may have been the cause of DNA hypermethylation following PET-MNP exposure. This study demonstrated that in order to combat these xenobiotic stressors, the freshwater species was either shielding its DNA or developing new epialleles (Dainelli et al., 2024).

Pflugmacher et al. performed a study to investigate if the accumulation of MP activates the different degrees of phytotoxicity in *Triticum aestivum* developed from various bottle caps which were extracted from different environment in two different cities irrespective of their age. Distinct collection of caps was consumed including three sets collected in Lahti, Finland, four batches of fresh caps, four cap collection in Singapore and two sets of synthetically seasoned fresh caps. Traditional media was developed for the propagation of the plant (Pflugmacher et al., 2021) (Walters & Kingham, 1990). This has been revealed that the proportion of MP can surpass 7% if the soil is extremely contaminated. The evaluation of the concentration of Malondialdehyde (MDA), an outcome of unsteady and deteriorating lipid peroxides, was used as a measure of lipid peroxidase for the estimation of oxidative response status. N-methyl-2-phenylindole, the chromogenic reagent, and MDA react to produce a stable chromophore with a maximum absorbance at 586 nm. Plants respond to oxidative stress by upregulating the production and activity of antioxidative enzymes such as catalase (Pflugmacher et al., 2021). The result in the concentration of MDA is shown in table no.4. Based on the information provided, it can be concluded that wheat seedlings exposed to MP from fresh and Lahti caps experienced oxidative stress, and to a lesser degree, caps from Singapore that were not artificially aged. ROS signaling in plants can be seriously disrupted by exogenous oxidative stress.

Moreover, the impact of polystyrene nano-plastics on *Cucumis sativus L.* leaves was investigated by Z. Li et al. provided PSNPs in various particle sizes. Five treatments totalling four distinct PSNP varied particle sizes with 500 nm and 700 nm PSNPs were applied to cucumber plants, the amount of MDA in the leaves significantly increased. The Superoxide dismutase enzyme activity of cucumber plants was shown to be considerably reduced upon exposure to 700 nm PSNPs. With the exposure of PSNPs of the diameter ranging 100 nm, 500 nm, and 700 nm, the relative peroxidase expression of genes in the leaves of cucumber plants increased. While the enzyme activity of catalase decreased when the size of PSNPs was changed to the diameter of 500 nm and 700 nm in the cucumber leaves. With the comparison of samples with the control, the increment of the enzyme activity of ascorbate peroxidase has been observed when exposed to 100 nm PSNPs. This study suggested that in the deterioration of ROS mechanism and the decrease in level of MDA, the enzyme activity of antioxidants played an important part (Z. Li et al., 2020).

Table 4. Alterations in the concentration of MDA and catalase at different locations

Exposure of MP to the wheat plant	Change in MDA concentration
For seven days	Increased by 91%
From Lahti city	Increased by 47%
From Singapore	Not affected
Exposure of MP to the wheat plant	**Catalase activity**
For seven days	Enhanced by 139%
From Lahti city	By 91%
From Singapore	Increased by 51%

(Pflugmacher et al., 2021)

6. CHALLENGES AND FUTURE DIRECTIONS

Currently, the major limitation the researchers are facing is the unknown distribution of the MPs/NPs and its concentration due to limitation of technology also known as limits of detection (LOD) (Schwaferts et al., 2019). Compared with aquatic ecosystems, the environmental behaviors of MPs/NPs in terrestrial ecosystems are much more complex due to the heterogeneity of the environmental media (e.g., soil) and intensive anthropogenic activities. To address the complex issues of microplastic and nano-plastic contamination on land, a concerted effort on several fronts is needed. The most important thing to do is look into workable ways to find, separate, measure, and describe these tiny particles in soil and living organisms. Secondly, to find valuable information on the potential ecotoxicity of synthetic micro- and nano-plastics that could be obtained by employing certain model systems for better understanding. Furthermore, a comprehensive outlook of the behaviour of these particles within soil profiles and their concentrations, as well as their dynamics and fate in soils, is necessary. It's critical to examine the interactions between plastic particles and agrochemicals in order to assess their impacts more fully. Additionally, determining the traits of plants and soil organisms that are relevant to the bioavailability of plastic particles is also crucial.

It is essential to comprehend how plastic wastes affect plants and soil organisms on a physical, physiological, and biochemical level as well as how they affect the microbiome and how those organisms react to other stresses. In the end, researching how plastic pollution affects the ability of the agroecosystem to produce biomass is essential for creating practical mitigation plans.

7. CONCLUSION

Microplastics and Nano-plastics are harmful contaminants that should be considered a contributing element to global warming due to their ubiquitous presence. We have a obscure understanding about how this aspect of global change affects plants. We provide a number of potential pathways via which these compounds might influence plant performance. While some of these methods have a good impact on the growth of roots and plants, while others have unfavourable impacts. The consequences differ depending on the kind of plant, which means they might result in modifications to the makeup of plant communities and even primary production. Determining the magnitude and direction of these impacts at the level of individual plants to ecosystems, depending on the kind of ecosystem and the extent and kind of pollution, will be challenging. Given the significance of plants in the climate system, it is imperative to test for these impacts. Widespread effects, even with relatively moderate effect sizes as one might anticipate for plant performance, might have significant implications on ecosystem processes and climate feedbacks.

REFERENCES

Adamczyk, S., Chojak-Koźniewska, J., Oleszczuk, S., Michalski, K., Velmala, S., Zantis, L. J., Bosker, T., Zimny, J., Adamczyk, B., & Sowa, S.L. J. (2023). Polystyrene nanoparticles induce concerted response of plant defense mechanisms in plant cells. *Scientific Reports*, 13(1), 22423. DOI: 10.1038/s41598-023-50104-5

Awasthi, J. P., Saha, B., Chowardhara, B., Devi, S. S., Borgohain, P., & Panda, S. K. (2018). Qualitative Analysis of Lipid Peroxidation in Plants under Multiple Stress Through Schiff's Reagent: A Histochemical Approach. *Bio-Protocol*, 8(8). Advance online publication. DOI: 10.21769/BioProtoc.2807

Azeem, I., Adeel, M., Ahmad, M. A., Shakoor, N., Jiangcuo, G. D., Azeem, K., Ishfaq, M., Shakoor, A., Ayaz, M., Xu, M., & Rui, Y. (2021). Uptake and Accumulation of Nano/Microplastics in Plants: A Critical Review. *Nanomaterials (Basel, Switzerland)*, 11(11), 2935. DOI: 10.3390/nano11112935

Bosker, T., Bouwman, L. J., Brun, N. R., Behrens, P., & Vijver, M. G. (2019, July). Microplastics Accumulate on Pores in Seed Capsule and Delay Germination and Root Growth of the Terrestrial Vascular Plant Lepidium Sativum. *Chemosphere*, 226, 774–781. DOI: 10.1016/j.chemosphere.2019.03.163

Castro, B., Citterico, M., Kimura, S., Stevens, D. M., Wrzaczek, M., & Coaker, G. (2021). Stress-induced reactive oxygen species compartmentalization, perception and signalling. *Nature Plants*, 7(4), 403–412. DOI: 10.1038/s41477-021-00887-0

Chen, G., Li, Y., Liu, S., Junaid, M., & Wang, J. (2022, February). Effects of Micro (Nano)Plastics on Higher Plants and the Rhizosphere Environment. *The Science of the Total Environment*, 807, 150841. DOI: 10.1016/j.scitotenv.2021.150841

Cui, Y., Zhang, Q., Liu, P., & Zhang, Y. (2022). Effects of Polyethylene and Heavy Metal Cadmium on the Growth and Development of Brassica chinensis var. Chinensis. *Water, Air, and Soil Pollution*, 233(10), 426. DOI: 10.1007/s11270-022-05888-z

Dainelli, M., Castellani, M. B., Pignattelli, S., Falsini, S., Ristori, S., Papini, A., Colzi, I., Coppi, A., & Gonnelli, C. (2024). Growth, physiological parameters and DNA methylation in Spirodela polyrhiza (L.) Schleid exposed to PET micro-nanoplastic contaminated waters. *Plant Physiology and Biochemistry*, 207, 108403. DOI: 10.1016/j.plaphy.2024.108403

De Souza Machado, A. A., Kloas, W., Zarfl, C., Hempel, S., & Rillig, M. C. (2018). Microplastics as an emerging threat to terrestrial ecosystems. *Global Change Biology*, 24(4), 1405–1416. DOI: 10.1111/gcb.14020

De Souza Machado, A. A., Lau, C. W., Kloas, W., Bergmann, J., Bachelier, J. B., Faltin, E., Becker, R., Görlich, A. S., & Rillig, M. C. (2019). Microplastics Can Change Soil Properties and Affect Plant Performance. *Environmental Science & Technology*, 53(10), 6044–6052. DOI: 10.1021/acs.est.9b01339

Diao, T., Liu, R., Meng, Q., & Sun, Y. (2023). Microplastics derived from polymer-coated fertilizer altered soil properties and bacterial community in a Cd-contaminated soil. *Applied Soil Ecology*, 183, 104694. DOI: 10.1016/j.apsoil.2022.104694

Dong, Y., Gao, M., Qiu, W., & Song, Z. (2021). Uptake of microplastics by carrots in presence of as (III): Combined toxic effects. *Journal of Hazardous Materials*, 411, 125055. DOI: 10.1016/j.jhazmat.2021.125055

Dovidat, L. C., Brinkmann, B. W., Vijver, M. G., & Bosker, T. (2020). Plastic particles adsorb to the roots of freshwater vascular plant *Spirodela polyrhiza* but do not impair growth. *Limnology and Oceanography Letters*, 5(1), 37–45. DOI: 10.1002/lol2.10118

Ekner-Grzyb, A., Duka, A., Grzyb, T., Lopes, I., & Chmielowska-Bąk, J. (2022). Plants oxidative response to nanoplastic. *Frontiers in Plant Science*, 13, 1027608. DOI: 10.3389/fpls.2022.1027608

Ekvall, M. T., Lundqvist, M., Kelpsiene, E., Šileikis, E., Gunnarsson, S. B., & Cedervall, T. (2019). Nano-plastics formed during the mechanical breakdown of daily-use polystyrene products. *Nanoscale Advances*, 1(3), 1055–1061. DOI: 10.1039/C8NA00210J

Ferrante, M. C., Monnolo, A., Del Piano, F., Mattace Raso, G., & Meli, R. (2022). The pressing issue of micro-and nanoplastic contamination: Profiling the reproductive alterations mediated by oxidative stress. *Antioxidants*, 11(2), 193. DOI: 10.3390/antiox11020193

Gao, M., Liu, Y., Dong, Y., & Song, Z. (2021). Effect of polyethylene particles on dibutyl phthalate toxicity in lettuce (Lactuca sativa L.). *Journal of Hazardous Materials*, 401, 123422. DOI: 10.1016/j.jhazmat.2020.123422

Gao, M., Wang, Z., Jia, Z., Zhang, H., & Wang, T. (2023). Brassinosteroids alleviate nanoplastic toxicity in edible plants by activating antioxidant defense systems and suppressing nanoplastic uptake. *Environment International*, 174, 107901. DOI: 10.1016/j.envint.2023.107901

Gao, Y., & Collins, C. D. (2009). Uptake Pathways of Polycyclic Aromatic Hydrocarbons in White Clover. *Environmental Science & Technology*, 43(16), 6190–6195. DOI: 10.1021/es900662d

Geyer, R., Jambeck, J. R., & Law, K. L. (2017). Production, use, and fate of all plastics ever made. *Science Advances*, 3(7), e1700782. DOI: 10.1126/sciadv.1700782

Giorgetti, L., Spanò, C., Muccifora, S., Bottega, S., Barbieri, F., Bellani, L., & Ruffini Castiglione, M. (2020). Exploring the interaction between polystyrene nanoplastics and Allium cepa during germination: Internalization in root cells, induction of toxicity and oxidative stress. *Plant Physiology and Biochemistry*, 149, 170–177. DOI: 10.1016/j.plaphy.2020.02.014

Gu, J., Chen, L., Wan, Y., Teng, Y., Yan, S., & Hu, L. (2022). Experimental Investigation of Water-Retaining and Unsaturated Infiltration Characteristics of Loess Soils Imbued with Microplastics. *Sustainability (Basel)*, 15(1), 62. DOI: 10.3390/su15010062

Hahladakis, J. N., Velis, C. A., Weber, R., Iacovidou, E., & Purnell, P. (2018). An overview of chemical additives present in plastics: Migration, release, fate and environmental impact during their use, disposal and recycling. *Journal of Hazardous Materials*, 344, 179–199. DOI: 10.1016/j.jhazmat.2017.10.014

Jia, L., Liu, L., Zhang, Y., Fu, W., Liu, X., Wang, Q., Tanveer, M., & Huang, L. (2023). Microplastic stress in plants: Effects on plant growth and their remediations. *Frontiers in Plant Science*, 14, 1226484. DOI: 10.3389/fpls.2023.1226484

Jiang, X., Chen, H., Liao, Y., Ye, Z., Li, M., & Klobučar, G. (2019). Ecotoxicity and genotoxicity of polystyrene microplastics on higher plant Vicia faba. *Environmental Pollution*, 250, 831–838. DOI: 10.1016/j.envpol.2019.04.055

Jing, M., Zhang, H., Wei, M., Tang, Y., Xia, Y., Chen, Y., Shen, Z., & Chen, C. (2022). Reactive Oxygen Species Partly Mediate DNA Methylation in Responses to Different Heavy Metals in Pokeweed. *Frontiers in Plant Science*, 13, 845108. DOI: 10.3389/fpls.2022.845108

Kalčíková, G., Gotvajn, A. Ž., Kladnik, A., & Jemec, A. (2017, November). Impact of Polyethylene Microbeads on the Floating Freshwater Plant Duckweed Lemna Minor. *Environmental Pollution*, 230, 1108–1115. DOI: 10.1016/j.envpol.2017.07.050

Kanwar, M. K., Yu, J., & Zhou, J. (2018). Phytomelatonin: Recent advances and future prospects. *Journal of Pineal Research*, 65(4), e12526. DOI: 10.1111/jpi.12526

Kaur, M., Xu, M., & Wang, L. (2022). Cyto–Genotoxic Effect Causing Potential of Polystyrene Micro-Plastics in Terrestrial Plants. *Nanomaterials (Basel, Switzerland)*, 12(12), 2024. DOI: 10.3390/nano12122024

Li, S., Guo, J., Wang, T., Gong, L., Liu, F., Brestic, M., Liu, S., Song, F., & Li, X. (2021). Melatonin reduces nanoplastic uptake, translocation, and toxicity in wheat. *Journal of Pineal Research*, 71(3), e12761. DOI: 10.1111/jpi.12761

Li, T., Cao, X., Zhao, R., & Cui, Z. (2023). Stress response to nano-plastics with different charges in Brassica napus L. during seed germination and seedling growth stages. *Frontiers of Environmental Science & Engineering*, 17(4), 43. DOI: 10.1007/s11783-023-1643-y

Li, X., Wang, R., Dai, W., Luan, Y., & Li, J. (2023). Impacts of Micro(nano)plastics on Terrestrial Plants: Germination, Growth, and Litter. *Plants*, 12(20), 3554. DOI: 10.3390/plants12203554

Li, X., Wang, R., Dai, W., Luan, Y., & Li, J. (2023, October 12). Impacts of Micro (Nano)Plastics on Terrestrial Plants: Germination, Growth, and Litter. *Plants*, 12(20), 3554. DOI: 10.3390/plants12203554

Li, Z., Li, Q., Li, R., Zhou, J., & Wang, G. (2021). The distribution and impact of polystyrene nano-plastics on cucumber plants. *Environmental Science and Pollution Research International*, 28(13), 16042–16053. DOI: 10.1007/s11356-020-11702-2

Li, Z., Li, R., Li, Q., Zhou, J., & Wang, G. (2020). Physiological response of cucumber (Cucumis sativus L.) leaves to polystyrene nano-plastics pollution. *Chemosphere*, 255, 127041. DOI: 10.1016/j.chemosphere.2020.127041

Liu, P., Zhan, X., Wu, X., Li, J., Wang, H., & Gao, S. (2020). Effect of weathering on environmental behavior of microplastics: Properties, sorption and potential risks. *Chemosphere*, 242, 125193. DOI: 10.1016/j.chemosphere.2019.125193

Liu, W., Ye, T., Jägermeyr, J., Müller, C., Chen, S., Liu, X., & Shi, P. (2021). Future climate change significantly alters interannual wheat yield variability over half of harvested areas. *Environmental Research Letters*, 16(9), 094045. DOI: 10.1088/1748-9326/ac1fbb

Lu, S., Chen, J., Wang, J., Wu, D., Bian, H., Jiang, H., Sheng, L., & He, C. (2023). Toxicological effects and transcriptome mechanisms of rice (Oryza sativa L.) under stress of quinclorac and polystyrene nanoplastics. *Ecotoxicology and Environmental Safety*, 249, 114380. DOI: 10.1016/j.ecoenv.2022.114380

Maity, S., Chatterjee, A., Guchhait, R., De, S., & Pramanick, K. (2020). Cytogenotoxic potential of a hazardous material, polystyrene microparticles on Allium cepa L. *Journal of Hazardous Materials*, 385, 121560. DOI: 10.1016/j.jhazmat.2019.121560

Meng, F., Yang, X., Riksen, M., Xu, M., & Geissen, V. (2021). Response of common bean (Phaseolus vulgaris L.) growth to soil contaminated with microplastics. *The Science of the Total Environment*, 755, 142516. DOI: 10.1016/j.scitotenv.2020.142516

Mittler, R. (2017). ROS Are Good. *Trends in Plant Science*, 22(1), 11–19. DOI: 10.1016/j.tplants.2016.08.002

Muyle, A. M., Seymour, D. K., Lv, Y., Huettel, B., & Gaut, B. S. (2022). Gene Body Methylation in Plants: Mechanisms, Functions, and Important Implications for Understanding Evolutionary Processes. *Genome Biology and Evolution*, 14(4), evac038. Advance online publication. DOI: 10.1093/gbe/evac038

Ng, E. L., Lwanga, E. H., Eldridge, S. M., Johnston, P., Hu, H. W., Geissen, V., & Chen, D. (2018). An overview of microplastic and nanoplastic pollution in agroecosystems. *The Science of the Total Environment*, 627, 1377–1388. DOI: 10.1016/j.scitotenv.2018.01.341

Pasaribu, B., Acosta, K., Aylward, A., Liang, Y., Abramson, B. W., Colt, K., Hartwick, N. T., Shanklin, J., Michael, T. P., & Lam, E. (2023). Genomics of turions from the Greater Duckweed reveal its pathways for dormancy and re-emergence strategy. *The New Phytologist*, 239(1), 116–131. DOI: 10.1111/nph.18941

Pflugmacher, S., Tallinen, S., Mitrovic, S. M., Penttinen, O.-P., Kim, Y.-J., Kim, S., & Esterhuizen, M. (2021). Case Study Comparing Effects of Microplastic Derived from Bottle Caps Collected in Two Cities on Triticum aestivum (Wheat). *Environments (Basel, Switzerland)*, 8(7), 64. DOI: 10.3390/environments8070064

Poma, A. M. G., Morciano, P., & Aloisi, M. (2023). Beyond genetics: Can micro and nano-plastics induce epigenetic and gene-expression modifications? *Frontiers in Epigenetics and Epigenomics*, 1, 1241583. DOI: 10.3389/freae.2023.1241583

Qiu, G., Han, Z., Wang, Q., Wang, T., Sun, Z., Yu, Y., Han, X., & Yu, H. (2023). Toxicity effects of nanoplastics on soybean (Glycine max L.): Mechanisms and transcriptomic analysis. *Chemosphere*, 313, 137571. DOI: 10.1016/j.chemosphere.2022.137571

Ren, F., Huang, J., & Yang, Y. (2024). Unveiling the impact of microplastics and nanoplastics on vascular plants: A cellular metabolomic and transcriptomic review. *Ecotoxicology and Environmental Safety*, 279, 116490. DOI: 10.1016/j.ecoenv.2024.116490

Ren, X., Tang, J., Wang, L., & Liu, Q. (2021). Microplastics in soil-plant system: Effects of nano/microplastics on plant photosynthesis, rhizosphere microbes and soil properties in soil with different residues. *Plant and Soil*, 462(1–2), 561–576. DOI: 10.1007/s11104-021-04869-1

Rillig, M. C., Lehmann, A., De Souza Machado, A. A., & Yang, G. (2019). Microplastic effects on plants. *The New Phytologist*, 223(3), 1066–1070. DOI: 10.1111/nph.15794

Sun, C., Yang, X., Gu, Q., Jiang, G., Shen, L., Zhou, J., Li, L., Chen, H., Zhang, G., & Zhang, Y. (2023). Comprehensive analysis of nanoplastic effects on growth phenotype, nanoplastic accumulation, oxidative stress response, gene expression, and metabolite accumulation in multiple strawberry cultivars. *The Science of the Total Environment*, 897, 165432. DOI: 10.1016/j.scitotenv.2023.165432

Sun, X.-D., Yuan, X.-Z., Jia, Y., Feng, L.-J., Zhu, F.-P., Dong, S.-S., Liu, J., Kong, X., Tian, H., Duan, J.-L., Ding, Z., Wang, S.-G., & Xing, B. (2020). Differentially charged nano-plastics demonstrate distinct accumulation in Arabidopsis thaliana. *Nature Nanotechnology*, 15(9), 755–760. DOI: 10.1038/s41565-020-0707-4

Surgun-Acar, Y. (2022). Response of soybean (Glycine max L.) seedlings to polystyrene nanoplastics: Physiological, biochemical, and molecular perspectives. *Environmental Pollution*, 314, 120262. DOI: 10.1016/j.envpol.2022.120262

Thompson, J. R., Wilder, L. M., & Crooks, R. M. (2021). Filtering and continuously separating microplastics from water using electric field gradients formed electrochemically in the absence of buffer. *Chemical Science (Cambridge)*, 12(41), 13744–13755. DOI: 10.1039/D1SC03192A

Tripathy, B. C., & Oelmüller, R. (2012). Reactive oxygen species generation and signaling in plants. *Plant Signaling & Behavior*, 7(12), 1621–1633. DOI: 10.4161/psb.22455

Ullah, R., Tsui, M. T.-K., Chow, A., Chen, H., Williams, C., & Ligaba-Osena, A. (2023). Micro(nano)plastic pollution in terrestrial ecosystem: Emphasis on impacts of polystyrene on soil biota, plants, animals, and humans. *Environmental Monitoring and Assessment*, 195(1), 252. DOI: 10.1007/s10661-022-10769-3

Vos, P., Hogers, R., Bleeker, M., Reijans, M., Lee, T. V. D., Hornes, M., Friters, A., Pot, J., Paleman, J., Kuiper, M., & Zabeau, M. (1995). AFLP: A new technique for DNA fingerprinting. *Nucleic Acids Research*, 23(21), 4407–4414. DOI: 10.1093/nar/23.21.4407

Walters, D. R., & Kingham, G. (1990). Uptake and translocation of α-difluoromethylornithine, a polyamine biosynthesis inhibitor, by barley seedlings: Effects on mildew infection. *The New Phytologist*, 114(4), 659–665. DOI: 10.1111/j.1469-8137.1990.tb00437.x

Wang, J., Lu, S., Bian, H., Xu, M., Zhu, W., Wang, H., He, C., & Sheng, L. (2022, October). Effects of individual and combined polystyrene nanoplastics and phenanthrene on the enzymology, physiology, and transcriptome parameters of rice (Oryza sativa L.). *Chemosphere*, 304, 135341. DOI: 10.1016/j.chemosphere.2022.135341

Wang, L., Wu, W. M., Bolan, N. S., Tsang, D. C., Li, Y., Qin, M., & Hou, D. (2021). Environmental fate, toxicity and risk management strategies of nano-plastics in the environment: Current status and future perspectives. *Journal of Hazardous Materials*, 401, 123415. DOI: 10.1016/j.jhazmat.2020.123415

Wang, Z., Li, W., Li, W., Yang, W., & Jing, S. (2023). Effects of microplastics on the water characteristic curve of soils with different textures. *Chemosphere*, 317, 137762. DOI: 10.1016/j.chemosphere.2023.137762

Xu, C., Wang, H., Zhou, L., & Yan, B. (2023). Phenotypic and transcriptomic shifts in roots and leaves of rice under the joint stress from microplastic and arsenic. *Journal of Hazardous Materials*, 447, 130770. DOI: 10.1016/j.jhazmat.2023.130770

Xu, Z., Zhang, Y., Lin, L., Wang, L., Sun, W., Liu, C., Yu, G., Yu, J., Lv, Y., Chen, J., Chen, X., Fu, L., & Wang, Y. (2022). Toxic effects of microplastics in plants depend more by their surface functional groups than just accumulation contents. *The Science of the Total Environment*, 833, 155097. DOI: 10.1016/j.scitotenv.2022.155097

Yang, C., & Gao, X. (2022). Impact of microplastics from polyethylene and biodegradable mulch films on rice (Oryza sativa L.). *The Science of the Total Environment*, 828, 154579. DOI: 10.1016/j.scitotenv.2022.154579

Younes, N. A., Dawood, M. F. A., & Wardany, A. A. (2019). Biosafety assessment of graphene nanosheets on leaf ultrastructure, physiological and yield traits of Capsicum annuum L. and Solanum melongena L. *Chemosphere*, 228, 318–327. DOI: 10.1016/j.chemosphere.2019.04.097

Yu, H., Zhang, Z., Zhang, Y., Song, Q., Fan, P., Xi, B., & Tan, W. (2021). Effects of microplastics on soil organic carbon and greenhouse gas emissions in the context of straw incorporation: A comparison with different types of soil. *Environmental Pollution*, 288, 117733. DOI: 10.1016/j.envpol.2021.117733

Yu, Z., Xu, X., Guo, L., Jin, R., & Lu, Y. (2024). Uptake and transport of micro/nano-plastics in terrestrial plants: Detection, mechanisms, and influencing factors. *The Science of the Total Environment*, 907, 168155. DOI: 10.1016/j.scitotenv.2023.168155

Yuan, W., Zhou, Y., Liu, X., & Wang, J. (2019). New Perspective on the Nanoplastics Distrupting the Reproduction of an Endangered Fern in Australian Freshwater. *Environmental Science & Technology*, 53(21), 12715–12724. DOI: 10.1021/acs.est.9b02882

Zhang, H., Lang, Z., & Zhu, J.-K. (2018). Dynamics and function of DNA methylation in plants. *Nature Reviews. Molecular Cell Biology*, 19(8), 489–506. DOI: 10.1038/s41580-018-0016-z

Zhou, C.-Q., Lu, C.-H., Mai, L., Bao, L.-J., Liu, L.-Y., & Zeng, E. Y. (2021). Response of rice (Oryza sativa L.) roots to nanoplastic treatment at seedling stage. *Journal of Hazardous Materials*, 401, 123412. DOI: 10.1016/j.jhazmat.2020.123412

Zhou, Q., Lian, J., Liu, W., Men, S., Wu, J., Sun, Y., Zeb, A., Yang, T., & Ma, Q. (2020). Transcriptome mechanisms underlying interaction of polystyrene nanoplastics and wheat Triticum aestivum L. DOI: 10.21203/rs.3.rs-98748/v1

Zuo, Z., Sun, L., Wang, T., Miao, P., Zhu, X., Liu, S., Song, F., Mao, H., & Li, X. (2017). Melatonin Improves the Photosynthetic Carbon Assimilation and Antioxidant Capacity in Wheat Exposed to Nano-ZnO Stress. *Molecules (Basel, Switzerland)*, 22(10), 1727. DOI: 10.3390/molecules22101727

Chapter 8
Effect of Micro/Nano-Plastics Accumulation on Soil Nutrient Cycling

Divya Kumari
https://orcid.org/0009-0002-0709-572X
Banasthali University, India

Pracheta Janmeda
https://orcid.org/0000-0003-0500-4636
Banasthali University, India

Nidhi Varshney
Banasthali University, India

Poornima Pandey
https://orcid.org/0000-0002-7381-3358
Banasthali University, India

ABSTRACT

The micro/nanoplastics (M/NPs) have attracted attention from around the world regarding their effects on the environment due to their broad distribution, potential ecological risks, and persistence. M/NPs, which are present in soil, water, and the atmosphere, are minute pieces of both organic and inorganic plastic trash. M/NPs are broadly recognized as a serious global ecological concern because of their widespread use and improper management of waste. The use of M/NPs in agriculture has its origins in different kinds of agricultural management practices, including as composting, mulching, and sewage sludge, affecting soil and plant properties. Polluting substances, notably plastic trash, are beginning to have a significant impact on crucial soil ecosystem activities, such as soil microbial interactions and

DOI: 10.4018/979-8-3693-3447-8.ch008

nitrogen cycling. The goal in presenting the evidence currently available is to show how M/NPs affect soil nutrient cycling by modulating soil nutrient availability, microbial communities that are functional, and soil enzyme activities that may have ecological significance.

1. INTRODUCTION

The impact of plastic waste on marine ecosystems has garnered substantial attention, making it an emerging global environmental problem. But new research has shown that plastic pollution also exists in terrestrial ecosystems, especially in the soil, which is a worrisome finding. Nanoplastics (NPs) are moretiny plastic particles, usually <1 μm, whereas Microplastics (MPs) are plastic particles, usually<5 mm. Because of their small size, soil organisms—from microbes to invertebrates that live in the soil—can readily reach and consume these particles by accident. Pollution from soil micro- and nanoplastics (M/NPs) has become a serious problem that could have a significant impact on ecosystem health, biodiversity, and soil health. The pervasiveness of plastic waste in terrestrial ecosystems emphasizes the seriousness of soil M/NPs pollution as an ecological issue. They can come from a variety of places, such as industrial processes, agricultural practices, and the fragmentation of bigger polymers. There are several ways that plastic particles might end up in soil, including when plastic-based mulches are used in agriculture, when M/NPs are deposited in the air, and when plastic debris is disposed of in landfills. These M/NPs have the ability to linger and accumulate in the soil, endangering soil organisms and possibly making their way up the food chain (Liu *et al.*, 2023).

In ecosystems, soil is a crucial component that facilitates the growth of plants, the cycling of water and nutrients, and other necessary processes. However, these functions could be seriously impacted by M/NPs soil contamination. It is essential to comprehend the toxic effects of M/NPs on soil biological species for a number of reasons. First, plant development, carbon sequestration, and nutrient cycling are all significantly influenced by the soil environment. Earthly ecosystems may experience a domino effect if this delicate balance is upset. Second, M/NPs have the ability to infiltrate the food chain through species that live in the soil, which could endanger animals at higher trophic levels, such as humans who eat crops that have been raised in soil polluted with M/NPs (Allen *et al.*, 2023; Li *et al.*, 2023). Toxic effects on soil organisms may be introduced by or amplified by the presence of M/NPs in the soil, according to toxicological theory. Assessments of soil health may face substantial difficulties due to the complicated, compound-specific toxicity dynamics that may result from the interaction between these pollutants and M/NPs. To be more precise, (i) M/NPs pollution in soil can negatively impact soil fauna, microbes, and biodiversity in general, upsetting the equilibrium as well as soil ecosystem stability (Lin *et al.*, 2020; Wang *et al.*, 2022);(ii) M/NP accumulation results in a deterioration of soil quality, which influences crop yield and sustainable soil utility by altering the soil structure, water-holding capacity, and nutrient cycling (Liu *et al.*, 2023; Yu *et al.*, 2021); (iii) M/NPs are absorbed via plants in soil and enter to the food chain, where they may endanger the health of people and other animals, especially those that eat crops and meat-eating creatures (Huerta Lwanga *et al.*, 2017). Moreover, the effects of soil M/NPs contamination go beyond the ecology of the soil. According to Gouin (2021), Horton and Dixon (2018), and Horton *et al.* (2017), these particles have the ability to contaminate freshwater and marine habitats by movement and discharge into water bodies. Furthermore, the ecological concerns associated with the presence of soil M/NPs are increased because they serve as transporters for another pollutant substances, mainly include heavy metals (Fu *et al.*, 2021; Yang *et al.*, 2022).

In conclusion, the results of this research will not only advance our understanding of soil M/NPs contamination and its toxicological effects, but they will also lay the framework for future investigations and formulation of policies intended to lessen the negative effects of plastic pollution on the environment in soil ecosystems. We offer a thorough analysis of the body of research in this unique review, exploring the effects of M/NPs pollution on food, plants, soil, as well as human health. Additionally, it examines the strategies for locating and controlling M/NPs, assesses the significant difficulties in eradicating M/NPs from soil, and identifies the top research priorities.

2. CLASSIFICATION OF PLASTICS

Basically, the plastic particles are mainly classified into four forms such as macroplastics (>200 mm), microplastics (5–200 mm), MPs (0.1 to 5 mm), and NPs (0.001 to 0.1 μm). The MPs can be further classified as follows:

2.1 Primary MPs

Small particles known as primary MPs are discharged into the environment directly. This class of MPs is thought to make up 15–31% of all MPs in the ocean. As a result of tires abrasion in case of driving, which releases trashes into the air that we breathe every day, and synthetic-textile washing (35% of primary MPs) are the main sources, in descending order of importance; MPs are also purposefully added into the body-care goods at a rate of 2% (Russo *et al.*, 2023). Plastic particles <5 mm that are discharged into surrounding are known as primary MPs.

The plastic manufacturing sectors have the ability to purposefully produce primary MPs (either a micro or nano size) for certain commercial applications, after which they may end up in the environment. MPs are useful as exfoliants, to extend product longevity, and to achieve sustainable release of main active components (Ziani *et al.*, 2023). MPs are used in both nanomedicine and agriculture (fertilizer, nutrients supply, and other active substances) due to their capacity to encapsulate and release items gradually.

Dalela and colleagues (2015) created poly(styrene-co-maleic acid)-paclitaxel nanoparticles, which they used as a nanomedicine to administer paclitaxel to solid tumors. This is an example of nanomedicine in action. Concurrently, Tian *et al.* (2022) coated urea in agriculture using three polymers including vegetable oil-based PU, epoxy resin, and liquid starch-based PUto allow for urea to release gradually.

Microbeads are eventually washed off to treatment facilities for wastewater or the environment directly after these goods are used. However, some MPs are too small to be eliminated by wastewater treatment methods, meaning they stay in the water. According to Bashir *et al.* (2021), wastewater treatment plants in Macao City, China, might discharge around 37 billion microbeads into the environment each year from personal care and cosmetic items.

2.2 Secondary MPs

Secondary MPs, which are created after the disintegration of plastics (for instance, bags, plastic bottles, bottles of detergent and soap, fishing nets, etc.). These make up between 68 and 81% of all MPs found in the oceans as well as seas. Thus, it is not unexpected that sea salt serves as a major source of food for secondary MPs. Although MPs are the end product of plastic destruction, the additives employed in plastics also carry a variety of other contaminants. Some chemicals that are categorized as "endocrine disruptors," including phthalates, used to enhance plastic's flexibility, and bisphenol A, which is used to increase the resistance of detergent bottles and some disposable tableware, are particularly significant. Furthermore, it is important to take into account additional contaminants, such as ambient toxins, that may have adsorbed on the surface that was exposed (Russo *et al.*, 2023). When big plastic particles are subjected to chemical, physical, and biological stresses, they break down and disintegrate, producing secondary MPs, which are polymers smaller than 5 mm (Boucher and Friot, 2017).

In the natural world, breakdown (degradation) mostly takes place by mechanical, water-soluble, photolytic, an oxidative, biological in nature, and thermal processes.

According to Chamas *et al.* (2020), hydrolytic degradation is the process by which plastic polymer bonds react with water molecules, breaking one or more of the polymer bonds and generating smaller plastic fragments. This may be a major factor in the MPs found in aquatic habitats. Tamayo-Belda *et al.* demonstrated that plastic can break down into NPs (PCL-NPs) through abiotic hydrolysis utilizing polycaprolactone (PCL) (Tamayo-Belda *et al.*, 2022). Microorganisms release extracellular enzymes during the biodegradation of plastics, which bind to the surface of plastics and cause hydrolysis to these polymeric smaller intermediates (Mohanan *et al.*, 2020). When a polymer of plastic framework reacts with oxygen, it forms carbonoxygen interactive bonds that cause polymeric chains to be shorten. This process is known as oxidative degradation. When plastics absorb photons in an existence of a chemical oxidant (like air), polymer links break, a process known as photo degradation occurs.

Thermal degradation results from temperature-induced changes in the properties of polymers. Climate change-related heat waves and warm, humid climates with high UV radiation levels speed up photolysis, thermo-oxidative, and photooxidative deterioration (Chamas *et al.*, 2020). Meides and colleagues demonstrated that polypropylene can undergo degradation under accelerated weathering circumstances (3200 hours) of thermal and photographic oxidation, resulting in the formation of 100,000 particles with a 192 μm diameter (Meides *et al.*, 2022). According to Egger *et al.*, there has been a documented fallout of plastic particles (ranging in size from 500 μm to 5 cm) from the North Pacific Garbage Patch to the deep sea. This indicates that photochemical oxidation on surface waters is most likely the cause of deterioration (Egger *et al.*, 2020). Plastics break down mechanically when they are subjected to outside pressures in their surroundings. Plastics can become fragmented when they collide and abrade with hard surfaces like rocks and sand due to wind, flowing water, or waves (Ziani *et al.*, 2023).

3. DIFFERENT SOURCES OF SOIL M/NPS

The primary method of producing food, particularly in developing nations, is farming. Since most people in developing nations depend on subsistence farming, the land and its quality of soil are valuable resources (Giller *et al.*, 2021). All living things rely on the soil to perform crucial ecological tasks. Regrettably, plastic garbage is now stored in the earth. Numerous processes, including street runoff, air deposition, landfill dumping, and agricultural activities, can release plastic pieces into the soil (Yang *et al.*, 2021). Different processes that led to M/NPs contamination in soil is shown in Figure 1.

3.1 M/NPs From Landfill Dumping

Plastic waste is among the many materials buried in landfills. 49% of the material produced worldwide is dumped in these landfills, making them the main repository for waste plastic (OECD, 2022). Microbes, moisture, air, heat, and other environmental factors enable plastics to break down into secondary M/NPs in landfills. Moreover, primary M/NPs may be disposed of in landfills. Therefore, landfills are important soil sinks for M/NPs. Mahesh and colleagues (2023) documented the existence of microplastics (MPs) at an open urban waste site, with a range of 180-1120 MP particles per kg of the soil. Similarly, Afrin and colleagues (2020) noted MPs derived from cellulose acetate (CA), HDPE, and LDPE in a landfill site in Bangladesh. These M/NPs can be moved from landfills to nearby new settings by agents like air, water, or other living things. Invertebrates living in the soil can deposit MNPs in a different environment by pushing or ingesting them. Earthworms, for instance, have been observed to consume M/NPs in soil on the surface and excrete them in soils that are deeper (Rillig *et al.*, 2017). Additionally, light M/NPs can be carried away by water runoff or blown by the wind, landing in wastewater treatment facilities, remote locations, or aquatic systems (Bullard *et al.*, 2021).

3.2 M/NPs From Sewage Sludge

M/NPs out of industry, dumps, urban runoff, and home wastewater are received by wastewater treatment plants. (Di Bella *et al.*, 2022; Bretas *et al.*, 2020; Franco *et al.*, 2023; Hassan *et al.*, 2023). Approximately 90% of these M/NPs are retained in the wastewater sludge. Di Bella *et al.* found that a waste sludge contained a variety of plastics, including fibers and fragments of PE, PP, and polybutadiene (PB)

Figure 1. Different sources leading to the M/NPs contamination in soil

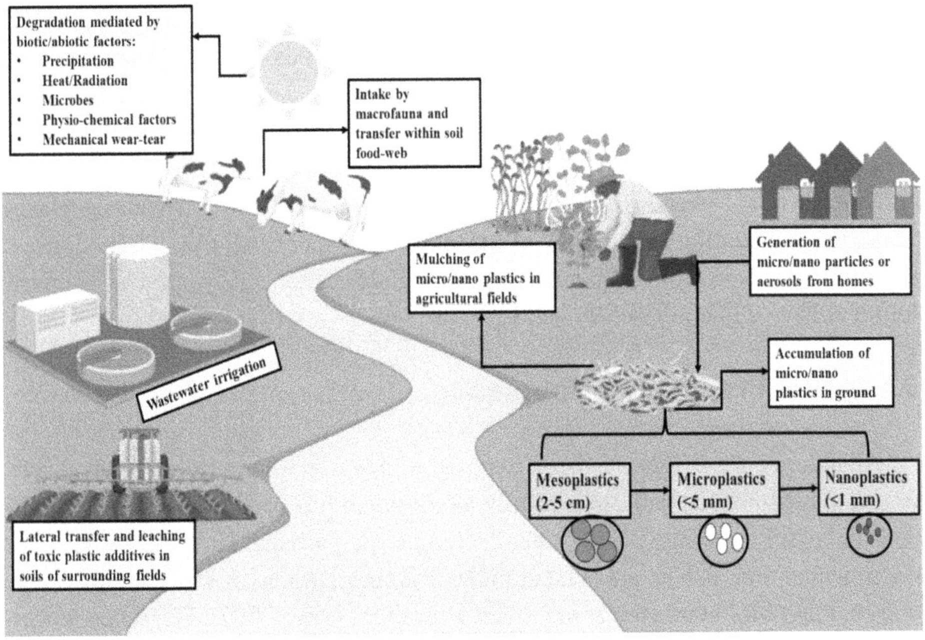

Harley-Nyang *et al.* (2022) revealed that, each month, sludge from particular wastewater-treatment plant retains 1.02 × 1010 to 1.61 × 1010 M/NPs in UK (Di Bella *et al.*, 2022). The sludge eventually contaminates the soil with M/NPs when it is either dumped in landfills or used as fertilizer in farmlands. Thirty-four years after the last sewage sludge application, Weber *et al.* found M/NPs in farmlands down to a depth of ninety centimeters. The highest concentrations were found in periodically ploughed topsoils (Weber *et al.*, 2022), indicating the function sewage sludge plays in introducing M/NPs into soils. Regrettably, there have been reports of M/NPs added to farmed soil through sewage as a biofertilizer spreading beyond the treated locations and contaminating other areas (Tagg *et al.*, 2022).

3.3 M/NPs From Agricultural Practices

Most M/NPs in agricultural soils are a result of agricultural operations. Plastic encapsulated (slow release) fertilizers, insecticides, as well as seed coatings are some of the methods by which primary M/NPs penetrate soils because they stay in the soil after the active component is released (Katsumi *et al.*, 2021; Wang *et al.*, 2020). M/NPs found in the soil are also a result of composting waste sludge that contains plastic pieces and irrigation with wastewater contaminated with M/NPs.

Composted samples obtained from certain municipal (organic) waste collections in the Netherlands included 2800 ± 616 MPs/kg, according to work by van Schothorst *et al.* (2021). Compost in China was also observed to have 2400 ± 358 MPs/kg, mainly fibers /films (Gui *et al.*, 2021). Other plastics used in agriculture, like protective nets, irrigation pipes, mulch, plastic irrigation, and greenhouse films, and drip irrigation systems, break down into secondary M/NPs within the soil (Lwanga *et al.*, 2022). For example, Li *et al.* (2022) found that topsoil ranging from 0–10 cm, collected MPs (7183 and 10,586 particles per kg) after nearly three decades of plastic mulching, accounting for 33–56% of the total MPs. According to van Schothorst *et al.* (2021), top soil from a Spanish farm that mulched for >12 years was similarly found to have 2242 ± 984 MPs per kg.

3.4 M/NPs From Other Sources

Abandoned trash that has been exposed to adverse weather conditions deteriorates into M/NPs as well. Regretfully, no research has been done on soil M/NPs near unlawful dumping sites or uncollected plastic garbage. However, it has been documented that M/NPs are present in the industrial, urban, as well as recreational soil regions (Mokhtarzadeh *et al.*, 2022; Fernandes *et al.*, 2022). Degraded, uncollected plastic trashes additionally add to these M/NPs, even if it's possible that they result from the breakdown of other materials in these locations. Uncollected garbage is frequently burned. MPs are reportedly found in the bottom ash left behind from burning (Yang *et al.*, 2021).

4. M/NPS FATE IN SOIL

The eco-corona qualities influence the consequences and fate of M/NPs in soil. These features can impact the way plastics interact with minerals from organic matter and clay, as well as the toxicity and ingestion of plastics by soil feeders like earthworms. Further information is needed to understand: (i) how significant soil biomolecules, like enzymes, DNA, or root exudates, are adsorbed on the surfaces of MPs and also what impact this has on eco-corona attributes; (ii) how various properties of eco-corona influence the interactions between M/NPs and the soil, that in turn affects the MPs' persistence, mobility, bioavailability, and toxicity. Soil pH can have an impact on the surface charge of plastics that interact with interface-reactive soil components, like clay or soil organic matter (SOM). However, uncharged polymers with hydrophobic surfaces are not affected by soil pH. The downward shift of microplastics (MPs) in natural soils that have not been disturbed should be encouraged by the existence of macropores and preferential route flows, including biopores and cracks, and restricted by microporosity due to MP accumulation in the soil's surface layer. Naturally, since plastic fragments can combine with these components of soil, dissolved organic matter (DOM) or clay minerals also display a role in plastic mobility in soil.

Tillage techniques improve leaching by increasing the porosity and aggregation of the topsoil. the high percentage of plastic trash (72%) found in aggregates and the presence of MP fibers in a cropped soil's micro-aggregates. The addition of plastic waste to aggregates may encourage the buildup of plastic waste in the soil. This may have an impact on the turnover of aggregates as well as the interactions between soil ingredients and the biota that resides in the aggregates. Because different MPs have different effects—polyethylene and polypropylene, for instance, increased aggregate formation—the impacts vary. MPs can therefore have an effect on functioning and structure of soil. Tillage, however, may also increase the amount of plastics in the topsoil by limiting the movement of soil M/NPs due to the plough pan development (Pathan *et al.*, 2020; Zhang and Liu, 2018; Wang *et al.*, 2018).

5. EFFECTS OF M/NPS ON THE SOIL CHARACTERISTICS

5.1 Impacts on Physical/Chemical Properties of Soil

M/NPs pollution is becoming a bigger problem because of its effects on soils. Nonetheless, there was a great deal of variance in the impacts of MP/NP in soil, suggesting a strong reliance on contextual factors. This review focuses on the effects of M/NP form, polymer type, and duration of incubation on soil parameters to understand the specific conditions affecting MP/NP-related impacts on soil. Soil is thought to be the main source of M/NPs; it has a larger storage capacity than the aquatic environment. Plastics can combine with soil particles to produce retention over time. According to Hurley and Nizzetto (2018), M/NPs in soil undergo a variety of fate processes, including storage, transport, erosion, destruction, and leaching into groundwater. MHigher density NPs tend to stay in the soil longer, moving downhill and possibly contaminating groundwater as well as infiltrating plants and the food chain. Less dense M/NPs, on the other hand, tend to stay near the surface and can be carried via water and wind erosion (Wu *et al.*, 2019), contaminating soils that are farther away. M/NPs can be preserved over time by being buried due to various processes, accumulation, floods, and other events. While certain techniques such as'tilling' might brought out buried substances to the soil surface, other soil properties like pH and microbial communities can also have an impact on the protective process. M/NPs are thought to be present on farmlands in America and Europe with amounts ranging between 63.000 and 430.000 tons, partly due to the ageing of plastic mulch and the addition of fertilizers like composts along with sewage sludge (which are not tested for plastic content).

5.2 Impact on Soil Microflora and Invertebrates

It appears that MPs raise the amount of nutrients, like carbon, nitrogen, or phosphorus, in soil dissolved organic matter (DOM). According to Liu *et al.* (2017), this implies that MPs might have a role in the accumulation of nutrients and organic molecules in soil. MPs have the capacity to drastically alter the DOM composition of soil. They have the ability to raise the levels of carbohydrates and aromatic compounds in the organic matter that is dissolved (Chen *et al.*, 2022). According to Meng *et al.* (2022), MPs appeared to increase microbial activity of soil. Increased extracellular enzyme activity could be a consequence of this increased microbial activity, suggesting improved organic chemical breakdown and conversion in the soil. The molecular chain position as well as functional groups of M/NPs can affect their capacity to adsorb other materials, such as antibiotics or heavy metals, depending on their chemical configuration (Fred-Ahmadu *et al.*, 2020). As a result, microbial activity and soil characteristics may be impacted by this interaction (Pathan *et al.*, 2020). For instance, it has been reported that polyethylene (PE) has a significant capacity to absorb phenanthrene and nitrogen rich heterocyclic counterparts, which can prevent soil microbial activity. Furthermore, studies have demonstrated that distinct kinds of polymers, including PE, PVC, and PP having distinct sorption abilities for particular substances (Anyanwu and Semple, 2016; Brennecke *et al.*, 2016; Wang *et al.*, 2018).

According to Wright and Thompson (2013), plastic pollution is a serious worldwide environmental problem. The occurrence of plastics in soil may possess profound biological effects, leading to develop a plastisphere inside the soil environment. Polymers like polypropylene, polyethylene, and polyvinyl chloride that are found in soil act as substrates for the microbial colonization. The plastic surfaces can be attached to and colonized by bacteria, fungus, and other microorganisms, creating a different microbe-based community termed as 'plastisphere' (Zettler *et al.*, 2013). This kind of plastic, the surroundings, and the length of exposure are some of the variables that might affect the microbial community within plastispheres. Certain bacteria have the ability to break down or alter plastic, while others could interact with it differently.

While some bacteria are capable of degrading specific forms of plastic, not all of them can break down plastics. For instance, it has been shown that certain bacteria, such as *Pseudomonas* and *Ideonella sakaiensis*, are capable of degrading PET and polyethylene, respectively. The plastisphere's growth in soil and its abundance of plastics can have a variety of ecological effects. Our understanding of how M/NPs affect soil respiration is constantly developing. For suitability of the overall microbial activity in the soil, soil respiration is very sensitive to the pH, moisture content, porosity, and texture of the soil (Luo and Zhou *et al.*, 2006). The introduction of M/NPs may have an impact on certain soil qualities. According to recent research, M/NPs may have a direct or indirect impact on the microbial population in the soil, which may then have an impact on the soil respiratory processes (Lozano *et al.*, 2021; Fei *et al.*, 2020; Lozano *et al.*, 2021). Additionally, MPs have ability to change the abundant variety of different microbial species by affecting the composition and diversity of soil microbial communities. In an appearance of the M/NPs, certain bacteria may flourish while others may suffer harm or even be repressed. An essential function of soil microbe-produced enzymes is to aid in the disintegration and decay of organic materials. These enzymes' activities can be changed by M/NPs pollution, which may have an impact on how quickly organic matter breaks down and how quickly nutrients cycle through the soil.

In a soil incubation experiment, Liang *et al.* (2021) investigated the impacts of MPs fibers by analyzing how the amount of water stable accumulates and the processes of β-glucosidase, N-acetyl-b-glucosaminidase, phosphatase, and β-D-celluliosidase enzymes changed with or without organic substances. Data showed that soil agglomeration and enzyme activity were affected by microplastic fibers in relation to organic content. According to additional research (Fei *et al.*, 2020; Wiedner and Polifka, 2020; Yi *et al.*, 2021; Li *et al.*, 2023), M/NPs can alter the microbial colonies in the soil, which can then impact the enzymatic activities of the soil (Hargreaves and Hofmockel, 2014). In their microplastic modification experiment, Lin *et al.* (2020) found substantial effects on the composition and abundance of microarthropod as well as nematode groups when polyethylene with low-density fragments were placed in the field. Interestingly, they found that MP had very little effect on the number of microbial communities in the soil. The findings of research investigations that have been published on the impacts of soil M/NPs on soil microbiota are recent and differ greatly from one another. Ya *et al.* (2022) found that exposure to PE and PP significantly altered the variety and richness of soil microbial communities. In particular, they discovered that the abundance of Bacteroidetes and *Acidobacteria* had increased, whilst *Deinococcus thermus* along with *Chloroflexi* had decreased simultaneously. M/NPs also affect the plants that are discussed in Table 1 and shown in Figure 2.

Figure 2. Effect of M/NPs on soil environment

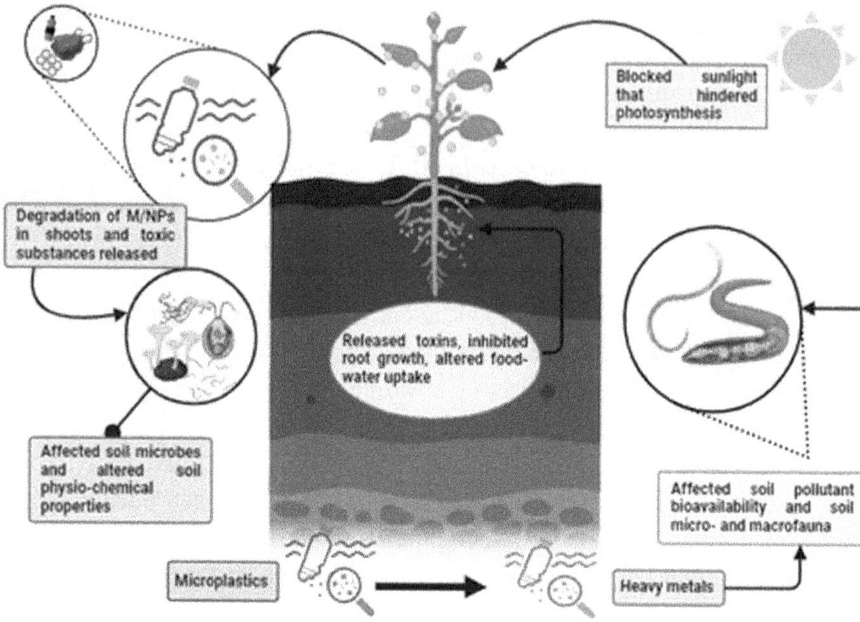

Table 1. Impacts of M/NPs on the plants

Plant species + size	M/NPs		Effects
	Type	Concentration	
A. thaliana	PS-NH_2: 71 nm, PS-SO_3H: 55 nm	0.3,1.0 g $kg^{-1;}$ 10, 50, 100 μg ml^{-1}	Reduced root elongation, seedling growth, and above-ground biomass Arabidopsis is able to absorb and distribute PS
Allium cepa (1.70 μm)	PES fibers	0.4% (w/w)	PES fibers raised the biomass above ground.
Allium cepa (20– 190 nm)	PS	0.01–1.0 g/l	PS particles entered root cells and gathered in the cytoplasm and vacuoles; PS particles aggregated in the epidermis, cortical cylinder, and central cylinder of the root
Allium fistulosum(8 μm in diameter)	PA beads: 15–20 μm PES fibers: 5000 μm length, 8 μm in diameter PEHD, PP: 2–3 mm spheres PS, PET: 2–3 mm cylinders	PES: 0.2%; others: 2.0%	Depending on the type of particle, demonstrated varying impacts on plant activity, included plant biomass, tissues elemental content, root characteristics, and soil microbial activity.

continued on following page

Table 1. Continued

Plant species + size	M/NPs		Effects
	Type	Concentration	
Arabidopsis thaliana(70–200 nm)	PS-COOH, PS-NH_2	10-100 mg/l; 10-50 mg/l	Plants absorbed PS-COOH more preferentially than PS-NH2, and it mostly accumulated in the root maturation zone's stele.
Arabidopsis thaliana and *T. aestivum* (40 nm, 1 μm)	PS	8.3 × 1011 n/ml; 5.3 × 107 n/ml	PS accumulated in the root cap cells of wheat and Arabidopsis.
Brassica rapa (70 nm, 5 μm)	PS	10 mg/kg	Depending on the magnitude of MP, PS affected the photosynthesis and growth of the plant.
Cucumis sativus (100, 300, 500, 700 nm)	PS	50 mg/l	Prominently increasing the amount of protein that is soluble in the fruits of cucumber and decreasing the levels of Mg, Fe, and Ca, PS at 300 nm also increased root activity,proline, and MDA contents. The impact of PS particle sizes was also seen.
Cucumis sativus (100-700 nm)	PS	50 mg/l	PS particles entered plant leaves and stems after being absorbed by plant roots
Cucurbita pepo(40–50 μm)	PE, PVC, PP, PET	0.02/0.10/0.20%	Impacted leaf size, the amount of chlorophyll, and photosynthetic efficiency, as well as hindered root and shoot growth.
Daucus carota (0.1–5 μm)	PS	10– 20 mg/l	Carrot roots may allow the ≤1 μm particle size of PS to penetrate and gather in the intercellular region.
Glycine max (2 × 2 cm,1 × 1 cm 0.5 × 0.5 cm)	(Bio) mulch film, PE	0/0.1/0.5/1%	PE decreased the length of the plant, diameter of culm, ratio of roots to shoots, and leaf area whereas bio-detritus had a negative impact on the viability of germination and biomass of root.
L. sativa (93.6 nm)	PS	0, 0.1, 1 mg/l	Reduced the nutritional quality, height, dry weight, as well as leaf area of the plant, and caused oxidative stress.
L. sativum (<0.125 mm)	PE, PP, PVC, PE+PVC	184 mg/kg	Adversely affected biometric features, according on the types of MPs and the duration of exposure
Lactuca sativa (a: 100 nm–18 μm and b: 18–150 μm)	PVC	0.5/1/2%	PVC (linked to photosynthesis),PVC-b (linked with root morphology). The total length, total surface area, and width of the roots increased by 0.5% and 1% of a; 1% of an increased the SOD activity of b.

continued on following page

Table 1. Continued

Plant species + size	M/NPs		Effects
	Type	Concentration	
Lepidium sativum (50, 500, 4800 nm)	Green fluorescent plastic	103-107 particles m/l	Impacts on root growth and germination rate that are both temporary and short-lived
Lycopersicon esculentum (52–368 μm)	PE, PP, PS	10, 100, 500, 1000 mg/l	Seed germination was inhibited by MPs (≤500 mg L−1), although under 1000 mg L−1 conditions, the effects were mitigated; PE was more hazardous to the development of seedlings than PS and PP.
Murraya exotica(12 ± 4.5 nm)	SMA	55 mg/l	Plant stems contained SMA nanoparticles, which were continuously enhanced throughout time.
Oryza sativa(50 μm)	BM, PE mulch film	1% (w/w)	Produced oxidative stress, which decreased the length and dry weight of the rice plant and had a detrimental impact on photosynthesis and nitrogen metabolism in the rice plant.
Oryza sativa(19 ± 0.16 nm)	PS	10– 100 mg/l	The rice roots took in PS particles, which gathered in the intercellular gaps.
Phaseolus vulgaris (250–500 μm, 500–1000 μm)	(Bio) mulch film, LDPE	0.5/1.0/1.5/2.0/2.5% (w/w)	The concentration of the LDPE-MP determines its effect; a concentration of less than 1.0% resulted in significantly larger specified root nodules, 2.5% in significantly longer specific roots, 1.0% in increased leaf area, and 0.5% in decreased relative chlorophyll amount in leaves. Bio-MP treatments demonstrated a considerable reduction in shoot, fruit, and root biomass and an increase of specific root height and nodules.
T. aestivum	MacroLDPE: 6.92 × 6.10 mm MacroBio: 6.98 mm × 6.01 mm Micro: 50 μm–1 mm	1% (w/w)	Both aboveand belowground areas were more negatively impacted by biodegradable plastic wastes than by PE.
T. aestivum (125 μm)	PE, PVC	1/5/10/20%	MPs had an adverse, dose-dependent effect on plant growth that reduced productivity both above and below ground.
Tobacco BY-2 cells (20–1000 nm)	PS	1:1000 v/v	Tobacco BY-2 protoplast cells were able to swallow PS particles up to 1000 nm, but tobacco BY-2 cells were only able to quickly ingest 20-40 nm PS nanobeads through endocytosis.

continued on following page

Table 1. Continued

Plant species + size	M/NPs		Effects
	Type	Concentration	
Triticum aestivum (100 nm)	PS	0.01-10 mg/l	Increased growth factors, chlorophyll amount, and shoot versus root biomass ratio; decreased concentrations of micronutrients; and modified metabolic activities in wheat seedlings.
Triticum aestivum (100 nm)	PS	0.01–10 mg/l	PS particles were found in xylems of the plant, both in the roots and shoots.
Triticum aestivum and *Lactuca Sativa* (0.2–10 μm)	PMMA, PS	0.5-50 mg/l; 150-500 mg/kg	The exposed plants' roots, stem, leaves, and xylem sap contained the 0.2 μm or 2 μm beads.
Vicia faba(0.1–5 μm)	PS	10-100 mg/l	While the 5 μm PS particles were rarely found in roots, the 0.1 μm PS molecules were able to penetrate the tips of broad bean roots.
Vicia faba(100 nm, 5 μm)	PS	10, 50, 100 mg/l	5 μm of PS reduced biomass and CAT enzyme production while increasing POD and SOD enzyme levels; 100 μm of PS (100 mg/l) reduced growth; 100 μm of PS caused more oxidative and genotoxic damage than 5 μm of PS; 100 μm of PS accumulated in root
Zea mays (3 μm)	PE	100 mg/l	Root contrasting carbon concentrations showed a considerable rise in those exposed to PE.
Zea mays(3 μm)	PE microbeads	0.0125 mg/l; 100 mg/l	PE may build up in the rhizosphere, affecting water and nutrient intake and finally reaching root eaters. PE bioaccumulation in the rhizosphere reduced transpiration, nitrogen availability, and growth.

(Wang *et al.*, 2022; Yue *et al.*, 2023)

Abbreviations: (Bio) mulch film: biodegradable plastic mulch film; Bio: biodegradable plastic; BM: PBAT based biodegradable mulch film; CAT: catalase; LDPE: low-density polyethylene; MDA: malondialdehyde; PA: polyamide; PBAT: butyleneadipate-co-terephthalate; PE: polyethylene; PEHD: polyethylene high density; PES: polyester; PET: polyethylene terephthalate; PMMA: polymethylmethacrylate; POD: peroxidase; PP: polypropylene; PS: polystyrene; PS-NH2: amino-modified polystyrene nanoparticles; PS-SO3H: sulfonic-acid-modified polystyrene nanoparticles; PVC: polyvinyl chloride; SMA: poly(styrene-co-maleic anhydride); SOD: superoxide dismutase;

According to Yu *et al.* (2020), M/NPs compete with microbial communities living in soil for the physicochemical niches, which lowers their efficacy and, in turn, lowers the extracellular enzyme processes. It is interesting to note that different aggregate-size fractions had diverse effects from M/NP exposure on enzyme activity. The different ways that each fraction responded to the presence of M/NPs demonstrated how complex the interactions are among MPs and soil microbial populations. The kind and form of the microplastic and the polymer may also be crucial factors in soil enzyme property. The effects of polyvinyl chloride and polyethylene MPs on enzyme activity varied. They had observed to increase enzymes such as acid phosphatase and urease. Due to their capacity to stick to or penetrate M/NPs and enhance degradation by establishing chemical interactions (for instance, carboxyl, carbonyl, and ester group), fungal species are effective degraders of M/NPs, eventually decreasing their hydrophobic nature (Russo *et al.*, 2023). Furthermore, particular microbial enzymes have the ability to alter M/NPs. Notably, a significant association was discovered between vegetable biomass and micro-eukaryotic community diversity, suggesting that crop growth may be aided by a more diversified micro-eukaryotic population. Note that micro-eukaryotes including a broad range of species, like algae or fungi. Likewise, there was a prominent correlation observed among the functional variety of bacterial communities as well as vegetable biomass, indicating that increased functional variety among the bacteria groups could be associated with enhanced crop growth. This kind of range of activities with their functions carried out by various bacterial groups within the colony is termed as functional diversity (Li *et al.*, 2023; Shah *et al.*, 2023).

5.3 Impact on Soil Pedogenesis

As a result of their extended duration of residence and strong reactivity, MPs and NPs have been shown to affect soil properties. One intriguing consequence of this is that they may have an impact on the soil pedologic mechanisms. It is conceivable to propose that the existence of M/NPs serves as a distinguishing characteristic for the subsurface and surface soil horizon classification. Furthermore, to the best of our knowledge, the literature has not yet addressed how this debris can alter the pedagogical processes. This is a very likely scenario that could lead to some fascinating discoveries. Within this framework, it is crucial to take into account the newly identified pyroplastics, which result from the widely used method of burning waste materials. These plastic types may enter the soil's geological cycle because of their resistance to deterioration (Pathan *et al.*, 2020).

6. CONCLUSION AND FUTURE PROSPECTS

The growing manufacturing and usage of plastics, together with inadequate management of plastic waste, were the main points of emphasis in this chapter. As a result, M/NPs are present or occur anywhere (the atmosphere, the hydrosphere, and the lithosphere). The detrimental effects of M/NPs on soil are explored along with their impact. Despite their tiny size, M/NPs showed a significant detrimental impact on soil ecosystem. Methods like incineration and landfilling merely convert the contaminant from one kind to another. Regretfully, the plastic shape that is created occasionally disappears from view and is more harmful to the ecosystem than the original form. To prevent the production of significant amounts of hazardous materials, plastic trash must be completely exposed to radiation. Regular disposal of plastic trash for conversion via circular-plastic economy efforts (reduce, reuse, and recycle) should be the first step towards effective plastic waste management. Minimizing plastic pollution is crucial, and it may be accomplished if communities receive the education necessary to become empowered and engaged in the process of reducing plastic pollution. Instead of using plastics, eco-friendly materials ought to be used. The breakdown of plastic trash is a global issue because microplastic pollutants (M/NPs) endanger aquatic life, including humans and birds. This project brought to light how plastic garbage degrades and takes on environmentally harmful forms.

High concentrations of M/NPs are continually discharged into the environment as a result of growing plastic production and consumption, endangering all living things. As a result, extreme efforts must be implemented to reduce plastic trash. M/NPs are a worldwide issue needing international cooperation since they can be carried over great distances, even to distant locations, via water and air. International cooperation is needed to reduce plastic waste, and the first step in this effort should be to develop research techniques for tracking, regulating, and assessing the breakdown and transformation of microplastics in diverse settings. In order to create innovative technology for the management of plastic waste, low and middle income nations should collaborate with high income countries. Waste management regulations need also be enacted at the regional as well as global levels. For low and middle income nations, garbage sorting and collection could be a good place to start because it can reduce the amount of rubbish dumped in water systems, on roadways, and in easily accessible areas. However, before being released into water systems, waste-traps in drainage systems can catch these. Consumer awareness is an additional instance that can be put into practice. Customers who receive management training and education regarding the effects of this trash will engage in all efforts to reduce plastic usage, including ceasing to litter. On the other hand, low- and middle-income nations, which are beset by unemployment or poverty, may find that recycling plastic trash into value-added items presents a chance for business development. This may only be accomplished by a shift in perspective and environmental education.

REFERENCES

Afrin, S., Uddin, M. K., & Rahman, M. M. (2020). Microplastics contamination in the soil from Urban Landfill site, Dhaka, Bangladesh. *Heliyon*, 6(11), e05572. DOI: 10.1016/j.heliyon.2020.e05572

Allen, S., Allen, D., Karbalaei, S., Maselli, V., & Walker, T. R. (2022). Micro(nano)plastics sources, fate, and effects: What we know after ten years of research. *Journal of Hazardous Materials Advances*, 6, 100057. DOI: 10.1016/j.hazadv.2022.100057

Anyanwu, I. N., & Semple, K. T. (2016). Assessment of the effects of phenanthrene and its nitrogen heterocyclic analogues on microbial activity in soil. *SpringerPlus*, 5(1), 279. DOI: 10.1186/s40064-016-1918-x

Bashir, S. M., Kimiko, S., Mak, C. W., Fang, J. K. H., & Gonçalves, D. (2021). Personal Care and Cosmetic Products as a Potential Source of Environmental Contamination by MPs in a Densely Populated Asian City. *Frontiers in Marine Science*, 8, 683482. DOI: 10.3389/fmars.2021.683482

Boucher, J., & Friot, D. (2017). Primary MPs in the Oceans: A Global Evaluation of Sources; IUCN: Gland, Switzerland. Environmental Science. *Geology*. Advance online publication. DOI: 10.2305/IUCN.CH.2017.01.en

Brennecke, D., Duarte, B., Paiva, F., Caçador, I., & Canning-Clode, J. (2016). Microplastics as vector for heavy metal contamination from the marine environment. *Estuarine, Coastal and Shelf Science*, 178, 189–195. DOI: 10.1016/j.ecss.2015.12.003

Bretas Alvim, C., Mendoza-Roca, J. A., & Bes-Piá, A. (2020). Wastewater treatment plant as microplastics release source - Quantification and identification techniques. *Journal of Environmental Management*, 255, 109739. DOI: 10.1016/j.jenvman.2019.109739

Bullard, J. E., Ockelford, A., O'Brien, P., & McKenna Neuman, C. (2021). Preferential Transport of MPs by Wind. *Atmospheric Environment*, 245, 118038. DOI: 10.1016/j.atmosenv.2020.118038

Chamas, A., Moon, H., Zheng, J., Qiu, Y., Tabassum, T., Jang, J. H., Abu-Omar, M., Scott, S. L., & Suh, S. (2020). Degradation Rates of Plastics in the Environment. *ACS Sustainable Chemistry & Engineering*, 8(9), 3494–3511. DOI: 10.1021/acssuschemeng.9b06635

Chen, M., Zhao, X., Wu, D., Peng, L., Fan, C., Zhang, W., Li, Q., & Ge, C. (2022). Addition of biodegradable microplastics alters the quantity and chemodiversity of dissolved organic matter in latosol. *The Science of the Total Environment*, 816, 151960. DOI: 10.1016/j.scitotenv.2021.151960

Dalela, M., Shrivastav, T. G., Kharbanda, S., & Singh, H. (2015). pH-Sensitive Biocompatible Nanoparticles of Paclitaxel-Conjugated Poly(styrene-co-maleic acid) for Anticancer Drug Delivery in Solid Tumors of Syngeneic Mice. *ACS Applied Materials & Interfaces*, 7(48), 26530–26548. DOI: 10.1021/acsami.5b07764

Di Bella, G., Corsino, S. F., De Marines, F., Lopresti, F., La Carrubba, V., Torregrossa, M., & Viviani, G. (2022). Occurrence of Microplastics in Waste Sludge of Wastewater Treatment Plants: Comparison between Membrane Bioreactor (MBR) and Conventional Activated Sludge (CAS) Technologies. *Membranes (Basel)*, 12(4), 371. DOI: 10.3390/membranes12040371

Dube, E., & Okuthe, G. E. (2023). Plastics and Micro/Nano-Plastics (MNPs) in the Environment: Occurrence, Impact, and Toxicity. *International Journal of Environmental Research and Public Health*, 20(17), 6667. DOI: 10.3390/ijerph20176667

Egger, M., Sulu-Gambari, F., & Lebreton, L. (2020). First evidence of plastic fallout from the North Pacific Garbage Patch. *Scientific Reports*, 10(1), 7495. DOI: 10.1038/s41598-020-64465-8

Fei, Y., Huang, S., Zhang, H., Tong, Y., Wen, D., Xia, X., Wang, H., Luo, Y., & Barceló, D. (2020). Response of soil enzyme activities and bacterial communities to the accumulation of microplastics in an acid cropped soil. *The Science of the Total Environment*, 707, 135634. DOI: 10.1016/j.scitotenv.2019.135634

Fernandes, E. M. S., de Souza, A. G., Barbosa, R. F. da S., & Rosa, D. dos S. (2022). Municipal Park Grounds and MPs Contamination. *Journal of Polymers and the Environment*, 30(12), 5202–5210. DOI: 10.1007/s10924-022-02580-5

Franco, A. A., Martín-García, A. P., Egea-Corbacho, A., Arellano, J. M., Albendín, G., Rodríguez-Barroso, R., Quiroga, J. M., & Coello, M. D. (2023). Assessment and accumulation of microplastics in sewage sludge at wastewater treatment plants located in Cádiz, Spain. *Environmental pollution, 317*, 120689. DOI: 10.1016/j.envpol.2022.120689

Fred-Ahmadu, O. H., Bhagwat, G., Oluyoye, I., Benson, N. U., Ayejuyo, O. O., & Palanisami, T. (2020). Interaction of chemical contaminants with microplastics: Principles and perspectives. *The Science of the Total Environment*, 706, 135978. DOI: 10.1016/j.scitotenv.2019.135978

Fu, L., Li, J., Wang, G., Luan, Y., & Dai, W. (2021). Adsorption behavior of organic pollutants on MPs. *Ecotoxicology and Environmental Safety*, 217, 112207. DOI: 10.1016/j.ecoenv.2021.112207

Giller, K. E., Delaune, T., Silva, J. V., van Wijk, M., Hammond, J., Descheemaeker, K., van de Ven, G., Schut, A. G. T., Taulya, G., Chikowo, R., & Andersson, J. A. (2021). Small farms and development in sub-Saharan Africa: Farming for food, for income or for lack of better options? *Food Security*, 13(6), 1431–1454. DOI: 10.1007/s12571-021-01209-0

Gouin, T. (2021). Addressing the importance of microplastic particles as vectors for long range transport of chemical contaminants: Perspective in relation to prioritizing research and regulatory actions. *Microplastics and Nanoplastics*, 1(1), 14. DOI: 10.1186/s43591-021-00016-w

Gui, J., Sun, Y., Wang, J., Chen, X., Zhang, S., & Wu, D. (2021). Microplastics in composting of rural domestic waste: abundance, characteristics, and release from the surface of macroplastics. *Environmental pollution, 274*, 116553. DOI: 10.1016/j.envpol.2021.116553

Hargreaves, S. K., & Hofmockel, K. S. (2014). Physiological shifts in the microbial community drive changes in enzyme activity in a perennial agroecosystem. *Biogeochemistry*, 117(1), 67–79. DOI: 10.1007/s10533-013-9893-6

Harley-Nyang, D., Memon, F. A., Jones, N., & Galloway, T. (2022). Investigation and analysis of microplastics in sewage sludge and biosolids: A case study from one wastewater treatment works in the UK. *The Science of the Total Environment*, 823, 153735. DOI: 10.1016/j.scitotenv.2022.153735

Hassan, F., Daffa, K., Nabilah, J., & Manh, H. (2023). Microplastic Contamination in Sewage Sludge: Abundance, Characteristics, and Impacts on the Environment and Human Health. *Environmental Technology & Innovation*, 31, 103176. DOI: 10.1016/j.eti.2023.103176

Horton, A. A. (2022). Plastic pollution: When do we know enough? *Journal of Hazardous Materials*, 422, 126885. DOI: 10.1016/j.jhazmat.2021.126885

Horton, A. A., & Dixon, S. J. (2018). MPs: An introduction to environmental transport processes. *WIREs. Water*, 5(2), e1268. DOI: 10.1002/wat2.1268

Huerta Lwanga, E., Mendoza Vega, J., Ku Quej, V., Chi, J. L. A., Sanchez Del Cid, L., Chi, C., Escalona Segura, G., Gertsen, H., Salánki, T., van der Ploeg, M., Koelmans, A. A., & Geissen, V. (2017). Field evidence for transfer of plastic debris along a terrestrial food chain. *Scientific Reports*, 7(1), 14071. DOI: 10.1038/s41598-017-14588-2

Hurley, R., & Nizzetto, L. (2017). Fate and occurrence of micro (nano) plastics in soils: Knowledge gaps and possible risks. *Current Opinion in Environmental Science & Health*, 1, 6–11. DOI: 10.1016/j.coesh.2017.10.006

Katsumi, N., Kusube, T., Nagao, S., & Okochi, H. (2021). The input-output balance of microplastics derived from coated fertilizer in paddy fields and the timing of their discharge during the irrigation season. *Chemosphere*, 279, 130574. DOI: 10.1016/j.chemosphere.2021.130574

Li, H., Luo, Q. P., Zhao, S., Zhou, Y. Y., Huang, F. Y., Yang, X. R., & Su, J. Q. (2023). Effect of phenol formaldehyde-associated microplastics on soil microbial community, assembly, and functioning. *Journal of hazardous materials, 443*(Pt B), 130288. DOI: 10.1016/j.jhazmat.2022.130288

Li, K., Jia, W., Xu, L., Zhang, M., & Huang, Y. (2023). The plastisphere of biodegradable and conventional microplastics from residues exhibit distinct microbial structure, network and function in plastic-mulching farmland. *Journal of Hazardous Materials*, 442, 130011. DOI: 10.1016/j.jhazmat.2022.130011

Li, S., Ding, F., Flury, M., Wang, Z., Xu, L., Li, S., Jones, D. L., & Wang, J. (2022). Macro- and microplastic accumulation in soil after 32 years of plastic film mulching. *Environmental pollution, 300*, 118945. DOI: 10.1016/j.envpol.2022.118945

Li, T., Cui, L., Xu, Z., Liu, H., Cui, X., & Fantke, P. (2023). Micro- and nanoplastics in soil: Linking sources to damage on soil ecosystem services in life cycle assessment. *The Science of the Total Environment*, 904, 166925. DOI: 10.1016/j.scitotenv.2023.166925

Liang, Y., Lehmann, A., Yang, G., Leifheit, E. F., & Rillig, M. C. (2021). Effects of Microplastic Fibers on Soil Aggregation and Enzyme Activities Are Organic Matter Dependent. *Frontiers in Environmental Science*, 9, 650155. DOI: 10.3389/fenvs.2021.650155

Lin, D., Yang, G., Dou, P., Qian, S., Zhao, L., Yang, Y., Fanin, N. (2020). MPs negatively affect soil fauna but stimulate microbial activity: Insights from a field-based microplastic addition experiment. *Proc. r. Soc*, 287(1934), 20201268.

Lin, D., Yang, G., Dou, P., Qian, S., Zhao, L., Yang, Y., & Fanin, N. (2020). Microplastics negatively affect soil fauna but stimulate microbial activity: insights from a field-based microplastic addition experiment. *Proceedings. Biological sciences, 287*(1934), 20201268. https://doi.org/DOI: 10.1098/rspb.2020.1268

Liu, H., Yang, X., Liu, G., Liang, C., Xue, S., Chen, H., Ritsema, C. J., & Geissen, V. (2017). Response of soil dissolved organic matter to microplastic addition in Chinese loess soil. *Chemosphere*, 185, 907–917. DOI: 10.1016/j.chemosphere.2017.07.064

Lozano, Y. M., Aguilar-Trigueros, C. A., Onandia, G., Maaß, S., Zhao, T., & Rillig, M. C. (2021). Effects of microplastics and drought on soil ecosystem functions and multifunctionality. *Journal of Applied Ecology*, 58(5), 988–996. DOI: 10.1111/1365-2664.13839

Lozano, Y. M., Lehnert, T., Linck, L. T., Lehmann, A., & Rillig, M. C. (2021). Microplastic Shape, Polymer Type, and Concentration Affect Soil Properties and Plant Biomass. *Frontiers in Plant Science*, 12, 616645. DOI: 10.3389/fpls.2021.616645

Luo, Y., & Zhou, X. (2006). *Soil Respiration and the Environment.* Academic Press., DOI: 10.1016/B978-0-12-088782-8.X5000-1

Lwanga, E. H., Beriot, N., Corradini, F., Silva, V., Yang, X., Baartman, J., Rezaei, M., van Schaik, L., Riksen, M., & Geissen, V. (2022). Review of Microplastic Sources, Transport Pathways and Correlations with Other Soil Stressors: A Journey from Agricultural Sites into the Environment. *Chemical and Biological Technologies in Agriculture*, 9(1), 20. DOI: 10.1186/s40538-021-00278-9

Mahesh, S., Gowda, N. K., & Mahesh, S. (2023). Identification of microplastics from urban informal solid waste landfill soil; MP associations with COD and chloride. *Water science and technology: a journal of the International Association on Water Pollution Research, 87*(1), 115–129. https://doi.org/DOI: 10.2166/wst.2022.412

Meides, N., Mauel, A., Menzel, T., Altstädt, V., Ruckdäschel, H., Senker, J., & Strohriegl, P. (2022). Quantifying the Fragmentation of Polypropylene upon Exposure to Accelerated Weathering. *Macroplastics and Nanoplastics*, 2, 1–13.

Meides, N., Menzel, T., Poetzschner, B., Löder, M. G. J., Mansfeld, U., Strohriegl, P., Altstaedt, V., & Senker, J. (2021). Reconstructing the Environmental Degradation of Polystyrene by Accelerated Weathering. *Environmental Science & Technology*, 55(12), 7930–7938. DOI: 10.1021/acs.est.0c07718

Meng, X., Zhang, J., Wang, W., Gonzalez-Gil, G., Vrouwenvelder, J. S., & Li, Z. (2022). Effects of nano- and microplastics on kidney: Physicochemical properties, bioaccumulation, oxidative stress and immunoreaction. *Chemosphere*, 288(Pt 3), 132631. DOI: 10.1016/j.chemosphere.2021.132631

Mohanan, N., Montazer, Z., Sharma, P. K., & Levin, D. B. (2020). Microbial and Enzymatic Degradation of Synthetic Plastics. *Frontiers in Microbiology*, 11, 580709. DOI: 10.3389/fmicb.2020.580709

Mokhtarzadeh, Z., Keshavarzi, B., Moore, F., Busquets, R., Rezaei, M., Padoan, E., & Ajmone-Marsan, F. (2022). Microplastics in industrial and urban areas in South-West Iran. *International Journal of Environmental Science and Technology*, 19(10), 10199–10210. DOI: 10.1007/s13762-022-04223-7

OECD. (2022). Plastic Pollution Is Growing Relentlessly as Waste Management and Recycling Fall Short. OECD Rep. Available online: https://www.oecd.org/environment/plastics/

Pathan, S., Arfaioli, P., Bardelli, T., Ceccherini, M., Nannipieri, P., & Pietramellara, G. (2020). Soil pollution from micro-and nanoplastic debris: A hidden and unknown biohazard. *Sustainability (Basel)*, 12(18), 7255. DOI: 10.3390/su12187255

Pathan, S. I., Arfaioli, P., Bardelli, T., Ceccherini, M. T., Nannipieri, P., & Pietramellara, G. (2020). Soil Pollution from Micro- and Nanoplastic Debris: A Hidden and Unknown Biohazard. *Sustainability (Basel)*, 12(18), 7255. DOI: 10.3390/su12187255

Rillig, M. C., Ziersch, L., & Hempel, S. (2017). Microplastic transport in soil by earthworms. *Scientific Reports*, 7(1), 1362. DOI: 10.1038/s41598-017-01594-7

Russo, M., Oliva, M., Hussain, M. I., & Muscolo, A. (2023). The hidden impacts of micro/nanoplastics on soil, crop and human health. *Journal of Agriculture and Food Research*, 14, 100870. DOI: 10.1016/j.jafr.2023.100870

Shah, T., Ali, A., Haider, G., Asad, M., & Munsif, F. (2023). Microplastics alter soil enzyme activities and microbial community structure without negatively affecting plant growth in an agroecosystem. *Chemosphere*, 322, 138188. DOI: 10.1016/j.chemosphere.2023.138188

Tagg, A. S., Brandes, E., Fischer, F., Fischer, D., Brandt, J., & Labrenz, M. (2022). Agricultural application of microplastic-rich sewage sludge leads to further uncontrolled contamination. *The Science of the Total Environment*, 806(Pt 4), 150611. DOI: 10.1016/j.scitotenv.2021.150611

Tamayo-Belda, M., Pulido-Reyes, G., González-Pleiter, M., Martín-Betancor, K., Leganés, F., Rosal, R., & Fernández-Piñas, F. (2022). Identification and toxicity towards aquatic primary producers of the smallest fractions released from hydrolytic degradation of polycaprolactone microplastics. *Chemosphere*, 303(Pt 1), 134966. DOI: 10.1016/j.chemosphere.2022.134966

Tian, H., Zhang, L., Dong, J., Wu, L., Fang, F., Wang, Y., Li, H., Xie, C., Li, W., Wei, Z., Liu, Z., & Zhang, M. (2022). A One-Step Surface Modification Technique Improved the Nutrient Release Characteristics of Controlled-Release Fertilizers and Reduced the Use of Coating Materials. *Journal of Cleaner Production*, 369, 133331. DOI: 10.1016/j.jclepro.2022.133331

van Schothorst, B., Beriot, N., Huerta Lwanga, E., & Geissen, V. (2021). Sources of Light Density Microplastic Related to Two Agricultural Practices: The Use of Compost and Plastic Mulch. *Environments (Basel, Switzerland)*, 8(4), 36. DOI: 10.3390/environments8040036

Wang, F., Gao, J., Zhai, W., Liu, D., Zhou, Z., & Wang, P. (2020). The influence of polyethylene microplastics on pesticide residue and degradation in the aquatic environment. *Journal of Hazardous Materials*, 394, 122517. DOI: 10.1016/j.jhazmat.2020.122517

Wang, J., Zheng, L., & Li, J. (2018). A critical review on the sources and instruments of marine microplastics and prospects on the relevant management in China. *Waste management & research: The journal of the International Solid Wastes and Public Cleansing Association. Waste Management & Research*, 36(10), 898–911. DOI: 10.1177/0734242X18793504

Wang, W., Yuan, W., Xu, E. G., Li, L., Zhang, H., & Yang, Y. (2022). Uptake, translocation, and biological impacts of micro (nano) plastics in terrestrial plants: Progress and prospects. *Environmental Research*, 203, 111867. DOI: 10.1016/j.envres.2021.111867

Wang, W., Zhao, Y., Bai, H., Zhang, T., Ibarra-Galvan, V., & Song, S. (2018). Methylene blue removal from water using the hydrogel beads of poly(vinyl alcohol)-sodium alginate-chitosan-montmorillonite. *Carbohydrate Polymers*, 198, 518–528. DOI: 10.1016/j.carbpol.2018.06.124

Weber, C. J., Santowski, A., & Chifflard, P. (2022). Investigating the dispersal of macro- and microplastics on agricultural fields 30 years after sewage sludge application. *Scientific Reports*, 12(1), 6401. DOI: 10.1038/s41598-022-10294-w

Wiedner, K., & Polifka, S. (2020). Effects of microplastic and microglass particles on soil microbial community structure in an arable soil (Chernozem). *Soil (Göttingen)*, 6(2), 315–324. DOI: 10.5194/soil-6-315-2020

Wright, S. L., Thompson, R. C., & Galloway, T. S. (2013). The physical impacts of microplastics on marine organisms: a review. *Environmental pollution, 178*, 483–492. DOI: 10.1016/j.envpol.2013.02.031

Wu, Y., Guo, P., Zhang, X., Zhang, Y., Xie, S., & Deng, J. (2019). Effect of microplastics exposure on the photosynthesis system of freshwater algae. *Journal of Hazardous Materials*, 374, 219–227. DOI: 10.1016/j.jhazmat.2019.04.039

Ya, H., Xing, Y., Zhang, T., Lv, M., & Jiang, B. (2022). LDPE microplastics affect soil microbial community and form a unique plastisphere on microplastics. *Applied Soil Ecology*, 180, 104623. DOI: 10.1016/j.apsoil.2022.104623

Yang, H., Dong, H., Huang, Y., Chen, G., & Wang, J. (2022). Interactions of MPs and main pollutants and environmental behavior in soils. *The Science of the Total Environment*, 821, 153511. DOI: 10.1016/j.scitotenv.2022.153511

Yang, L., Zhang, Y., Kang, S., Wang, Z., & Wu, C. (2021). Microplastics in soil: A review on methods, occurrence, sources, and potential risk. *The Science of the Total Environment*, 780, 146546. DOI: 10.1016/j.scitotenv.2021.146546

Yang, Z., Lü, F., Zhang, H., Wang, W., Shao, L., Ye, J., & He, P. (2021). Is incineration the terminator of plastics and microplastics? *Journal of Hazardous Materials*, 401, 123429. DOI: 10.1016/j.jhazmat.2020.123429

Yi, M., Zhou, S., Zhang, L., & Ding, S. (2021). The effects of three different microplastics on enzyme activities and microbial communities in soil. *Water environment research: a research publication of the Water Environment Federation, 93*(1), 24–32. DOI: 10.1002/wer.1327

Yu, H., Fan, P., Hou, J., Dang, Q., Cui, D., Xi, B., & Tan, W. (2020). Inhibitory effect of microplastics on soil extracellular enzymatic activities by changing soil properties and direct adsorption: An investigation at the aggregate-fraction level. *Environmental pollution, 267*, 115544. DOI: 10.1016/j.envpol.2020.115544

Yu, Z. F., Song, S., Xu, X. L., Ma, Q., & Lu, Y. (2021). Sources, migration, accumulation and influence of MPs in terrestrial plant communities. *Environmental and Experimental Botany*, 192, 104635. DOI: 10.1016/j.envexpbot.2021.104635

Yue, Y., Li, X., Wei, Z., Zhang, T., Wang, H., Huang, X., & Tang, S. (2023). Recent Advances on Multilevel Effects of Micro(Nano)Plastics and Coexisting Pollutants on Terrestrial Soil-Plants System. *Sustainability (Basel)*, 15(5), 4504. DOI: 10.3390/su15054504

Zettler, E. R., Mincer, T. J., & Amaral-Zettler, L. A. (2013). Life in the "plastisphere": Microbial communities on plastic marine debris. *Environmental Science & Technology*, 47(13), 7137–7146. DOI: 10.1021/es401288x

Zhang, G. S., & Liu, Y. F. (2018). The distribution of microplastics in soil aggregate fractions in southwestern China. *The Science of the Total Environment*, 642, 12–20. DOI: 10.1016/j.scitotenv.2018.06.004

Chapter 9
Micro/Nano-Plastics Pollution:
Challenges to Agriculture Productivity

Nandini Arya
Uttaranchal University, Dehradun, India

Atin Kumar
https://orcid.org/0000-0002-1653-2146
Uttaranchal University, Dehradun, India

ABSTRACT

Plastic pollution, particularly in the form of microplastics and nanoplastics, has emerged as a significant environmental concern with far-reaching implications. This chapter delves into the origins, characteristics, effects, and identification and methods of removal of these micro/nano plastic particles that are prevailing in our ecosystems. The properties of these plastics make them versatile and widely used materials in the commercial world, but this has led to their extensive production and disposal, contributing to the formation and accumulation of more and more microplastics and nanoplastics in various natural habitats. These smaller plastic fragments, categorized as primary and secondary micro/nano plastic, pose a greater threat to the environment due to their ability to carry harmful substances and disrupt ecosystems. Collaborative efforts among researchers, policymakers, and industries are essential to developing effective strategies for detecting, mitigating, and preventing further plastic pollution.

DOI: 10.4018/979-8-3693-3447-8.ch009

INTRODUCTION

Plastic started popularizing in the market on a commercial scale in 1950 used for Industrial use, packaging products, cosmetic products, mulch nets, storage bins, and nets in agriculture. Because they are lightweight, robust, corrosion-resistant, durable, non-biodegradable, and cost less in the pocket. They are produced by the polymerization of various monomers and other additives, petroleum being a major ingredient (Yadav et al. 2022). About 86% of plastic goes into landfills, and about 4,900 million metric tons of plastic waste ended up in landfills or openly lying in environment between 1950 and 2015 (Rogers K. 2024), because of its non-biodegradable properties, contributing as a major source of land water, and air pollution. Currently, global production of plastic has increased to 320 million tons per year and is expected to rise continuously. Yet, Macro plastic isn't a major concern these larger fragments of plastic break down into smaller fragments through mechanical wave action, hydrolysis, photo-degradation, abrasion, action of UV radiation (Alimi et al. 2018) & action of different chemicals present in the environment leading to the formation of Microplastic and Nano plastic which are turning out to be much vicious enemy of biosphere. They are not visible to the naked eye being >5mm in size. Microplastic is 0.1μm-5mm and nanoplastic is less than 0.1μm in size. Due to improper waste management system (Huang et al. 2023) & their small size they are easily carried away by air and water mixing well into the ecosystem.They are excellent vectors carrying heavy metals and other pathogenic/invading substances with them as they travel through water or air making their way into organisms hindering their metabolisms and introducing harmful diseaeses. They have ability to accumulate in sensitive habitats such as sea grass beds and mangroves hindering their biological processes and leading to physical damage and reduced biodiversity (Machado et al. 2019). Microplastic has two types of origin firstly, primary source which are found as microbeads, microplastic fibers in cosmetic products, or produced during the laundering of synthetic fabric reaches in the environment through wastewater system from households (Rogers 2024). They are usually intentionally produced in industries to be added in the products increasing their efficiency (Frias et al. 2019). Then comes the secondary type of microplastic which are formed due to the abrasion of larger pieces of plastic due to environmental weathering factors. The percentage of microplastic released in the environment from secondary sources is much higher than from primary sources (Frias et. al 2019). In agriculture Plastic in fields has many sources they can be due to washout water from landfills reaching the field during rain or mulches or fertilizers used. Microplastic in agricultural fields present due to various sources into the soil leach further contaminating the groundwater and polluting it for any further uses. Micro/nano plastic is also reaching the natural soils away from urban areas or places of plastic contamination through atmospheric deposition. They are posing a major threat to agriculture productivity by ultimately altering the soil's physical properties and transforming soil density, nutrient cycle mechanism, water retention quality, soil aeration, aggregation, and intoxicating soil with heavy metals& other toxicants which are likely to stick to the plant roots absorbed by the plant, they also interfere with the number of soil microorganisms by introducing invasive species leading to the extinction of natural species or by bioaccumulation in which a high concentration of micro/nano plastic gets build in tissues of organisms over time disrupting their physiological process and eventually killing them eventually deteriorating the whole soil ecosystem (Pandey et al. 2024). The plant absorbing such harmful substances leads to the indirect consumption of plastic by predators and one such major predator is humankind only leading to the introduction of poison into the human body, such cultivation would either be hard to flourish or deprived of many nutrients just a bunch of plastic accumulated plants (Alimi et al. 2018). The fact that makes this much more concerning is that according to estimations by different studies plastic waste in landfills will be increased by 12,000 million metric tons in the next 2 decades (Rogers 2024). Both of the versions of plastic are harmful to the environment but still smaller size of nano plastic than microplastic which makes it much easier to travel and penetrate through tissues and skin of the organisms or travel in plants, which can go undetected. Nanoplastics are much easier to uptake as compared to microplastic which can adhere to the surface of leaves, making the penetration easier lead to a higher rate of uptake by plants. The rate of accumulation of nanoplastic is much higher than microplastic, they exhibit greater mobility which leads them to being transported throughout the plant parts. They are potential bioaccumulators of plastic disrupting the germination rate, causing physiological changes, and hindering the metabolism of plants making them potentially more harmful for the plants (Bratovic, 2023).

Sources of Microplastic and Nanoplastic Pollution

Microplastic and nanoplastic pollution is rising at a very high rate in metropolitan cities as compared to rural areas due to excessive usage of plastic but the microplastic and nanoplastic are also being detected in rural or areas very far from such cities' land and water in large number. Few pathways are being identified as the most prevalent pathways through which agricultural land of different areas is contaminated.

- **Agriculture waste:** The nearest and most common sources of microplastic and nanoplastic pollution, Mulching film greenhouse covering (Xu et al. 2020), and polymer-coated fertilizers or wear and tear of agriculture machinery which are used to enhance productivity are some of the pollutant sources

(Moeck et al. 2023). Fragments of microplastic are the most common shape seen in microplastic released in the soil by such material (Xu et al. 2020)

- **Industrial dump or landfill runoff water:** Waste water from industries like textile, sugar, and cosmetics or effluent of wastewater treatment plants being released into agriculture fields (Moeck et al. 2023) or runoff water from landfill areas usually reaches fields during the rainy season, such water contains a high amount of microplastic usually in the shape of microfibers and polyester. (Xu et al. 2020).
- **Urban influence:** The microplastic from tire abrasion, agriculture vehicles or construction site-released waste can get deposited in soil because of atmospheric deposition, and airborne MPs (Micro Plastics) from sources like household or textile industry dust can settle down in the field or on the plant leaves. (Moeck et. al. 2023)

Polymers like PP, PE, and PS type of microplastic are prevalent but, there are many shapes and sizes of microplastic like Microfibers are a notable source of primary microplastics in the environment, originating from shedding synthetic textiles during washing processes. Heterogeneous fragments and films, often derived from plastic film and compost applications in agriculture, are widespread in terrestrial environments. Plastic films utilized in agriculture, like mulching films, play a role in the prevalence of films as a common form of microplastic pollution in fields. Although less common than fibers, fragments, and films, granules, and foams are also detected in atmospheric deposition samples in urban areas. Microplastic in soil are easily detected in the soil but less technology available to detect sizes less than 1mm is the main cause of less detection of nanoplastic detection in soil (Xu et al. 2020).

Effect of Different Shapes, Sizes, and Concentrations of Micro/Nanoplastics

These tiny winy particles are a major threat to the environment and as they are a just a smaller version of plastic, they inherit the properties of plastic such as hydrophobic and other surface properties that naturally lead them to repeal water instead, which makes them highly dangerous as they absorb and retain a range of heavy metals, organic pollutants, pathogens, and other contaminants which makes them a vector to carry harmful substances as they travel. Micro/Nano plastics are chemically stable which means they can be in the environment for a prolonged period without losing any of their properties. Due to their high surface area to volume ratio because of their tiny size (Figure 1), they provide a large surface area for the interaction with other pollutants this interaction potential makes them much more harmful to the agro-ecosystem and further organisms as these can be transferred from one organism to another in the food chain. Once they enter the soil they alter the soil structure by blocking the soil pore which leads to a reduction in the poor soil aeration and water absorption capacity. Higher concentrations of Micro/Nano plastic can intensely harm the soil-plant interaction and ecosystem as well as marine ecosystem as they can easily penetrate through the gills of fish and create blockages in their organs by traveling through their blood circulation system, it harmful to all the poultry as well as fish farming practices (Bratovic, 2023).

Figure 1. Size comparison of microplastic particles with normal-sized plastic sheet

Micro/nanoplastics have a variety of different shapes, sizes, and densities detected in the environment, ranging from fibers to pellets and films, granules in different size ratios (Rosal et al. 2021). Different types of microplastics, their sizes, and their concentration affect the plant, not all microplastic can disrupt the functioning of the plant. A few such examples are shown in Table 1.

Table 1. Effect of types of microplastic on plants

NAME OF PLANT	TYPE OF MICRO PLASTIC	SIZE OF MP	CONCENTRATION OF MICRO PLASTIC	EFFECT OF MICRO PLASTIC ON PLANT FUNCTIONING
Lactuca sativa (lettuce)	Polystyrene microbeads	0.2µm-1.0µm	10^3,10^5&10^7 particle/ mL	Polystyrene Microbeads are transported to the stem via transpiration, a decline in overall chlorophyll content and a reduced germination rate (Tang 2020, Sandhya 2024)
*Lolium Perenne(*perennial ryegrass*)*	High-density polyethylene	102.6µm	1g/kg dry soil	Decreased rate of germination and total root biomass is reduced. (Tang 2020)
*Vigna mungo L.(*Blackgram)	Polyethylene	60µm	0.25%,0.5%,0.75%, 1%	Reduction in germination rate, root and shoot length, due to physical blockage of the pores by microplastics disrupted the rate of nutrients and water. (Sandhya 2024)
Lemna minor (duckweed)	Polyethylene microsphere	10-45µm	50,000 microplastic/mL	There was no significant change in chlorophyll content but root growth is been retarded due to mechanical blocking. (Tang 2020)
Solanum lycopersicum L. (Tomato)	Polyethylene terephthalate & low-density polyethylene	60µm	0.25%,0.5%,0.75%, 1%	A significant decline in the growth rate of plants is noted, reduced rate of germination alongside root and shoot length is noted. (Sandhya 2024)
*Oryza sativa (*Rice*)*	Polystyrene & polytetrafluoroethylene	10µm	0.04,0.1,0.2g/L	A decline in growth rate in lower parts of plant is noted. (Sandhya 2024)
Lepidium sativum (garden cress)	Green fluorescent plastic particles	50nm, 500nm, 4800nm	10^3,10^7 particles	Initial exposure showed a decline in the growth rate of the plant, but as the exposure prolonged the impact was negligible. (Tang 2020)
Ipomoea batatas (sweet potato)	Polyethylene	5µm	1% to 5%	Increases in Cd accumulation and glutathione levels in tissues of sweet potatoes, while rises in soil electrical conductivity are also noted. (Sandhya 2024)

Low-density polyethylene (LDPE) and PVC (Poly Vinyl Chloride) with different shapes are very common in the agriculture field which negatively impacts the growth of plants, the studies show a decline in the total fresh weight of the plant in the presence of these microplastic concentrations (Hasan & Jho 2023). Polystyrene (PS) Nano plastic with a concentration of 20 μg L^{-1} (1.1 × 108 particles/mL) can cause local infection in the liver resulting in a metabolism disorder, they can be capable of passing through the blood-brain barrier which has a lethal effect on organisms, these PS nano plastics can be transported through food chain. Nanoplastic of size and density (39.4 nm, 10 mg L^{-1}) accumulates in the body of soil and marine fauna by entering their blood circulation or cell membrane (Shen et al. 2019). These results show they not only hinder the crops but can lead to a decline in the growth of other agriculture sectors such as fishery or poultry, the presence of 1% (w/w) and 10% (w/w) polylactic acid (PLA) MPs and high-density polyethylene (HDPE) had a significant increase of soil pH (Wang et al. 2021)

Figure 2. Depicting the impact of microplastic on plants

Changes in plant stress levels are noticed due to their exposure to micro/nanoplastics, This upregulation typically manifests as heightened levels of reactive oxygen species (ROS), malondialdehyde (MDA), and thiobarbituric acid reactive substances (TBARS), which serve as indicators of stress. These increased secretions of antioxidant enzymes have triggered the defense mechanisms to mitigate the stress induced by nano- and microplastics (NMPs). The elevation in free radicals, including hydrogen peroxide and hydroxyl radicals, can lead to oxidative harm within the plant, impacting functions such as the synthesis of secondary metabolites. But, the concentration of these particles matters as for different types of plants different concentrations are required in the soil for the plant to get affected.Hence, it is very important to conduct studies under different field conditions for a comprehensive understanding of the impacts of nano and microplastics on plants (Figure 2). This gives a level of realism that mirrors natural environmental conditions, encompassing factors like soil diversity, and microbial communities.This helps in developing real-world management strategies and policies aimed at mitigating the impacts of NMP pollution on plants and ecosystems. Field research plays a crucial role in translating scientific findings into actionable solutions for environmental sustainability by generating data directly applicable to environmental conservation efforts (Zantis et al. 2022).

Soil Microbiome Disruption

Soil microbiome plays a vital role in enhancing the growth and healthy lifecycle of plants, they provide environmental aid such as fixing nitrogen, regulating the carbon cycle, water retention, plant defense, and maintaining even composition of plant nutrients throughout the soil which helps in balancing the environmental order and can help the humanity in taking a victorious step towards sustainable agriculture. The plant microbiome is a beneficial factor for growth. It can also be marked by the presence of organic compounds secreted by the roots of the plant which supports microbial activities near the plant. (Vincze et al. 2024). But, day-by-day increasing concentrations of microplastic and finer nanoplastic, estimation from many studies these pollutants collect up to 0.5 megatons annually in agricultural land (Maddela et al. 2023), causing disruptions that lead to changes in the diversity and structure of the microbial community, this can be indicated by variation of enzymatic activities in the soil, disruptions in nutrient cycling (Iqbal et al. 2020) and substrate availability which are influenced by microbes. In many cases, the microplastics and nanoplastic carry heavy metals or organic pollutants such as cyclic aromatic hydrocarbon and polychlorinated biphenyls with them as they travel from various places or absorb pesticides that are sprayed in the field accumulation of such additives gives them a high potential of phytotoxicity impairing the soil structure and plant growth as well as inducing oxidative damage and intestinal stress in the nematodes and earthworms decreasing their survival and reproductive ability (Yu et al. 2022). Although, the studies have shown different concentration ranges have different impacts on the agro eco-system such as at the concentration range of 0.01-1% there is no effect on the soil bacterial colony, even a small concentration of 0.10% will reduce the growth and survival of earthworm. There are specific concentrations at which the microplastic can actually either promote the soil activites or stay neutral for example according to US regulations the maximum potential rate of microplastic should range from 0.5 to 3.2 t/ha/year, this usually reaches to 9 to 63 t/ha/year which is a matter of great concern (Ng et. al 2018). The fact that it is very difficult to control the amount of microplastic is threatening.

Micro/Nano Plastic Interaction With Fertilizer and Pesticides

Fertilizers and pesticides are some of the most used chemicals and artificial agents to regulate agriculture productivity according to the needs of the farmers and to control the pests and diseases of the field, but it is noticed that their interaction with Micro/Nano plastic has led to a dangerous effect which was supposed to have a beneficial impact. Such as Micro/Nanoplastics can adsorb agrochemicals such as fertilizers and pesticides. This adsorption of agrochemicals impacts the transport and distribution of these chemicals in the soil, such as iron (Fe), manganese (Mn), copper (Cu), and zinc (Zn) in plants. This interaction between micro/nanoplastic with fertilizer has the potential to disturb the nutrient equilibrium in the plants, affecting their overall health and plant productivity, and potentially changing their efficacy and environmental destiny. They also serve as carriers for agrochemicals, aiding in the release of heavy metals into the environment. This phenomenon results in unintended contamination of soil and water in agro eco-systems. Their presence has amplified the phytotoxicity of agrochemicals by increasing their bioavailability and uptake by plants. This interaction often leads to adverse effects on plant health and ecosystem functioning. The phytotoxicity of microplastics varies depending on the type of plastic, size, and concentration. High concentrations of microplastics and nanoplastics increases the stress responses in plants, affecting their ability to uptake and utilize nutrients effectively. Micro/nanoplastics exhibit a good sorption capacity, leading to the accumulation of agrochemicals on their surfaces. When microplastics and nanoplastics are present in the soil, they get attached to plant roots and they hinder their absorption capacity as they create blockage, creating difficulties for plants in absorbing water and other essential nutrients (Figure 3). This change in root structure can impede plants' uptake of vital nutrients. This accumulation may extend the persistence of these chemicals in the environment. Other, than changing all the physical properties their presence has also impacted the soil microbiota, potentially influencing the degradation and transformation of fertilizers and pesticides. This interaction has the potential to modify soil nutrient cycling and the dynamics of microbial communities. All these factors showcases that This interaction has the potential to modify soil nutrient cycling and the dynamics of microbial communities (Verma et al. 2023). In many cases, the fertilizer in the market is available with a polymer coating one such study case study on maize showcases the experiment research focused on two maize cultivars, examining how PCF-MPs (polymer-coated fertilizer-micro plastics) influenced plant growth, rhizosphere metabolites, and soil enzyme activities. Results demonstrated that PCF-MPs had distinct effects on maize growth specific to each cultivar, showcasing variations in shoot biomass and root diameter. Analysis of rhizosphere metabolomics highlighted differences in metabolite profiles attributed more to maize cultivars than the presence of MPs. Moreover, soil enzyme activities exhibited slight increases with the introduction of PCF-MPs, notably affecting phytase activity in one maize cultivar. Significantly, PCF-MPs did not display phytotoxicity on maize cultivars and could even enhance the growth of select cultivars at higher concentrations. The study underscored the importance of considering cultivar-specific responses and soil properties when evaluating the environmental risks of phytotoxicity and soil chemical, physical, and microbial health deterioration associated with MPs in agricultural settings (Lian et. al. 2021). The different concentrations have different impacts on plant growth and maintaining a positive concentration of microplastic and nanoplastic can be very challenging.

Figure 3. Illustration of how microplastics are evenly distributed around roots and in soil disturbing its physical structure

CASE STUDY

There have been many cases around the globe that showcase how big of a problem microplastic and nanoplastic in agroecosystems are becoming few of them are:

Case 1: Mangrove Ecosystem Threatened Due to Microplastic in Kerala

BASIS: Mangroves are vital ecosystems that provide numerous ecological benefits, including shoreline protection, carbon sequestration, and habitat for diverse flora and fauna. However, these ecosystems are facing a severe threat from microplastic pollution, as highlighted in a recent study conducted in Mangalavanam, Kerala, India. The study, led by a team of researchers from Cochin University of Science and Technology, aimed to assess the impact of microplastic pollution on urban mangrove ecosystems. Microplastics, tiny plastic particles less than 5 mm in size, have become a major environmental and soil productivity concern due to their widespread presence in marine and terrestrial biosphere. Samples were conducted in February 2021, with a focus on accessibility and maintaining a distance between sampling spots ensuring that the collected sample covered a wide area, it gave a better understanding of the concentration variations and their effects. Soil, sediment, and water samples were collected and prepared for analysis following established protocols from relevant research papers. Quality control measures were implemented to prevent environmental contamination during sample collection and analysis.

Findigs: The investigation unveiled a notable abundance of microplastics in the soil, sediment, and water samples obtained from Mangalavanam. Examination displayed a diverse range of microplastic varieties and sizes, signaling a substantial level of pollution poses a serious threat to the biodiversity and ecological balance of these fragile habitats urban mangrove ecosystem. Statistical analysis validated the disparities in microplastic composition across the three environmental compartments. need for conservation initiatives to protect urban mangrove ecosystems from the detrimental effects of microplastic pollution. By raising awareness, implementing sustainable practices, and fostering community engagement, we can work towards safeguarding these valuable ecosystems for future generations (Kannankai et al. 2022).

Case Study 2: Effect of Microplastic Pollution on Watermelon and Tomato Cultivation

BASIS: Agricultural soils used for cultivating watermelons and canning tomatoes were investigated. Aimed to identify and quantify MPs in soil samples collected from these agricultural fields in Ilia County, Western Greece. Soil samples were collected from five watermelon fields (WSS_1 to WSS_5) and five canning tomato fields (TSS_1 to TSS_5). Additionally, 10 soil samples were collected from non-cultivated lands as blanks for comparison surface soil was collected from a depth of 0-30cm, and a quantity of 1.5 kg was collected from different locations. The samples were air-dried, sieved, and subjected to a floating extraction method to isolate MPs without destroying or fragmenting them. The MPs were then counted manually, and the procedure was repeated for each sample. The fields had been cultivated with watermelons and canning tomatoes for over ten years.

FINDING: The study has raised concerns about the potential impact of MPs on food safety and environmental sustainability. The accumulation of microplastics in agricultural soils could lead to contamination of crops, affecting human health throughout the food chain. Moreover, the long-term presence of MPs in soils may harm soil quality and ecosystem health. varying levels of microplastic abundance in the soil samples, with higher concentrations observed in soils cultivated with watermelons compared to canning tomatoes. The presence of black polyethylene (PE) covers used in agricultural practices is identified as a significant source of microplastics in the soil. It can potentially transfer these contaminants to crops cultivated in these soils raising alarm regarding the contamination of food items with microplastics. This scenario threatens human health as consuming contaminated produce may introduce risks associated with microplastic ingestion. The long-term accumulation has raised doubts about soil quality, biodiversity, and ecosystem functioning, raising the need for proactive measures to mitigate the spread of microplastics (Isari et al. 2021).

Case Study 3: Microplastic Pollution In Terrestrial Ecosystem

BASIS: The fate and interactions of microplastics with existing co-contaminants in the soil were investigated. The researchers highlighted the potential risks posed by microplastic pollution to soil ecosystems and emphasized the need for sustainable soil management practices. It is emerging as a significant environmental concern, with potential threats to terrestrial ecosystems and soil sustainability. This case study delves into the interactions of microplastics with other soil pollutants and their impacts on soil ecosystem health. The study utilized a multidisciplinary approach, starting with an extensive literature review to investigate the sources, distribution, and impacts of microplastic pollution in terrestrial environments. Researchers analyzed previous studies focusing on how interactions between microplastics and soil contaminants influence soil properties, microbial communities, and overall ecosystem health. Field observations and laboratory experiments were conducted to assess the presence of microplastics in soil samples collected from diverse terrestrial locations. Techniques such as microscopy, spectroscopy, and chemical analysis were employed to detect and quantify microplastics in the soil samples accurately. Moreover, statistical analyses were performed to assess the correlations between microplastic contamination, soil characteristics, and microbial diversity. The results of these analyses were crucial in concluding the potential risks of microplastic pollution to soil sustainability and biodiversity in terrestrial ecosystems.

Findings: The study identified various sources of microplastic pollution in soil, including plastic mulch, pharmaceuticals, cosmetics, tire abrasions, textile industries, sewage sludge, and plastic dumping. These sources contribute to the widespread distribution of microplastics in terrestrial environments as well as The presence of microplastics in soil poses a significant threat to soil sustainability and biodiversity in terrestrial ecosystems. Microplastic contamination can lead to adverse effects on soil health, fertility, pH levels, water-holding capacity, and soil microbial enzymatic activities, ultimately disrupting the balance of the soil ecosystem (Rai et al. 2023).

Removal Methods of Micro/Nano Plastics in Agro-Ecosystem

There have been a lot of studies going on to deal with the removal of micro/nano plastics as they are not only hard to identify but very difficult to separate from the environment. Few of the treatment

1. Centrifugation

This is a method used to remove microplastic by using a solution with a specific density, mainly a solution of $CaCl_2$ solution with a density of 1.4g/cm^3 is used to separate microplastic from soil. This involves various key steps which follow as first sampling and sieving of the soil sample is done which is the removal of larger debris and particles and only a fine fraction of soil is used, the extracted sample is treated with Fenton reagent which destroys all the non-plastic organic matter that could interfere in the further process, the final treated product is then centrifuged with $CaCl_2$ solution of density 1.4 g/cm^3. Centrifugation separates the microplastics from the soil matrix based on density differences. The microplastics, being less dense than the soil particles, will move to the top of the centrifuge tube during centrifugation. The separated microplastics are stained with Nile Red, a fluorescent dye that binds to hydrophobic substances, including certain types of plastics. This staining process helps to visualize and identify the microplastics under fluorescence microscopy. Then the stained microplastics are then examined using fluorescence microscopy. By exciting the Nile Red-stained microplastics with specific wavelengths of light, the plastic particles emit fluorescence, making them visible under the microscope, allowing the visualization and identification of the microplastics in the sample, and Image processing techniques are applied to the microscopic images obtained during fluorescence microscopy. All These techniques help quantify the number of microplastics present in the soil sample and analyze their characteristics. Software tools like Fiji and Python may be used for image analysis and quantification. This approach enables the effective extraction and examination of microplastics, thereby aiding in the evaluation of plastic contamination in environmental samples. Shown in fig 2. (Grause et al 2021).

2. Leachate Treatment System

It is the process designed to treat leachate, which is the liquid that drains or "leaches" from a landfill. This liquid typically contains various contaminants, such as micro/nano plastics, that can pose environmental risks if not properly managed. The leachate treatment system aims to remove pollutants and contaminants from the leachate before it is released into the environment. As, this results in one of the major causes of soil and water pollution. There are multiple stages of filtration from which the leachate goes through, firstly through the Adjustment Tank stage this may involve initial adjustments to the leachate's pH, temperature, or other parameters to optimize subsequent treatment processes, after this stage around 16.67% of micro/nano plastics is removed. Then comes the Membrane Bioreactor (MBR) which contains biological catalyst-supported catalysis that combines biological treatment processes with membrane filtration to remove contaminants, including microplastics, from the leachate, reducing 50% more microplastic. Two-stage anionic/Oxic (AO) Treatment is the next process involving two stages - an anoxic stage followed by an oxic stage - to promote the removal of organic matter and contaminants from the leachate 20% reduction in microplastic number is noticed. Further, Ultrafiltration (UF): UF is a membrane-based separation process that can effectively remove 75% of microplastics and other particles from the leachate. Nanofiltration (NF) and Reverse Osmosis (RO) are the last processes that may not be as effective for microplastic removal, they are commonly used in leachate treatment systems for removing other contaminants (Kundu et al. 2021)

3. Dynamic Membrane Technology

It is a very innovative approach process that is used in wastewater treatment processes, including urban wastewater treatment, surface water treatment, industrial wastewater treatment, and sludge treatment. The technology involves the formation of a dynamic membrane (DM) that acts as a secondary barrier within the treatment system to enhance the removal of contaminants, including microplastics. The dynamic membrane is created by the accumulation of a cake layer on supporting membrane filters within the wastewater treatment system. This cake layer effectively traps and removes contaminants, including microplastics, from the liquid stream. But, Various parameters, such as the characteristics of the supporting membrane, the composition of the matter deposited on the membrane, operating pressure, and cross-flow velocity, influence the formation and performance of the dynamic membrane, offering versatility in its application across different wastewater treatment processes and systems, making it a promising solution for enhancing the removal of microplastics and other contaminants from liquid streams. Various tests and studies showed that around 90% of retained microplastic particles and cellulose acetate with a size of less than 90 μm were successfully removed using dynamic membrane technology & and it is found to be an ideal membrane for the implementation of a domestic household system (Kundu et al. 2021)

Impact of Microplastic in Different Agroecosystem

Microplastic/ Nanoplastic impact does not only impact the productivity and health of crops but there other sectors such as livestock, fisheries, and aquaculture also. Microplastic and nanoplastic can enter the bodies of livestock in three primary ways ingestion, inhalation, and skin contact, once they enter the body of the animal lead to various health complications such as oxidative stress, apoptosis, inflammation in the digestive tract, dysregulation of the endocrine system, accumulation in different organs, can increase bacterial contamination can lead to issues can lead to biological amplification along the food chain, posing risks to human health through the consumption of contaminated products such as milk, meat, and eggs (Khan et al. 2024). The long-term consequences of these pollutants on livestock health and the potential risks for disrupting reproductive and fertility (Urli et al. 2023). The sources can be the presence of microplastics and nanoplastic in animal feeds and forages, additives such as Bisphenol A (BPA) have high negative effects. The excreta or the manure produced from such contaminated poultry will also have microplastic remains in them leading to the futher contamination of soil as soon as it is mixed with soil. Looking at the sector of fisheries and agriculture major sources of microplastic are plastic used in fishing gear, nets, buckets, machines, aerators, water pumps, and bait. Soil erosion, flooding, and runoff water from agriculture fields can transport microplastic from land into the water bodies. Microplastics can accumulate in the tissues of aquatic organisms over time. As predators consume prey with microplastics, the concentration of these particles can increase up the food chain, potentially reaching harmful levels (Vandermeersch et al. 2015). They can adsorb and concentrate harmful chemicals and pollutants from the surrounding environment. When ingested by aquatic organisms, these contaminants can be transferred, leading to toxic effects developing immunity and fertility-related issues, blockages in the digestive system, and malnutrition. Zhou et al. (2020) addressing these challenges is crucial for developing a common monitoring protocol to keep track of and develop mitigation methods. Livestock animals, such as cows and sheep, may inadvertently consume microplastics present in their feed, water sources, or contaminated soil through grazing or contaminated fodder which can be transferred to the milk or meat produced by the cows and can futher be harmful for the calfs and human consuming. Such ingestion of microplastic lead to reduction in muscle mass and decreasing the amount of meat (Corte Pause et al. 2024).

CONCLUSION

The prevalence of plastic pollution, particularly in the form of microplastics and nanoplastics, poses a significant Challenge in the successful cultivation of crops, and rearing of animals and in various sectors of agro-ecosystems. Since the start of commercial production of plastic in the market, the issue has prevailed but came into notice very recently leading to its widespread destruction, causing harm to wildlife and potentially entering the food chain. There are cases study that mainly showcases the real-life destruction caused by microplastics and day-to-day increasing microplastic content highly damaging the biosphere. There are many sources of such pollution which are very difficult to control and the migration methods available are becoming very difficult to act. There is a high need of Continued research and collaboration among scientists, policymakers, and industries to address this pressing environmental issue and work towards a more sustainable future. Better technology for the detection of nanoplastic in different habit, need much development and all the challenges such as blocking pathways of MPs and NPs to enter in soil and water, Atmospheric deposition, and stopping crops and animals from consuming it need to taken into serious concern in order to avoid adversities in production and produce a safe cultivation.

REFERENCES

Alimi, O. S., Budarz, J. F., Hernandez, L. M., Tufenkji, N., & Sinton, D. (2018). Microplastics and Nanoplastics in Aquatic Environments: Aggregation, Deposition, and Enhanced Contaminant Transport. *Environmental Science & Technology*, 52(4), 1704–1724. DOI: 10.1021/acs.est.7b05559 PMID: 29265806

Bratovic A. (2023). European Journal of Advanced Chemistry Research. DOI: DOI: 0.24018/ejchem.2023.4.1.124

Corte Pause, F., Urli, S., Crociati, M., Stradaioli, G., & Baufeld, A. (2024). Connecting the Dots: Livestock Animals as Missing Links in the Chain of Microplastic Contamination and Human Health. *Animals (Basel)*, 14(2), 350. DOI: 10.3390/ani14020350 PMID: 38275809

Frias, J. P. G. L., & Nash, R. (2019). Microplastics: Finding a consensus on the definition. *Marine Pollution Bulletin*, 138, 145–147. DOI: 10.1016/j.marpolbul.2018.11.022 PMID: 30660255

Grause, G., Kuniyasu, Y., Chien, M.-F., & Inoue, C. (2021). Separation of microplastic from soil by centrifugation and its application to agricultural soil. *Chemosphere*, 132654. Advance online publication. DOI: 10.1016/j.chemosphere.2021.132654 PMID: 34718018

Huang, G., Quershi, M., Song, L., & Di, S. H. (2023, July 2). Yao, Sun w. *The Science of the Total Environment*, 896. Advance online publication. DOI: 10.1016/j.scitotenv.2023.165308 PMID: 37414186

Iqbal, S., Xu, J., Allen, S. D., Khan, S., Nadir, S., Arif, M. S., & Yasmeen, T. (2020). Unraveling consequences of soil micro- and nano-plastic pollution for soil-plant system with implications for nitrogen (N) cycling and soil microbial activity. .DOI: 10.1016/j.chemosphere.2020.127578

Isari, E. A., Papaioannou, D., Kalavrouziotis, I. K., & Karapanagioti, H. K. (2021). Microplastics in Agricultural Soils: A Case Study in Cultivation of Watermelons and Canning Tomatoes. *Water (Basel)*, 13(16), 2168. DOI: 10.3390/w13162168

Kannankai, M. P., Alex, R. K., Muralidharan, V. V., Nazeerkhan, N. P., Radhakrishnan, A., & Devipriya, S. P. (2022). Urban mangrove ecosystems are under severe threat from microplastic pollution.

Khan, A., Qadeer, A., Wajid, A., Ullah, Q., Rahman, S. U., Ullah, K., Safi, S. Z., Ticha, L., Skalickova, S., Chilala, P., Bernatova, S., Samek, O., & Horky, P. (2024). Microplastics in animal nutrition: Occurrence, spread, and hazard in animals. *Journal of Agriculture and Food Research*, 17, 101258. DOI: 10.1016/j.jafr.2024.101258

Lian, J., Liu, W., Meng, L., Wu, J., Zeb, A., Cheng, L., & Sun, H. (2021). Effects of microplastics derived from polymer-coated fertilizer on maize growth, rhizosphere, and soil properties. *Journal of Cleaner Production*, 318, 128571. DOI: 10.1016/j.jclepro.2021.128571

Maddela, N. R., Ramakrishnan, B., Kadiyala, T., Venkateswarlu, K., & Megharaj, M. (2023). Do Microplastic and Nanoplastics Pose Risks to Biota in Agricultural Ecosystems? *Soil Systems*, 7(1), 19. DOI: 10.3390/soilsystems7010019

Mariano, S., Tacconi, S., Fidaleo, M., Rossi, M., & Dini, L. (2021). Micro and Nanoplastics Identification: Classic Methods and Innovative Detection Techniques. *Frontiers in Toxicology*, 3, 636640. DOI: 10.3389/ftox.2021.636640 PMID: 35295124

Moeck, C., Davies, G., Krause, S., Schneidewind, U. (2023). Microplastics and nanoplastics in agriculture—A potential source of soil and groundwater contamination? Grundwasser - Zeitschrift der Fachsektion Hydrogeologie, 23-35. DOI: 10.1007/s00767-022-00533-2

Ng, E.-L., Johnston, P., Hu, H.-W., Geissen, V., & Chen, D. (2018). Microplastic in Agroecosystem. *The Science of the Total Environment*, 627, 1377–1388. DOI: 10.1016/j.scitotenv.2018.01.341 PMID: 30857101

Pandey, K. A. (2024). A Comprehensive Exploration of Soil, Water & Air Pollution in Agriculture, BFC Publications.

Rai, M., Pant, G., Pant, K., Aloo, B. N., Kumar, G., Singh, H. B., & Tripathi, V. (2023). Microplastic Pollution in Terrestrial Ecosystems and Its Interaction with Other Soil Pollutants: A Potential Threat to Soil Ecosystem Sustainability. *Resources*, 12(6), 67. DOI: 10.3390/resources12060067

Rogers, K. (2024). Microplastics. Encyclopedia Britannica. https://www.britannica.com/technology/microplastic

Sandhya. (2024). A Comprehensive Exploration of Soil, Water & Air Pollution in Agriculture, BFC Publications.

Tang D. (2020). Health, Safety and Environment Program, Curtin University. DOI: 10.9734/ajee/2020/v13i130170

Urli, S., Corte Pause, F., Crociati, M., Baufeld, A., Monaci, M., & Stradaioli, G. (2023). Impact of Microplastics and Nanoplastics on Livestock Health: An Emerging Risk for Reproductive Efficiency. *Animals (Basel)*, 13(7), 1132. DOI: 10.3390/ani13071132 PMID: 37048387

Vandermeersch, G., Van Cauwenberghe, L., Janssen, C. R., Marques, A., Granby, K., Fait, G., Kotterman, M. J. J., Diogène, J., Bekaert, K., Robbens, J., & Devriese, L. (2015). A critical view on microplastic quantification in aquatic organisms. *Environmental Research*, 143, 46–55. Advance online publication. DOI: 10.1016/j.envres.2015.07.016 PMID: 26249746

Verma, K. K., Song, X.-P., Xu, L., Huang, H.-R., Liang, Q., Seth, C. S., & Li, Y.-R. (2023). Nano-microplastic and agro-ecosystems: A mini-review. *Frontiers in Plant Science*, 14, 1283852. DOI: 10.3389/fpls.2023.1283852 PMID: 38053770

Vincze, É.-B., Becze, A., Laslo, É., & Mara, G. (2024). Beneficial Soil Microbiomes and Their Potential Role in Plant Growth and Soil Fertility. *Agriculture*, 14(1), 152. DOI: 10.3390/agriculture14010152

Wang, W. (2021, February 15). A. Adams, Sun, Zhang, Qingdao University of Science and Technology. *Journal of Hazardous Materials*, 424(Part C). Advance online publication. DOI: 10.1016/j.jhazmat.2021.127531 PMID: 34740160

Xu, C., Zhang, B., Gu, C., Shen, C., Yin, S., Aamir, M., & Li, F. (2020). Are we underestimating the sources of microplastic pollution in the terrestrial environment? *Journal of Hazardous Materials*, 400, 123228. Advance online publication. DOI: 10.1016/j.jhazmat.2020.123228 PMID: 32593024

Yadav V, Dhanger S, & Sharma J. (2022). Microplastics accumulation in agricultural soil: Evidence for the presence, potential effects, extraction, and current bioremediation approaches. DOI: 10.7324/JABB.2022.10s204

Yu J.R., Adingo S., Liu X.L., Li X.D., Sun J., & Zhang X.N. (2022). Micro plastics in soil ecosystem – A review of sources. DOI: 10.17221/242/2021-PSE

Zhou, A., Zhang, Y., Xie, S., Chen, Y., Li, X., Wang, J., & Zou, J. (2020). Microplastics and their potential effects on the aquaculture systems: A critical review. *Reviews in Aquaculture*, 1–15. DOI: 10.1111/raq.12496

Chapter 10
Micro/Nano-Plastic Pollution in Aquarium Systems:
Sources, Fate, Hazards, and Ecological Imbalances

Bahati Shabani Nzeyimana
https://orcid.org/0009-0007-3640-5471
Bishop Heber College, Bharathidasan University, India

Swagata Chakraborty
https://orcid.org/0000-0002-9898-5244
Bharathidasan University, India

R. Priyadharshini
Bishop Heber College, Bharathidasan University, India

Mariaselvam Sheela Mary
Bishop Heber College, Bharathidasan University, India

M. Govindaraju
Bharathidasan University, India

ABSTRACT

Microplastics and nanoplastics (MPN) pose a growing threat to aquatic ecosystems, including closed systems like aquariums. This chapter delves into the various sources of MNPs in aquariums ranging from synthetic decorations and fish food to dustfall and tap water. It explores the fate and transport of these particles, including settling, interaction with the substrate, and potential ingestion by aquatic organisms. The

DOI: 10.4018/979-8-3693-3447-8.ch010

chapter then dissects the hazards MNP poses to captive animals, encompassing physical harm, chemical toxicity, and disruptions in vital biological functions. The chapter proposes mitigation strategies such as MNP-free aquarium equipment, high-quality fish food, and efficient filtration systems. It emphasizes education and awareness among hobbyists alongside standardized protocols for MNP detection and monitoring recognizing the need for further research. The chapter calls for investigations into long-term effects and mitigation strategies.

1. INTRODUCTION

The Aquarium provides a fascinating view of the underwater world, fostering an appreciation for aquatic life and offering a source of relaxation. However, beneath this beauty lies a growing issue of uncontrolled micro and Nano plastics (MNP) pollution from various sources. (Thiounn & Smith, 2020). MNPs are comprehended as Particles, plastics that are less than millimeters in size diameter, are ubiquitous in aquarium environments and cause water to be unsafe for aquatic life(Nzeyimana & Mary, 2024). Studies are increasingly revealing their presence in the biased lens of ecosystems with aquariums being no exception(Raju et al., 2018). This chapter aims to offer a comprehensive of knowledge of micro and nano plastic Exploration within aquarium systems. We Begin by exploring various Pathways by which (MNPs) pass through into aquariums followed by an examination of the effect within the system (Andrady & Neal, 2009).

Microplastics and nanoplastics pose a growing threat to Aquatic ecosystems including closed-loop systems like the aquarium, this chapter delivered into the unique dynamics of the Environment, exploring its diverse Source, fate, hazards, and ecological consequences(Choi et al., 2022)(Choi et al., 2022). Primary sources include synthetic fiber from decorations, microplastics in fish food, and Biodegradation of large Plastics, Secondary sources involve reaching equipment external Sources like dustfalls and exploring the fate of transport of MNPs reveal processes like settling ingestion by organisms and the potential trophic transfer and influenced by water flow and the filtration system(Shirazimoghaddam et al., 2023).

We delve the life into the potential Hazard of MNPs to the health and well-being of aquatic organisms inhabiting aquariums, Finally, we will discuss the potential ecological imbalances of MNPs that might contribute to these closed systems(Maria Tsakona et al., 2021). Hazards to aquatic inhabitants range from physical harm and chemical toxicity to destruction in a fading behavior and potential impact on the production food web.MNP pollution can also disrupt an ecological balance impacting, the microbial Community, the Plankton population perlite, pre-dynamic, and introducing invasive species (Bridson et al., 2021). Therefore, the Mitigation strategy of MNP will focus on free Alternatives, high-quality fish food efficient filtration, and education for aquarium hobbyists haven standardized detection and maintenance protocols are crucial alongside continued research on long-term effects and effective mitigation strategies. This chapter emphasizes the need to address MNP prohibitions in the aquarium by highlighting potential risks, to the captive organism, and the broader ecological balance.

It is essential to remember that up to 90% of oceanic Plastic pollution comes from the Telestial River network, and understanding the process that controls their mobilization transport ultimately fate into the environment is a crucial challenge (Rummel, 2022). The widespread presence of microplastic within river sediment beds suggests that their movement is governed by the same physical processes that regulate sediment transport (Hermabessiere et al., 2017). In a particularly physical aspect of the flow, regimes are expected to have a primary influence on the transport of destiny Plastic particles. And investigating the unexplored sources and the critical imbalances associated with MNP pollutions in miniature of the aquatic environments by shedding right on these intricacies, we can gain variable insight into the potential risk to captive, aquatic organisms and the broader ecological balance within this artificial ecosystem.

This chapter also will contribute to the existing knowledge of MNP by focusing on the specific context of aquarium systems. Both freshwater aquariums and marine coral reef aquariums While research opens the aquatic environment, excessive Dynamic events across control, the system, the aquarium unique challenging and the tolerant investigation, understanding a source, fate, and the impact of them (Billings et al., 2023). This study is crucial for the sharing of information, best practices in aquarium management, and minimizing the potential harm to captive organisms and the broader environment.

1.1 Backgraound and Justfication of MNP Pollutions

MNP pollution causes harm to the fish and other aquatic animals in the aquariums as well as the state of the relevant ecosystem. These are tiny pieces of plastic usually shed from synthetic materials; they infiltrate the tank through gravel, fish chow, or water sources. The particles are vengeful ones—they are small but create significant problems to the underwater friends (Dhiman et al., 2023).

Some of these mitigation measures mentioned before include employing other substrates, the quality of the water to be treated, and the use of filtration systems that are designed to capture MNPs. That is why investigation and promotion programs are called for as ways to define the extent of the issue and to establish feasible solutions. It was found that aquarists could lessen the quantity of MNP in water that is within their sphere of influence and hence help maintain the aquatic ecosystem. These are the micro beads and synthetic fibers as well as particles from the breakdown of plastic products (Mcdevitt et al., 2017).

Although they are tiny, MNPs have significant impacts on water and aquatic species and general health. As for the variety of entry points into the ecosystem, let it be pointed out that MNPs in aquariums have a concept of the initial injection of items. Major nutrients in MNP can contaminate the tank water through top offs or water changes. Besides, small plastics are present in fish meals most of the time, which is another source of pollution. They could be deposited in the water as sand or gravel sometimes used in aquariums degrades with time (Pfohl et al., 2022).

MNPs can settle at the bottom of an aquarium once introduced into it and are taken in by other organisms such as fish, crustaceans and plants (Bellasi et al., 2021). Eating MNPs can lead to several adverse effects such as body harm reduced feeding efficiency, and disruption of physiological functions. Also, dangerous contaminants like organic compounds and heavy metals can move through MNPs and build up in the food chain, putting aquatic life at risk (Raamsdonk et al., 2020).

Pollution control in the aquarium system MNP needs to employ several methods. This entails putting strategies in minimizing the MNPs that go into aquariums like preferring fish food with no MNP contamination and using different substrates. Other measures that can be taken in order to minimize MNPs in aquariums include using electricity filtration systems that are competent in trapping the particles as well as evaluate water quality. As a result, scientific investigations and awareness raising aiming at minimizing the sources of MNP pollution in aquarium systems and protecting aquatic ecosystems' biological diversities and overall health are crucial undertakings (Zhou et al., 2021); Li et al., 2018; Maes et al., 2017; Gallowaya & Lewisa, 2016).

This is because studies carried out on open water environments and enclosed tanks of aquatic facilities give out different angles and perspectives on the complexity of aquatic systems. These different dynamics are due to the open space marine environment which is characterized by fluctuations as compared to the closed environments of an aquarium (Carr et al., 2003; Melo-Merino et al., 2020).

In such controlled environments with closed tank systems, researchers can effectively change and observe values of temperature, pH, salinities and nutrients among others. This controlled environment allows a highly regulated examination of particular phenomena or the effects of certain stressors on the targets of the investigation. Such benefits include flexibility where researchers can isolate factors, conduct the analysis again, and test different scenarios. In addition, closed aquarium systems are ideal for long-term studies and observation because they can provide a stable environment that allows researchers to control the conditions under which aquatic creatures grow, reproduce, and interact over relatively extended periods of time (Cole et al., 2009).

On the other hand, research conducted in the natural environments within aquatic systems establishes how actual ecosystems are complex and dynamic. These ecosystems are characterized by variation in the surroundings owing to factors such as seasonal fluctuations, trends in weather conditions, nutritional stimuli, as well as, interactivity with other organisms. Fieldwork is required when observing dense layers of interconnection in ecological food chains comprising predator-prey, competition, and mutualism, among others, but this is only possible when the aquariums are at their best. Further, research in the scope of environmental ecology and ecological physiology revealed trends of human activity such as pollution and climate change and the ability of ecosystems to open aquatic habitats to change (Curcuruto et al., 2023; Yazdian et al., 2014; Snelgrove et al., 2000).

Some wildlife environments might be as wild as closed tank systems, but the problem is that such environments might be 'loose' as compared to the accurate controlled environment provided by the closed tank systems. On the other hand the very stochastic and diverse character of natural systems can present difficulties for investigation in open water environments and it is sometimes hard to control for certain variables or replicate experiments with the precision that can be accomplished in a closed systems (Martin, 1975; Carr et al., 2003).

In an epilogue, it is possible to claim that the generalisation of the study outcome of the learning process could be complemented by the analysis of study findings from the open water conditions and aquarium facilities. When used together, the two methods will help create a better comprehension of the relations, intersections and phenomena that define the aquatic environment, and in turn aid in formulating adequate management and protection policies (Cózar et al., 2014); Mccormick, 2022).

Aquarium systems present certain issues with regard to MNP presence since such systems are designed to mimic natural environments necessary to support various forms of aquatic life (Chae & An, 2017). The places of origin and originations, life trajectories, and consequences of MNPs in and between aquariums are crucial knowledge and objects for effective MNP management and the welfare of captive species and their habitats.

It is important to state that MNPs that appear in aquarium systems originate from multiple sources (Windsor et al., 2019). Some of the commercial products may include synthetic decorations, filter media or the very fish foods could contain MNPs as a component. Also, the water used to fill the aquariums such as tap water, river water or seawater, may contain MNPs. Other activities in aquatic systems for example cleaning practices where plastic-based equipment is used also introduced MNP. Also, the degradation of dep negates larger plastic waste in the aquarium and the detachment of MNPs from organisms in contact with the system also contributes to MNP pollution (Granek et al., 2020).

After entering the aquarium systems MNPs take various fates. They can be assumed in the substrate as they are small in size and thus accumulate in the substrate over time. It is therefore possible to filter some MNPs but at the same time possible to have MNPs that pass through the filtration media or even originate from the media themselves. Feeding or filter-feeding on the particles in the water by the organisms in an aquarium is also a common means of MNPs intake. These MNPs once ingested can accumulate in the tissues of organisms and may cause health consequences and the phenomenon of biomagnification occurs in the food chain (Turner & Holmes, 2011; Koelmans et al., 2019; Carbery et al., 2018).

Depending on the type and magnitude of MNPs introduced into an aquarium system, the resulting effects may be positive, negative or mixed. Absorbed MNPs negatively impact the health of the digestive tracts of the aquarium organisms by causing physical harm, inflammation, and blockage leading to inefficient feeding and consequent health complications. Also, MNPs have the potential of adsorbing and accumulating hazardous elements in the environment, which in turn could be discharged within the body in case of consuming the MNPs, thus causing other health implications. Furthermore, MNPs can alter water parameters in a way that can prove hazardous to the stability of ecosystems of the aquarium, such as upset of nutrient cycling, microbial composition and the general species/ Genetic/Bio-diversity. In addition, the MNPs released from the aquarium systems are transported into the large environment through wastewater disposal, thereby increasing the pollution of water and endangering the lives of aquatic organisms (Rostami et al., 2021).

Thus, to manage the sources, fate, and impacts of MNPs in aquarium systems effectively, one has to consider key issues related to proper choice of products, routine practices, and waste disposal procedures. Thus, preventing the pollution of MNP within aquariums and reducing the possible negative influences on captured organisms will bring a positive impact on aquatic and different environments. Thus, the exploration of this issue and preventive actions should remain a priority in addressing this environmental problem.

2. SOURCES OF MNP IN THE AQUARIUM

The plastics that are less than 5 millimeters are called microplastics and that are less than 1 millimeter are called nanoplastics. They originate from various sources and accumulate in this environment causing threats. There are two sources of MNPs Incorporating primary and secondary sources from which microplastics originate(Sander et al., 2023). Small-sized particles that are released straight into the marine ecosystems and contribute mainly to synthetic fibers, substrates, cleaning tools, fish feed, cosmetic products, vehicle tires, and city dust are primary microplastics. Secondary microplastics

are microplastics that occur in larger plastic objects' degradation by UV and Weathering. The larger plastic objects like plastic bags, plastic bottles, and fishing nets.

2.1 Synthetic Fibers, Substrates, Cleaning Tools

According to IUCN (2017) over one-third (35%)of the annual total primary microplastics are contributed by synthetic textiles which enter the world marine environment and contribute to the microplastics problem to a greater extent(Schmidt et al., 2019). synthetic clothes are washed in aquariums. Synthetic clothes release microplastics due to chemical stresses that fabric undergoes during washing in which microfibers detach from the clothes that constitute the textile. The dust, composed of artificial polymers, contains more durability and a slow degradation rate it remains in the environment for the long term (Ahlawat, 2024). Many studies have shown that based on the fabric type the microplastic production is different. To reduce microplastic removal, it is necessary to understand the difference between the concentrations of microplastics released from various fabrics. fabric migration, friction, shape transformation, and restoration, fabric movements are caused during washing and drying. the movement of fabric can be affected by its physical properties, microbeads are mainly released from textiles and textile manufacturing industries and also by household washing. washing of clothes in the ponds in remote villages makes the aquarium to be contaminated with microbeads, due to this the plastics from the clothes detach and accumulate in the pond ecosystems. The household usage water and waste plastics are dumped into the aquarium by people surrounding the ecosystems

2.2 Microplastics in Fish Food and Additives

Plastic consists of polymers and is also composed of various materials known as supplements, by adding different kinds of supplements to the polymer, they can be changed and their range of applications extended. Microplastics in fish food cause damage to the fish, secondary consumers, and tertiary consumers and its additives such as paint, fillers, and flame retardants pose potential threats to the organism. Microplastics biodegrade by UV, thermal radiation, or leach and enter the aquarium or ground these additives serve the purpose of modifying the physical and chemical properties of plastic, and the selection of additives importantly influences both the cost and production of plastic materials(Pouech et al., 2014). Through this process, the fragmented microorganisms enter the primary consumer (small fish)and transfer to the secondary consumer(large fish) likewise they pass on to various Trophic levels.

It causes damage to the well-being of fish and consumers of fish. These additives change the size, and shape of the organism causing ecotoxicity. The specific type of additives plays an important role in detecting the physicochemical properties of MPs and their behaviors and interactions within the environment (Iftikhar et al., 2024).

Table 1. Source micro & nano plastic in aquarium

Sources	Description
Synthetic fibers	Found in decorations, substrates, and cleaning tools
Microplastics in fish food	Contamination from fish food and additives
Biodegradation of plastics	Breakdown of larger plastic components within the aquarium
Leaching from decorations	Plasticizers and additives leaching from decorations and equipment
Dustfall and tap water	External sources of microplastics entering the aquarium ecosystem

(Choi et al., 2022)

Polymer biodegradation is the process by which microorganisms completely digest the hydrocarbon in the polymer under which the carbon dioxide and microbial biomass are produced under oxic conditions, or carbon dioxide, methane, and microbial biomass are produced under anoxic conditions(Angnunavuri et al., 2020). The degradation takes place in two steps. In the first step, the plastic disintegrates into microbeads and in the second step, it is continuously consumed by the microorganisms.(Foltz et al., 2015). Inside the microbial cell, the plastic molecules are metabolically used leading to the $CO2$ formation and microbial biomass under aerobic conditions or $CO2$, $CH4$, and microbial biomass under anaerobic, methanogenic conditions. The plastic from various sources degrades by photooxidation and weathering. Due to UV light, the plastic divides into fragments and by weathering there is a change in the size, shape, and characteristics of the plastics.

2.3 Leaching From Plasticizer and Additives

plasticizer and supplements leach from microbeads. These plasticizers and supplements arise from industrial products during and after the manufacture of products. They generate from raw materials as well as from waste. The causes physical, chemical, and biological impacts to the environment. physical harm includes ingestion and rafting by aquatic organisms. Due to Leaching the organism has various biological effects including altered gene expression and oxidative stress. biological Leaching includes retarded behavior, survival, growth, development

2.4 The Process of Filtration of Additives From Plastics Consists of Four Stages

1. scattering the supplement towards the top of the polymer
2. Disappearance takes place as the supplement isolates from the polymer top.
3. Absorption happens as the supplement becomes soaked within the plastic matrix in the surrounding medium.4. Distribution and absorption take place within the matrix.

Moreover, the filtration of supplement substances is primarily related to the particle dimension. Small-sized substances are easy to discharge.

2.5 Dust Fall and Tap Water

Study shows that MPs were found in the tap water of Mosul city, which investigated the components providing their existence. There was proof of microplastic pollution at every sampling spot. It was noticed that most microplastics were fibers and fragments (93%) in transparent, white, blue, red, black, green, orange, yellow, and other colors (Sulpizio, 2019). Among the polymers, polyvinyl chloride (PVC) and poly amide were more abundant. Microplastic amounts in tap water were relatively greater in contrast to other studies, pointing out that DWTPs cannot remove all particles and MPs released from aged plastic pipes in districts.

3. FATE AND TRANSPORT OF MNP IN AQUARIUM

3.1 Settling and Deposition

The micro-nano plastics originate from various sources like primary, and secondary, go to household drainage then move to open aquariums like lakes and settle there(S. A. Carr et al., 2016) The micro-nano plastics being nonbiodegradable cause harm to aquatic organisms by settling in aquariums. over time it accumulates causing water to worsen the quality of water. The deposition of micro-nanoplastics changes the soil nutrient level. The micro-nano plastics settle and deposit for a long time.

3.2 Interaction With Substrates

By the types of substrates, habitat, season, and location biofilms are formed. In plastic debris, Microbial communities are found in marine ecosystems, on submerged surfaces biofilm forms which is the result of the selective attachment of microorganisms, and same species competition among microbial communities. For settling larger surface area availability makes the weathering processes favor biofilm growth which in turn covers plastic debris from UV light (Hahladakis et al., 2018). However, biofilms can biodegrade the polymer. The biological significance of biofilm emergence comprises effects on trophic transfer of MP and corresponding pollutants, community arrangements of microbial aggregation, and potentially toxic to Primary consumers.

3.3 Ingestion by Aquatic Organisms

Aquatic organisms like fish and aquatic birds ingest micro-nano plastics through the water. it causes harm to their body(Athey et al., 2020). A common phenomenon is the consumption of microplastics from the entry into the digestive system through eating or drinking either straight or indirectly and is exasperated across various trophical levels, habitats, and geographic areas. The aquatic organisms can ingest Micronanoplastics in water which may stuck inside the throat or cause damage to the intestinal layer.in several cases, death has been recorded Studies have shown that a lot of aquatic organisms are harmed due to MNP. recent studies show that MNP-harmed organisms have increased

3.4 Microplastic Consumption Mechanism

There are two methods of consumption of food 1. Direct(primary intake) 2. Indirect (secondary intake)Primary intake: In this, straightforward consumption of Microplastic by the organism in any form.(Carson, 2013). Bigger animals scratch or bite the plastic and plastic waste is found with animal traces which are Noticeable to the researcher. This signals the active feeding for microbeads or more complex signs of foraging on microbeads attempt by small-sized organisms. Secondary intake: secondary intake is those which predate upon primary consumers. consumption of microbeads and some other attached poisonous substances through feeding of biofilm by small invertebrates is a secondary consumption. Living beings are interested in vibrant items as kit resemble their prey. the presence of vibrant artificial items has been reported in gut analysis(Thushari & Senevirathna, 2020). Studies have shown that living beings actively feed due to producer confusion and substance signaling facilitated by substance attractants (algae-derived dimethyl sulfide) attached to the microbeads in seawater.

Table 2. Micro and nano plastics fate and transport in aquarium

Processes	Description
Settling and deposition	Microplastics settling at the bottom of the aquarium or adhering to surfaces
Interaction with substrates	Adherence of microplastics to substrate materials
Ingestion by organisms	Aquatic organisms consuming microplastics
Trophic transfer	Transfer of microplastics through the food chain
Influence of water flow	Impact of water flow patterns on microplastic distribution
Filtration and biological activity	Removal and breakdown of microplastics by filtration and organisms

(Foltz et al., 2015)

3.5 Potential Trophic Transfer

Microplastics are taken by lagoon species, in lagoon species the microplastics pass on from unicellular to larval killers have not yet been identified. Microplastic transfer at various Trophic levels in lake ecosystems. The microplastics in water are ingested by fishes like jilapi, Sendai, zooplankton, and other aquatic organisms. These microplastics are transferred from zooplankton to smaller fishes which then transferred to larger fishes. The larger fishes are taken by humans for commercial purposes(Wang et al., 2020). The larger fishes are eaten by humans which causes humans to ingest microplastics. This causes changes in the trophic levels of organisms. primary consumers (zooplankton) consume either straight-forward or secondarily taken. These microplastic substances proceed to the food chain and are passed on through various trophic levels.

4. HAZARDS OF MICRONANO PLASTIC TO AQUATIC ORGANISM

4.1 Physical Harm From Ingestion

intake and entanglement causes physical harm. suffocation, laceration, and increased energetic cost to movement are effects of entanglement and have been reported for hundreds of organisms, particularly sea turtles, seabirds, and marine mammals. Blockages of the intestinal tract, lacerations, and reduction in feeding activity are caused by consumption. Based on the species and the dimensions and form of the debris we know the type and extent of internal harm. A further physical mechanism has been hypothesized, where plastic debris may serve as an anthropogenic raft for non-native and significantly invasive species in aquatic environments.

4.2 Chemical Toxicity

Through the food cycle, microplastic is passed on to various trophic levels. (Juying et al., 2016). The composition of the plastic waste, as well as the enormous surface area of microplastics, makes them vulnerable to attaching watery organic contaminants and poisonous substances to filtrate. Intake of microplastics introduces poisonous substances to the bottom level of the food cycles in the marine ecosystem, where poisonous substances form in the tissues of aquatic living species (Fakolade, O. A & Atanda, 2015). Perhaps because plastics are commonly considered to be biochemically inert plastic additives, also known as "plasticizers," may be fully integrated into plastics during production and injection molding to improve their properties or extend their life by providing resistance to heat (e.g., polybrominated diphenyl ethers), oxidative damage (e.g., nonylphenol), and antimicrobials (triclosan) (Iheanacho et al., 2023). These supplements cause harm to the environment and the marine ecosystem. Because they both prolong in the environment and the decomposing period of Polymer slows down and releases potentially poisonous substances into marine aquatic life. A few chemicals can move away from the artificial matrix of plastic due to inadequate polymerization of polymers during manufacture. The chemical toxicity causes imbalance in the body of organisms which leads to various temporary or permanent morphological changes in the aquatic creatures during primitive phases as well as causes sexual disintegration in adults. The chemical toxicity leads to intersex abnormalities in aquatic organisms.

4.3 Disruptions In Feeding Behaviour and Nutrient Uptake

Microplastic disrupts feeding behavior and nutrient uptake. Persistent vulnerability has no potential effects on the development and existence of perch but the nutritional standard of fish is lower with low amounts of polypeptide and dust matter in the exposed fish. HDPE potentially alters nutrion digestion involves bile acid anabolism, levocarnitine genesis, and alpha-keto propanoic acid digestion of yellow perch, and disintegrates stomaic histology and microbiota diversity. studies point out that long-term risk to MNP proceeds to weakened nutrition usage and fish well-being.

4.4 Impacts on Immune Function and Reproduction

Plastic microbeads primarily cause harm to tissue and cell death in the sexual parts of living beings, greater ROS production. Microplastic consumption causes damage to immune function and reproduction. The progeny develops with reduced anatomical growth like changes in egg breeding practices, irregular body growth, and external appearance does not help in the survival of large fleas. microbeads rupture the intracellular signaling paths and change the immune balance.MNP exposure causes the sexual ability of organisms to worsen which is evident in C.elegans

4.5 Specific Vulnerabilities of Different Species

Studies demonstrated that long-term exposure to microplastic particles could influence potential alternations in the intestines of rare minnows (Gobiocypris rarus). The species are affected by intestinal damage which leads to oxidative stress. These changes were related to lipid metabolism and amino acid metabolism in the guts. In addition, on long-term exposure immune-related genes change and cause the inflammatory response and innate immune response of rare minnow at the gene level(Agbekpornu & Kevudo, 2023). The results of microbead intake consist of lower reproductive health, suffocation in water, inability to prevent consumers, damaged feeding capacity, and the significance of consuming and transporting poisonous substances stuck to the top of plastics.

Table 3. Hazard of Microplastic spollutions in aquatic organisms

Hazard	Description
Physical harm	Ingestion causing physical damage to the digestive system or blockages
Chemical toxicity	Toxic substances associated with microplastics impact organism health
Disrupted feeding behavior	Altered feeding patterns affecting nutrient intake and energy balance
Impaired immune function	Reduced ability to fight off diseases and infections
Reproductive impacts	Negative effects on reproductive success and population dynamics

(Thushari & Senevirathna, 2020)

5. ECOLOGICAL IMBALANCES CAUSED BY MICRO-NANO PLASTICS POLLUTION

5.1 Shifts In Microbial Communities

Due to ecological imbalances, there is a shift in microbial communities. If a particular species increases or decreases the microbial community shift occurs which affects the ecosystem and various trophic levels.

5.2 Impacts on Planktonic

Parents exhibit to microplastics at some stage can generate weak progeny, suggesting that microplastics comprise of oxidative stress that decreases parental interest in consequent stages of life (Issac & Kandasubramanian, 2021). Progeny character Decreases massively causing a reduction in the size of zooplankton populations for long drawn. Both zooplankton and phytoplankton are affected due to Microplastic consumption which causes ecological imbalances.

5.3 Alterations in Predator-Prey Dynamics

When a consumer intakes D. magna a larger amount of MNPs pass into various animals through trophic transfer, in which the biomagnification level is greater and produces a threat to the habitat. The producer lifeforms disrupt community form and ecological functions and are considered to be carriers and border agents of MNPs, A diversity of species, like radio pods, and silkworms intake the nano and microplastics which affect their reproductive Rate metabolism, and death rate. MNP exposure ended in damage to food intake and floating behavior of mysid shrimp larvae(Malinowski et al., 2023). The damaged floating practices including limiting trapping capabilities and moderate capacity to evade predators, would conclude changes in primary consumer-secondary consumer relationship and energy pass on among aquatic food cycles, thus proceeding to a significant threat. If a prey dies then there is no option for predators rather to die. If a predator dies then accumulations of prey species cause ecological imbalances.

6. MITIGATION STRATEGIES OF MICRO AND NANOPLASTIC POLLUTIONS

Table 4. Mitigation strategies for pollution in the aquarium

Strategy	Description
MNP-free alternatives for decorations	Use of natural materials or certified microplastic-free products
High-quality fish food with minimal MNP content	Selection of fish food with low microplastic contamination
Efficient filtration systems and maintenance	Installation of effective filtration systems and regular cleaning

continued on following page

Table 4. Continued

Strategy	Description
Education and awareness for aquarium hobbyists	Providing information on microplastic pollution and its impacts
Standardized protocols for MNP detection	Development of protocols for consistent monitoring and detection
Research on long-term effects and mitigation	Investigation into the long-term impacts and development of mitigation

(Issac & Kandasubramanian, 2021)

6.1 MNP-Free Alternatives and Future Directions

MNPs in aquarium systems can pose significant risks to aquatic life, including fish and invertebrates, as well as potential harm to water quality. Fortunately, several alternatives to traditional decorations and equipment can help mitigate these risks:

Utilizing natural substrates like sand or gravel formed from broken coral, limestone, or quartz is preferable to utilizing artificial gravel or sand. These substrates imitate natural environments and encourage the growth of good microorganisms in addition to offering a natural look.

By removing nitrates and adding oxygen, live aquatic plants enhance the aquarium's aesthetic appeal while also contributing to its improved water quality. To guarantee success, select native or low-maintenance species.

For ornamentation, use driftwood or naturally occurring rocks like slate or granite. These objects give fish hiding places and create ecosystems that appear natural without adding MNPs.

Seek for ornaments made of ceramic rather than plastic. Ceramic ornaments are harmless and won't contaminate the water with dangerous elements. Make sure they are paint- or glaze-free and safe for aquariums, as these could harm aquatic life.

Choose Glass or Stainless-Steel Equipment rather than plastic equivalents, go with glass or stainless-steel pumps, heaters, and filters. The likelihood of MNP contamination is decreased by these materials' strength, inertness, and slow rate of degradation. The use of Stainless Steel and Brass takes into account hardware such as tank stands, mounting brackets, or fasteners. Because of their longevity and resistance to corrosion, these metals minimize plastic waste and the need for replacement.

Polylactic acid (PLA) and polyhydroxyalkanoates (PHA) are examples of natural polymers that can be converted into biodegradable plastic substitutes if plastic is still required (Ju et al., 2021). These materials decompose more quickly than traditional plastics, yet they still have some impact on the environment (Aley & Pranjal, 2023), (Chen et al., 2022), (Prata et al., 2019).

6.2. High-Quality Fish Food With Minimal MNP Content

Keeping water homes for fish safe and strong needs top fish food with very little MNP. MNPs are tiny bits, under 5 mm, that dirty the water and harm fish and sea life. Fish-eating MNPs could get sick. If people eat these fish, they might also get sick. Fish are key in the water food chains. We must make better fish food that cuts down MNP. This food should also make sure fish get all the nutrients they need to deal with this big problem. Using other ingredients like algae, bug meal, and plant proteins without MNPs is one way to do this. These sources give sustainable protein and cut down on the risk of adding MNPs to water life. (De-la-Torre, 2020; Alberghini et al., 2023).

Moreover, cutting-edge processing methods can lessen the amount of MNP pollution in fish meal. Before being added to fish feed formulations, MNP particles can be efficiently extracted from raw materials using methods like density separation, filtration, and screening. To reduce the unintentional introduction of MNPs, strict quality control procedures must be implemented throughout the production process. The assessment of the effectiveness of these tactics in mitigating MNP pollution in fish meals is mostly dependent on scientific studies. The development of sustainable aquafeed solutions can benefit greatly from studies evaluating the MNP content of various feed ingredients and the effects of ingesting MNP on fish health and ecosystem dynamics. In conclusion, it is critical to design fish food that is high-quality and contains as little MNP as possible in order to protect aquatic ecosystems and maintain the viability of fisheries. Through the utilization of substitute components and inventive processing methods backed by scientific investigation, the aquaculture sector might proactively tackle the worldwide issue of MNP contamination in aquatic habitatAlso, new tech can lower how much MNP waste gets into fish food. Methods like filtering and sorting can pull MNP bits out of the basic materials before they're mixed into fish feed. We need to check quality at every step of making it. This will help stop MNPs from getting in by mistake. Checking how well different ways work to cut down microplastic (MNP) in fish food needs a lot of science research. Finding fish food that doesn't harm the planet can greatly gain from looking into MNP in feed parts. Also, knowing how eating MNP affects fish health and how they live with other sea life is key.

In the end, it's very important to make a fish feed that's good quality and has very little MNP. This is needed to keep our water life safe and make sure fishing stays a good choice. Through the utilization of substitute components and inventive processing methods backed by scientific investigation aquaculture sector might proactively tackle. The worldwide issue of MNP contamination in aquatic habitats (Chen et al., 2022); Dhiman et al., 2023; Prata et al., 2019).

6.3. Efficient Filtration Systems and Regular Maintenance

To sustain the water quality which is important for the fish and other life forms in an aquarium, good filtration is crucial. Thus, chemical, biological and mechanical methods of filtration guarantee the best possible purity. Solidified organic matter and debris are discharged through mechanical filtration with materials such as filter floss or sponges. The ammonia and nitrites that are toxic to the organisms are channeled through the useful bacteria in the process of biological filtration and are changed to nitrates which are relatively harmless. In chemical filtering, dissolved interferences are removed by means of Reagents or Activated carbon. This means then that some level of maintenance is required in order to keep up the filtering efficiency. These include replacing chemical filtration components especially those that are fixed, washing filter media, and siphoning materials from the substrate. Measuring and recording pH, ammonia, nitrite, and nitrate at least once a month with adjustments being made in the water formula forms part of the process. Also, verify the flow rate of water and look for faults if any in the concerned equipment. Often, the usage of toxic materials is controlled so that they do not accumulate, thus maintaining a balanced environment in the aquatic biome. Such aquarium conditions that some hobbyists maintain will also be helpful to contribute to the healthy living of the aquatic inhabitants (Carla et al., 2020); (Padervand et al., 2020); (Wolff et al., 2021); (Carolina et al., 2016).

6.4. Education and Awareness for Aquarium Hobbyists

Indeed, to entice aquarium hobbyists to apply standard approaches, the clients must be educated. Having accurate understanding of species compatibility, water chemistry, and maintenance of fish tank is also very essential (Chomiak et al., 2024). The changes of nitrogen cycle influence the quality of water and therefore the general health state of the fish that inhabit the lake. Therefore, it is possible to prevent the spreading of invasive species when one has a green lens and a conscious mind of their existence. In addition, the worst aspect is raising enthusiasts' awareness of the ethical approach to acquiring aquarium specimens and encouraging captive-generated offspring in preference to wild-harvested fish. This is made possible through advocacy for sustainable aquarium vocational embraces by supporting groups that encourage it and the conservation of species by carrying out the selection processes. Additionally, reducing the environmental effect is realised through stressing the importance of proper disposal methods of the fishing vessels and the aquatic animals. It is possible that enthusiasts of the water institution will be able to affect positively conservation of water bodies and other course by raising awareness (Maceda-veiga & Dom, 2016); Marchio, 2018; Williams et al., 2022).

6.5. Standardized Protocols for MNP Detection and Monitoring

Standardized protocols combining methods from both physical and chemical techniques are usually used in microplastic detection and monitoring in aquarium systems. These techniques frequently consist of:

One approach involves filtering microplastic particles from water samples with a fine mesh screen. Among these approaches is density separation which employs techniques such as gradient centrifugation for organic matter versus microplastics separation based on variations in densities (Hidalgo-ruz et al., 2012).

Contact angle/of Microplastic polymers can be identified by their characteristic Raman or Fourier-transform infrared spectroscopy vibrational frequencies. Microplastics are usually countable and are usually visually characterized and quantified using optical or SEM (Scanning electron microscopy) techniques (Marine Debris Program, 2015).

Analytical methods like gas chromatography coupled with mass spectrometry which is abbreviated as GC-MS are effective in measuring or identifying decomposition byproducts or additives of plastic. Organic material can be assisted in disintegrating through the use of acidic decompositions, or enzymatic processes that are used to reduce the amount of microplastics, for analysis (Müller et al., 2020; Reproducibility et al., 2020). That is why, finally, microplastic monitoring must be well-orchestrated and, therefore, based on sample collection and measurement protocols which include the mentioned aspects such as the sampling method, the sample treatment, the quality assurance measures, and the data analysis techniques. These procedures often possess guidelines set by agencies like the United States NOAA or the European Union MSFD including the GESAMP (Group of Experts on the Scientific Aspects of Marine Environmental Protection), and OSPAR (Oslo-Paris Convention) also published guidelines regarding Microplastics presence and interference in marine system. Several defined techniques are usually as follows, to be applied in the detection as well as monitoring of microplastics. Several defined techniques are usually applied to the detection and monitoring of microplastics.

Sampling: Therefore sites to be used in the sampling process must be chosen with regard to surrounding conditions. Possible sources of microplastics should also be considered. Methods of data collection include water filtration. Sediment grabs and surface trawls are also used (Hidalgo-ruz et al., 2012; Claessens et al., 2013).

Sample Processing: After collection, the samples are subjected to processes to wash the samples and get rid of organic materials and microplastics. The methods include chemical digestion, density separation, and sieving among others (Hidalgo-ruz et al., 2012; Claessens et al., 2013).

Microplastic Identification: Out of the two microscopy techniques, Fourier-transform infrared spectroscopy (FTIR) and optical microscopy are used in identifying microplastics. The process here involved is the separation of the plastic particles from other materials (Lenz et al., 2015; Frias et al., 2016).

Quantification: The sample, once identified, undergoes establishment of the concentration level of microplastics therein. This might be attained using batch analysis—for example, applying Raman spectroscopy or just counting with a microscope the number of particles in the sample (Song et al., 2015; Mintenig et al., 2017).

Data Analysis: The data to be gathered shall be analyzed with a vie to mapping the types, quantities, and locations of microplastics. This is where statistical analysis may be applied to conclude (Costa et al., 2019; Li et al., 2018).

Quality Control: The implementation of tighter controls regarding the quality of the output from the process can enhance its accuracy and reliability. Some of the methods used in this include the use of standards, duplicate samples, and blanks (Costa et al., 2019 ; Li et al., 2018).

Reporting: Technically formatted reports so that they look suitable to be presented as a paper before a scientific journal or to the regulators. Be specific on the processes conducted, the results achieved, and limitations or any factors that were not known during the case.

These actions are the basis of the best practices designed for microplastic identification and follow-up to ensure data comparability across different studies and sites (Cowger et al., 2021; Hermsen et al., 2018).

6.6. Research on Long-Term Effects and Mitigation Strategies

In the research related to the effects of microplastics in long-term exposure in aquarium systems, several concerning discoveries have been made (Guilhermino et al., 2021). Ecosystems of water are at risk in case of accumulation of microplastics in aquariums; microplastics are obtained from cosmetics, degradation of larger plastics, and synthetic fibres. Chemically to leach toxins, to adsorb on them, as well as to alter the chemistry of water through changes in the levels of pH and dissolved oxygen, they are capable of disturbing ecosystems. For that matter, ingesting microplastics may lead to abrasion or internal harm, inflammation, and disturbances in feeding and reproduction in the water inhabitants, which may influence the population status. The options that can be utilised in aquarium systems mirror equal elements of removal and precautionary strategies that have been mentioned. Some of these are as follows; avoiding use of plastic ornaments and tools, proper installation of filtering systems that will capture the microplastic substances, and constant processes such as; water exchange and gravel cleaning. Also, as far as ensuring proper consumption and disposal among aquarists, it may assist in lowering the level of entry of microplastics into relevant ecosystems by increasing awareness of the sources and consequences of microplastics. Both researchers and legislators as well as other interested parties need to collaborate to design and implement proper mitigation measures that will help safeguard the well-being of ecosystems in aquaria (Ogunola et al., 2018; Duan et al., 2021; A. S. Shafiuddin Ahmed, 2023; Katyal et al., 2020; Onyena et al., 2022).

7. EXPECTED OUTCOMES FROM THE STUDY

7.1 Inform Best Practices for Minimizing MNP Pollution

Several vital steps can be performed to decrease the levels of microplastic pollution in the aquarium systems. First of all, in the case of using non- or low-plastic materials in the aquarium equipment such as, stainless steel or glass, the risk of plastic degradation rises. Secondly, there is the possibility of preventing microplastics from entering the water column through proper filtering that harnesses appropriate methods such as filters with proper mesh elements, small diameter meshes. As it has been said, filters should be cleaned on a regular basis for the best results to be achieved. Moreover, keeping contact with substrates and decorations fabricated from plastic as decreased also decreases the chance of microplastic release. Non packaged foods and practices that aim at avoiding overfeeding are some of the feeding practices that must be uregarded since they are a cause of microplastics. Finally, adopting proper waste disposal processes, such as recycling and correctly disposing of plastics, eliminates microplastics' contamination of water systems. To effectively prevent the input of microplastics further into the environment and to study the effects of the already existing microplastics more research and continuous monitoring is required.

7.2 Raise Awareness Among Aquarium Hobbyists and Industry Stakeholders

People related to aquarium hobbies and industries have significant parts to play in the marine conservation process. This way, they are able to manage the adverse effects of ecological change within the context of aquarium systems on marine life. Some of the things that should be in the awareness include water quality, temperature, and nutrient concentration for the appropriate growth of various animals in the water body. This also decreases the dependence on wild birds and animals by promoting different forms of sustainability, including captive-bred birds and animals and sources of livestock. More so, publicizing the need to preserve biodiversity and the consequences of invasive species' presence raises awareness about the necessity of stewardship. Interactions with researchers and conservation organizations help in the exchange of knowledge and ideas and new and improved husbandry practices at aquariums. Thus, promoting scientific knowledge among hobbyists and interested individuals not only serves the interests of the well-being of aqua captures, but also of global ocean conceded objectives and is part of mankind's collective responsibility to guarantee the well-being of the oceans for future generations.

7.3 Contribute to the Broader Understanding of MNP Pollution and Its Ecological Impacts

Plastics that have been discarded and later degraded release micro-plastics which are very hazardous to the ecosystem. Ranging from a size of less than 5mm they are found in all major ecosystems of the world both on land and in the seas. Versatility and adaptability make them blend with ecosystems; therefore, they become a threat to different forms of life. Thus, end-users of these microplastics such as marine animals may be physically affected and poisoned as they ingest these particles. Besides, they alter the cycle of nutrients, disrupt the flow of ecological succession, and in many cases, are toxic to human beings through biomagnification. In light of these facts, efficient measures that shield ecosystems as well as people's health from this universal issue must acknowledge the existence of interactions, dispersion. To protect ecosystems and human health from this ubiquitous pollution, effective mitigation solutions must take into account an understanding of their dynamics, dispersion patterns, and effects.

8. CONCLUSION

Micro- and nanoplastic pollution in aquarium systems represents a major ecological challenge, disrupting the health and behavior of aquatic organisms and altering the balance in these engineered environments. This chapter has reviewed the sources, fate, and hazards of MNPs to the holobionts within home aquaria, thus underlining the critical need for effective mitigation strategies.

In this light, the major sources of the MNPs—synthetic decorations, fish food, and exterior pollutants such as dustfall and tap water—are responsible for their wide occurrence in the aquarium system. These particles, upon introduction, tend to interact with substrates, settle and are ingested by aquatic organisms, causing physical harm, chemical toxicity, and disturbances in biological functions. Ecological imbalances resulting from MNP pollution further highlight the importance of addressing this issue due to its impact on microbial communities, plankton populations, and predator-prey dynamics.

Mitigation strategies discussed in further detail include stocking MNP-free aquarium equipment, feeding high-quality fish food, and using efficient filtration systems. This shows that educating and raising awareness among hobbyists, alongside implementing standard protocols for detecting and monitoring MNPs, can help minimize pollution. Recognizing the need for further research, this chapter calls for investigations into long-term effects and effective mitigation strategies.

8.1 Future Prospects of Micro and Nano Plastic Pollution in the Aquarium

Dealing with MNP pollution in the aquarium environment needs to be multifaceted, utilizing cutting-edge research, the application of new technologies, and public education. The future outlook for improving our knowledge and control of MNP pollution will focus on the following areas:

Chronic Effects Research: Establish long-term studies investigating the chronic impacts of MNP exposure on different aquatic species. For instance, the focus of such studies should be on the physiological, behavioral, and reproductive impacts of long-term exposure to MNP, especially on species that are commonly kept in aquariums. This can provide a general understanding of what prolonged exposure to MNPs does to aquatic organisms and their ecosystems.

Advanced Filtration Technologies: New technologies for filtration should be developed and tested to capture even the smallest plastic particles without disturbing the aquarium ecosystem. Innovations such as nanofiber filters, electrostatic precipitators, and biofiltration systems could further increase the efficiency with which MNPs can be removed from aquarium water. It will be highly critical that technologies designed for this purpose are inexpensive and achievable in practice by an average hobbyist.

Alternative Materials: Alternative, eco-friendly materials that can be used to decorate the aquarium and act as substrates and equipment can reduce MNPs in the aquarium. Research on the development of sustainable, biodegradable, and nontoxic materials should be initiated, which do not break down into microplastics. This will involve co-venturing into production with manufacturers of these alternatives and promoting them.

Public Awareness Campaigns: Develop education campaigns targeted at aquarium hobbyists and the general public so as to raise awareness about sources of MNP pollution, its effects, and ways to mitigate it. Sample campaigns could feature content on selecting MNP-free goods, proper aquarium maintenance, and the environmental impact of MNP reduction. These messages can be disseminated further through social media, workshops, and partnerships with aquarium organizations.

Regulatory Policies: Advocate for and support the development of policies and regulations aimed at reducing MNP pollution at the source. This should include a reduction in plastic production, better management of waste, and guidelines controlling the use of plastics in aquarium products. Policy engagement is necessary with the lawmakers, industry, and environmental organizations to see the policies through.

Interdisciplinary Research: Encourage interdisciplinary research collaborations on the multifaceted issue of MNP pollution. Collaboration between the fields of environmental, materials, biological science, and engineering can be combined to reach innovative solutions and a deeper understanding of MNP dynamics. Therefore, research partnerships and funding opportunities in such formation have to be encouraged.

Taking these areas into consideration, something better can be achieved in overcoming MNP pollution in aquarium systems. This comprehensive approach is going to protect aquatic organisms' health and diversity and contribute to the overall efforts of sustainability. The continued long-term viability of public aquariums as an educational and recreational resource depends on our ability to manage and reduce MNP pollution.

REFERENCES

Agbekpornu, P., & Kevudo, I. (2023). The Risks of Microplastic Pollution in the Aquatic Ecosystem. *The Risks of Microplastic Pollution in the Aquatic Ecosystem.*, 8(January). Advance online publication. DOI: 10.5772/intechopen.108717

Ahlawat, J. (2024). *Nanoparticles characterization and bioremediation-A synergy to the potential environmental benefit.*

Alberghini, L., Truant, A., Santonicola, S., Colavita, G., & Giaccone, V. (2023). Microplastics in Fish and Fishery Products and Risks for Human Health: A Review. *International Journal of Environmental Research and Public Health*, 20(1), 789. Advance online publication. DOI: 10.3390/ijerph20010789 PMID: 36613111

Aley, S., & Pranjal, A. (2023). PHA - Based Bioplastic : a Potential Alternative to Address Microplastic Pollution. In *Water, Air, & Soil Pollution.* Springer International Publishing. DOI: 10.1007/s11270-022-06029-2

Andrady, A. L., & Neal, M. A. (2009). Applications and societal benefits of plastics. *Philosophical Transactions of the Royal Society of London. Series B, Biological Sciences*, 364(1526), 1977–1984. DOI: 10.1098/rstb.2008.0304 PMID: 19528050

Angnunavuri, P. N., Attiogbe, F., Dansie, A., & Mensah, B. (2020). Consideration of emerging environmental contaminants in africa: Review of occurrence, formation, fate, and toxicity of plastic particles. *Scientific African*, 9, e00546. DOI: 10.1016/j.sciaf.2020.e00546

Athey, S. N., Albotra, S. D., Gordon, C. A., Monteleone, B., Seaton, P., Andrady, A. L., Taylor, A. R., & Brander, S. M. (2020). Trophic transfer of microplastics in an estuarine food chain and the effects of a sorbed legacy pollutant. *Limnology and Oceanography Letters*, 5(1), 154–162. DOI: 10.1002/lol2.10130

Bellasi, A., Binda, G., Pozzi, A., Boldrocchi, G., & Bettinetti, R. (2021). Chemosphere The extraction of microplastics from sediments : An overview of existing methods and the proposal of a new and green alternative. *Chemosphere*, 278, 130357. DOI: 10.1016/j.chemosphere.2021.130357 PMID: 33823347

Billings, A., Carter, H., Cross, R. K., Jones, K. C., Pereira, M. G., & Spurgeon, D. J. (2023). Co-occurrence of macroplastics, microplastics, and legacy and emerging plasticisers in UK soils. *The Science of the Total Environment*, 880(April), 163258. DOI: 10.1016/j.scitotenv.2023.163258 PMID: 37019241

Bridson, J. H., Gaugler, E. C., Smith, D. A., Northcott, G. L., & Gaw, S. (2021). Leaching and extraction of additives from plastic pollution to inform environmental risk: A multidisciplinary review of analytical approaches. *Journal of Hazardous Materials*, 414(March), 125571. DOI: 10.1016/j.jhazmat.2021.125571 PMID: 34030416

Carbery, M., O'Connor, W., & Palanisami, T. (2018). Trophic transfer of microplastics and mixed contaminants in the marine food web and implications for human health. *Environment International*, 115(March), 400–409. DOI: 10.1016/j.envint.2018.03.007 PMID: 29653694

Carla, A., Tasso, C., Wambier, S., Luiz, A., Junior, F., & Nardes, C. (2020). *Case Studies in Chemical and Environmental Engineering Filtration, assimilation and elimination of microplastics by freshwater bivalves.* DOI: 10.1016/j.cscee.2020.100053

Carolina, N., Beljanski, A., Cole, C., Fuxa, F., Setiawan, E., Singh, H., Arbor, A., Advisor, F., & Alford, L. K. (2016). *Efficiency and Effectiveness of a Low-Cost, Self-Cleaning Microplastic Filtering System for Wastewater Treatment Plants The University of Michigan – Ann Arbor.*

Carr, M. H., Neigel, J. E., Estes, J. A., Andelman, S., Robert, R., Carr, M. H., Neigel, J. E., Estes, J. A., Andelman, S., Warner, R. R., & Largier, J. L. (2003). Comparing Marine and Terrestrial Ecosystems : Implications for the Design of Coastal Marine Reserves Warner and John L. Largier Source : Ecological Applications, Vol. 13, No. 1, Supplement : The Science of Marine Reserves Published by : Wiley Stable. *Ecological Applications*, 13(1), S90–S107. DOI: 10.1890/1051-0761(2003)013[0090:CMATEI]2.0.CO;2

Carr, S. A., Liu, J., & Tesoro, A. G. (2016). Transport and fate of microplastic particles in wastewater treatment plants. *Water Research*, 91, 174–182. DOI: 10.1016/j.watres.2016.01.002 PMID: 26795302

Carson, H. S. (2013). The incidence of plastic ingestion by fishes: From the prey's perspective. *Marine Pollution Bulletin*, 74(1), 170–174. DOI: 10.1016/j.marpolbul.2013.07.008 PMID: 23896402

Chae, Y., & An, Y. J. (2017). Effects of micro- and nanoplastics on aquatic ecosystems: Current research trends and perspectives. *Marine Pollution Bulletin*, 124(2), 624–632. DOI: 10.1016/j.marpolbul.2017.01.070 PMID: 28222864

Chen, J., Wu, J., Sherrell, P. C., Chen, J., Wang, H., Zhang, W., & Yang, J. (2022). How to Build a Microplastics-Free Environment. *Advanced Science (Weinheim, Baden-Wurttemberg, Germany)*, 2103764(6), 1–36. DOI: 10.1002/advs.202103764 PMID: 34989178

Choi, S., Kim, J., & Kwon, M. (2022). The Effect of the Physical and Chemical Properties of Synthetic Fabrics on the Release of Microplastics during Washing and Drying. *Polymers*, 14(16), 3384. Advance online publication. DOI: 10.3390/polym14163384 PMID: 36015640

Chomiak, K. M., Owens-rios, W. A., Bangkong, C. M., Day, S. W., Eddingsaas, N. C., Hoffman, M. J., Hudson, A. O., & Tyler, A. C. (2024). *Impact of Microplastic on Freshwater Sediment Biogeochemistry and Microbial Communities Is Polymer Specific.*

Claessens, M., Van Cauwenberghe, L., Vandegehuchte, M. B., & Janssen, C. R. (2013). New techniques for the detection of microplastics in sediments and field collected organisms. *Marine Pollution Bulletin*, 70(1–2), 227–233. DOI: 10.1016/j.marpolbul.2013.03.009 PMID: 23601693

Cole, D. W., Cole, R., Gaydos, S. J., Gray, J., Hyland, G., Jacques, M. L., Powell-Dunford, N., Sawhney, C., & Au, W. W. (2009). Aquaculture: Environmental, toxicological, and health issues. *International Journal of Hygiene and Environmental Health*, 212(4), 369–377. DOI: 10.1016/j.ijheh.2008.08.003 PMID: 18790671

Costa, P., Duarte, A. C., Rocha-santos, T., & Prata, J. C. (2019). *Trends in Analytical Chemistry Methods for sampling and detection of microplastics in water and sediment : A critical review Density separation.* DOI: 10.1016/j.trac.2018.10.029

Cowger, W., Steinmetz, Z., Gray, A., Munno, K., Lynch, J., Hapich, H., Primpke, S., De Frond, H., Rochman, C., & Herodotou, O. (2021). Microplastic Spectral Classification Needs an Open Source Community: Open Specy to the Rescue! *Analytical Chemistry*, 93(21), 7543–7548. DOI: 10.1021/acs.analchem.1c00123 PMID: 34009953

Cózar, A., Echevarría, F., González-Gordillo, J. I., Irigoien, X., Úbeda, B., Hernández-León, S., Palma, Á. T., Navarro, S., García-de-Lomas, J., Ruiz, A., Fernández-de-Puelles, M. L., & Duarte, C. M. (2014). Plastic debris in the open ocean. *Proceedings of the National Academy of Sciences of the United States of America*, 111(28), 10239–10244. DOI: 10.1073/pnas.1314705111 PMID: 24982135

Curcuruto, M., Williams, S., Brondino, M., & Bazzoli, A. (2023). Investigating the Impact of Occupational Technostress and Psychological Restorativeness of Natural Spaces on Work Engagement and Work–Life Balance Satisfaction. *International Journal of Environmental Research and Public Health*, 20(3), 2249. Advance online publication. DOI: 10.3390/ijerph20032249 PMID: 36767614

De-la-Torre, G. E. (2020). Microplastics: An emerging threat to food security and human health. *Journal of Food Science and Technology*, 57(5), 1601–1608. DOI: 10.1007/s13197-019-04138-1 PMID: 32327770

Dhiman, S., Sharma, C., Kumar, A., & Pathak, P. (2023). *Microplastics in Aquatic and Food Ecosystems : Remediation Coupled with Circular Economy Solutions to Create Resource from Waste*.

Duan, J., Bolan, N., Li, Y., Ding, S., Atugoda, T., Vithanage, M., Sarkar, B., Tsang, D. C. W., & Kirkham, M. B. (2021). Weathering of microplastics and interaction with other coexisting constituents in terrestrial and aquatic environments. *Water Research*, 196, 117011. DOI: 10.1016/j.watres.2021.117011 PMID: 33743325

Fakolade, O. A., & Atanda, A. I. (2015). Literature review Literature review. *Literature Review*, (November), 33–37.

Foltz, G. R., Schmid, C., & Lumpkin, R. (2015). Transport of surface freshwater from the equatorial to the subtropical North Atlantic Ocean. *Journal of Physical Oceanography*, 45(4), 1086–1102. DOI: 10.1175/JPO-D-14-0189.1

Frias, J. P. G. L., Gago, J., Otero, V., & Sobral, P. (2016). Microplastics in coastal sediments from Southern Portuguese shelf waters. *Marine Environmental Research*, 114, 24–30. DOI: 10.1016/j.marenvres.2015.12.006 PMID: 26748246

Gallowaya, T. S., & Lewisa, C. N. (2016). Marine microplastics spell big problems for future generations. *Proceedings of the National Academy of Sciences of the United States of America*, 113(9), 2331–2333. DOI: 10.1073/pnas.1600715113 PMID: 26903632

Granek, E. F., Brander, S. M., & Holland, E. B. (2020). Microplastics in aquatic organisms: Improving understanding and identifying research directions for the next decade. *Limnology and Oceanography Letters*, 5(1), 1–4. DOI: 10.1002/lol2.10145

Guilhermino, L., Martins, A., Cunha, S., & Fernandes, J. O. (2021). Science of the Total Environment Long-term adverse effects of microplastics on Daphnia magna reproduction and population growth rate at increased water temperature and light intensity : Combined effects of stressors and interactions. *The Science of the Total Environment*, 784, 147082. DOI: 10.1016/j.scitotenv.2021.147082 PMID: 33894603

Hahladakis, J. N., Velis, C. A., Weber, R., Iacovidou, E., & Purnell, P. (2018). An overview of chemical additives present in plastics: Migration, release, fate and environmental impact during their use, disposal and recycling. *Journal of Hazardous Materials*, 344, 179–199. DOI: 10.1016/j.jhazmat.2017.10.014 PMID: 29035713

Hermabessiere, L., Dehaut, A., Paul-Pont, I., Lacroix, C., Jezequel, R., Soudant, P., & Duflos, G. (2017). Occurrence and effects of plastic additives on marine environments and organisms: A review. *Chemosphere*, 182, 781–793. DOI: 10.1016/j.chemosphere.2017.05.096 PMID: 28545000

Hermsen, E., Mintenig, S. M., Besseling, E., & Koelmans, A. A. (2018). Quality Criteria for the Analysis of Microplastic in Biota Samples. *Critical Reviews in Environmental Science and Technology*, 52(18), 10230–10240. DOI: 10.1021/acs.est.8b01611 PMID: 30137965

Hidalgo-ruz, V., Gutow, L., Thompson, R. C., & Thiel, M. (2012). *Microplastics in the Marine Environment: A Review of the Methods Used for Identification and Quantification*. DOI: 10.1021/es2031505

Iheanacho, S., Ogbu, M., Bhuyan, M. S., & Ogunji, J. (2023). Microplastic pollution: An emerging contaminant in aquaculture. *Aquaculture and Fisheries*, 8(6), 603–616. DOI: 10.1016/j.aaf.2023.01.007

Issac, M. N., & Kandasubramanian, B. (2021). Effect of microplastics in water and aquatic systems. *Environmental Science and Pollution Research International*, 28(16), 19544–19562. DOI: 10.1007/s11356-021-13184-2 PMID: 33655475

Ju, S., Shin, G., Lee, M., Koo, J. M., Jeon, H., Ok, Y. S., Hwang, D. S., Hwang, S. Y., Oh, D. X., & Park, J. (2021). *Biodegradable chito-beads replacing non-biodegradable microplastics for cosmetics.* DOI: 10.1039/D1GC01588E

Juying, C., Lead, W., Kiho, K., Ofiara, D., Zhao, Y., Bera, A., Lohmann, R., & Baker, M. C. (2016). Part V. Assessment of Other Human Activities and the Marine Environment. [First World Ocean Assessment]. *First Global Integrated Marine Assessment*, 2011, 1–34.

Katyal, D., Kong, E., & Villanueva, J. (2020). *Microplastics in the environment : impact on human health and future mitigation strategies.* DOI: 10.5864/d2020-005

Koelmans, A. A., Mohamed Nor, N. H., Hermsen, E., Kooi, M., Mintenig, S. M., & De France, J. (2019). Microplastics in freshwaters and drinking water: Critical review and assessment of data quality. *Water Research*, 155, 410–422. DOI: 10.1016/j.watres.2019.02.054 PMID: 30861380

Lenz, R., Enders, K., Stedmon, C. A., MacKenzie, D. M. A., & Nielsen, T. G. (2015). A critical assessment of visual identification of marine microplastic using Raman spectroscopy for analysis improvement. *Marine Pollution Bulletin*, 100(1), 82–91. DOI: 10.1016/j.marpolbul.2015.09.026 PMID: 26455785

Li, J., Liu, H., & Paul Chen, J. (2018). Microplastics in freshwater systems: A review on occurrence, environmental effects, and methods for microplastics detection. *Water Research*, 137(May), 362–374. DOI: 10.1016/j.watres.2017.12.056 PMID: 29580559

Maceda-veiga, A., & Dom, O. (2016). *The aquarium hobby : can sinners become saints in freshwater fish conservation?* DOI: 10.1111/faf.12097

Maes, T., Jessop, R., Wellner, N., Haupt, K., & Mayes, A. G. (2017). A rapid-screening approach to detect and quantify microplastics based on fluorescent tagging with Nile Red. *Scientific Reports*, 7(March), 1–10. DOI: 10.1038/srep44501 PMID: 28300146

Malinowski, C. R., Searle, C. L., Schaber, J., & Höök, T. O. (2023). Microplastics impact simple aquatic food web dynamics through reduced zooplankton feeding and potentially releasing algae from consumer control. *The Science of the Total Environment*, 904(August), 166691. Advance online publication. DOI: 10.1016/j.scitotenv.2023.166691 PMID: 37659532

Marchio, E. A. (2018). The Art of Aquarium Keeping Communicates Science and Conservation. *Frontiers in Communication*, 3(April), 1–9. DOI: 10.3389/fcomm.2018.00017

Marine Debris Program. (2015). *Laboratory Methods for the Analysis of Microplastics in the Marine Environment: Recommendations for quantifying synthetic particles in waters and sediments.*

Martin, R. G. (1975). Sexual and Aggressive Behavior, Density and Social Structure in A Natural Population of Mosquitofish, Gambusia affinis holbrooki. *Copeia*, 1975(3), 445. DOI: 10.2307/1443641

Mccormick, A. A. (2022). *Microplastic contamination and possible sources in a small public aquarium.*

Mcdevitt, J. P., Criddle, C. S., Morse, M., Hale, R. C., Bott, C. B., & Rochman, C. M. (2017). *Addressing the Issue of Microplastics in the Wake of the Microbead-Free Waters Act: A New Standard Can Facilitate Improved Policy*. DOI: 10.1021/acs.est.6b05812

Melo-Merino, S. M., Reyes-Bonilla, H., & Lira-Noriega, A. (2020). Ecological niche models and species distribution models in marine environments: A literature review and spatial analysis of evidence. *Ecological Modelling, 415*(September), 108837. DOI: 10.1016/j.ecolmodel.2019.108837

Mintenig, S. M., Int-Veen, I., Löder, M. G. J., Primpke, S., & Gerdts, G. (2017). Identification of microplastic in effluents of waste water treatment plants using focal plane array-based micro-Fourier-transform infrared imaging. *Water Research*, 108, 365–372. DOI: 10.1016/j.watres.2016.11.015 PMID: 27838027

Müller, Y. K., Wernicke, T., Pittroff, M., Witzig, C. S., Storck, F. R., Klinger, J., & Zumbülte, N. (2020). *Microplastic analysis — are we measuring the same ? Results on the first global comparative study for microplastic analysis in a water sample.*

Nzeyimana, B. S., & Mary, A. D. C. (2024). Sustainable sewage water treatment based on natural plant coagulant: Moringa oleifera. *Discover Water*, 4(1), 15. Advance online publication. DOI: 10.1007/s43832-024-00069-x

Ogunola, O. S., Onada, O. A., & Falaye, A. E. (2018). *Mitigation measures to avert the impacts of plastics and microplastics in the marine environment (a review).* DOI: 10.1007/s11356-018-1499-z

Onyena, A. P., Aniche, D. C., Ogbolu, B. O., Rakib, R. J., Uddin, J., & Walker, T. R. (2022). Governance Strategies for Mitigating Microplastic Pollution in the Marine Environment. *RE:view*, 15–46.

Padervand, M., Lichtfouse, E., Robert, D., Wang, C., Padervand, M., Lichtfouse, E., Robert, D., & Wang, C. (2020). Removal of microplastics from the environment. A review To cite this version : HAL Id : hal-02562545 Removal of microplastics from the environment. A review. *Environmental Chemistry Letters*, 18(3), 807–828. DOI: 10.1007/s10311-020-00983-1

Pfohl, P., Wagner, M., Meyer, L., Domercq, P., Praetorius, A., Hu, T., Hofmann, T., & Wohlleben, W. (2022). *Environmental Degradation of Microplastics: How to Measure Fragmentation Rates to Secondary Micro- and Nanoplastic Fragments and Dissociation into Dissolved Organics.* DOI: 10.1021/acs.est.2c01228

Prata, J. C., Patr, A. L., Mouneyrac, C., Walker, T. R., Duarte, A. C., & Rocha-santos, T. (2019). *Solutions and Integrated Strategies for the Control and Mitigation of Plastic and Microplastic Pollution.*

Raamsdonk, L. W. D. Van, Zande, M. Van Der, & Koelmans, A. A. (2020). *and Potential Health E ff ects of Microplastics Present in the Food Chain.*

Raju, S., Carbery, M., Kuttykattil, A., Senathirajah, K., Subashchandrabose, S. R., Evans, G., & Thavamani, P. (2018). Transport and fate of microplastics in wastewater treatment plants: Implications to environmental health. *Reviews in Environmental Science and Biotechnology*, 17(4), 637–653. DOI: 10.1007/s11157-018-9480-3

Reproducibility, M., Al-azzawi, M. S. M., Kefer, S., Weißer, J., Reichel, J., Schwaller, C., Glas, K., Knoop, O., & Drewes, J. E. (2020). *Validation of Sample Preparation Methods for Microplastic Analysis in Wastewater.*

Rostami, S., Talaie, M. R., Talaiekhozani, A., & Sillanpää, M. (2021). *Evaluation of the available strategies to control the emission of microplastics into the aquatic environment.*

Rummel, C. (2022). *Ecotoxicological and Microbial Studies on Weathering Plastic.*

Sander, M., Weber, M., Lott, C., Zumstein, M., Künkel, A., & Battagliarin, G. (2023). Polymer Biodegradability 2.0: A Holistic View on Polymer Biodegradation in Natural and Engineered Environments. *Advances in Polymer Science*, 293, 65–110. DOI: 10.1007/12_2023_163

Schmidt, N., Fauvelle, V., Ody, A., Castro-Jiménez, J., Jouanno, J., Changeux, T., Thibaut, T., & Sempéré, R. (2019). The Amazon River: A Major Source of Organic Plastic Additives to the Tropical North Atlantic? *Environmental Science & Technology*, 53(13), 7513–7521. DOI: 10.1021/acs.est.9b01585 PMID: 31244083

Shafiuddin Ahmed. (2023). *Microplastics in aquatic environments: A comprehensive review of toxicity, removal, and remediation strategie.*

Shirazimoghaddam, S., Amin, I., Faria Albanese, J. A., & Shiju, N. R. (2023). Chemical Recycling of Used PET by Glycolysis Using Niobia-Based Catalysts. *ACS Engineering Au*, 3(1), 37–44. DOI: 10.1021/acsengineeringau.2c00029 PMID: 36820227

Snelgrove, P. V. R., Austen, M. C., Boucher, G., Heip, C., Hutchings, P. A., King, G. M., Koike, I., Lambshead, P. J. D., & Smith, C. R. (2000). Linking biodiversity above and below the marine sediment-water interface. *Bioscience*, 50(12), 1076–1088. DOI: 10.1641/0006-3568(2000)050[1076:LBAABT]2.0.CO;2

Song, Y. K., Hong, S. H., Jang, M., Han, G. M., Rani, M., Lee, J., & Shim, W. J. (2015). A comparison of microscopic and spectroscopic identification methods for analysis of microplastics in environmental samples. *Marine Pollution Bulletin*, 93(1–2), 202–209. DOI: 10.1016/j.marpolbul.2015.01.015 PMID: 25682567

Sulpizio, J. (2019). Microplastics in our waters, an unquestionable concern. *York Daily Record*, 24–26. https://www.proquest.com/newspapers/microplastics-our-waters-unquestionable-concern/docview/2322667200/se-2?accountid=206735

Thiounn, T., & Smith, R. C. (2020). Advances and approaches for chemical recycling of plastic waste. *Journal of Polymer Science*, 58(10), 1347–1364. DOI: 10.1002/pol.20190261

Thushari, G. G. N., & Senevirathna, J. D. M. (2020). Plastic pollution in the marine environment. *Heliyon*, 6(8), e04709. DOI: 10.1016/j.heliyon.2020.e04709 PMID: 32923712

Tsakona, M., Baker, E., Rucevska, I., Maes, T., Appelquist, L. R., Macmillan-Lawler, M., Harris, P., Raubenheimer, K., Langeard, R., Savelli-Soderberg, H., Woodall, K. O., Dittkrist, J., Zwimpfer, T. A., Aidis, R., Mafuta, C., & Schoolmeester, T. (2021). Marine Litter and Plastic Waste Vital Graphics. *UN environment programme.*

Turner, A., & Holmes, L. (2011). Occurrence, distribution and characteristics of beached plastic production pellets on the island of Malta (central Mediterranean). *Marine Pollution Bulletin*, 62(2), 377–381. DOI: 10.1016/j.marpolbul.2010.09.027 PMID: 21030052

Wang, X., Liu, L., Zheng, H., Wang, M., Fu, Y., Luo, X., Li, F., & Wang, Z. (2020). Polystyrene microplastics impaired the feeding and swimming behavior of mysid shrimp Neomysis japonica. *Marine Pollution Bulletin*, 150(September), 110660. DOI: 10.1016/j.marpolbul.2019.110660 PMID: 31727317

Williams, S., Stoskopf, M., Drive, C., City, M., Carolina, N., Francis-floyd, R., Koutsos, L., Dierenfeld, E., Dierenfeld, E. S., Drive, G., Louis, S., Cicotello, E., German, D., Semmen, K., Keaffaber, J., Olea-popelka, F., Livingston, S., Sullivan, K., & Valdes, E. (2022). *Recommendations and Action Plans to Improve Ex Situ Nutrition and Health of Marine Teleosts*. DOI: 10.1002/aah.10150

Windsor, F. M., Durance, I., Horton, A. A., Thompson, R. C., Tyler, C. R., & Ormerod, S. J. (2019). A catchment-scale perspective of plastic pollution. *Global Change Biology*, 25(4), 1207–1221. DOI: 10.1111/gcb.14572 PMID: 30663840

Wolff, S., Weber, F., Kerpen, J., Winklhofer, M., Engelhart, M., & Barkmann, L. (2021). *Elimination of Microplastics by Downstream Sand Filters in Wastewater Treatment.*

Yazdian, H., Jaafarzadeh, N., & Zahraie, B. (2014). Relationship between benthic macroinvertebrate bio-indices and physicochemical parameters of water: A tool for water resources managers. *Journal of Environmental Health Science & Engineering*, 12(1), 1–9. DOI: 10.1186/2052-336X-12-30 PMID: 24410768

Zhou, A., Zhang, Y., Xie, S., Chen, Y., Li, X., Wang, J., & Zou, J. (2021). Microplastics and their potential effects on the aquaculture systems: A critical review. *Reviews in Aquaculture*, 13(1), 719–733. DOI: 10.1111/raq.12496

Chapter 11
Physiological and Toxicological Effects of Nano/Microplastics on Marine Birds

Anubha Singh
Gautam Buddha University, India

Jyoti Upadhyay
Gautam Buddha University, India

ABSTRACT

Micro/nano plastics are emerging as a severe threat to marine birds/ environment worldwide due to anthropogenic litter. Increase in production of microplastics combined with inefficient waste management has led to its bioaccumulation in the marine environment. Marine sea birds are known bioindicators for plastic pollution as they get absorbed and accumulated in the tissues of these birds. Negative effects of microplastics on marine sea birds are based on toxicological consequences that can be observed due to their ingestion like starvation, suffocation, and entanglement. In the majority of cases, these microplastics are easily taken up by birds which results in disturbance in their physiology like skin lesions, diminished body weight, fledgling success, and reproductive output. In conclusion, this chapter demonstrates the current status and effect of microplastic on the marine ecosystem related to at-risk species of sea birds which will help create awareness in regard to waste management policies and advanced technologies present to reduce plastics in the marine environment.

DOI: 10.4018/979-8-3693-3447-8.ch011

1. INTRODUCTION

Plastics have become a major marine pollutant due to anthropogenic intervention causing potential harm to marine wildlife through ingestion and entanglement (NOAA., 2014). A study revealed, that approximately 56% of marine seabirds have interacted with marine debris, leading to health deterioration in several species (Lavers et al., 2014). The foraging trend of these seabirds depends upon the resemblance of marine debris and the scent of chemicals to the natural food (Robards et al., 1995). There is widespread distribution of plastics in the form of mega, macro, micro, and nano-plastics, out of which microplastics are dispersed and accumulated across the marine and coastal ecosystem impacting other ecosystems as well. Compared to other debris like paper, metal, glass, rubber, etc., plastic litter persists due to its long shelf life and becomes debris.

Plastics are made of various synthetic or semi-synthetic organic polymers and these polymers emanate from petroleum, natural gas, oil, and coal and are made of polymer chains of, silicon, hydrogen, carbon, oxygen, and chloride, and most prominent synthetic plastics are polyethylene terephthalate (PET), polypropylene (PP), polystyrene (PS), and polyethylene (PE). The size of primary microplastics manufactured is smaller than 5 mm, and secondary microplastics following the breakdown reach these sizes in an ecosystem (Arthur et al., 2009; Cole et al., 2011). Larger plastic items, reach the size of these microplastics via various processes including UV-degradation, weathering, and oxidation (Andrady, 2015). Plastics have high stability, versatility, and ease of manufacture, useful for a variety of products. Plastic production and consumption have significantly increased since the early 1950s (Geyer et al. 2017; MacLeod et al., 2021). The plastic products i.e. macro plastics with a diameter of > 5 mm, are non-biodegradable and the broken particles can spread into the environment due to waves or winds as they are lightweight with high extended stability and durability as well as small size (Susanti et al., 2020). In recent times, the presence of plastic particles, classified as microplastics (MPs) with a diameter of 5 mm or less, and nanoplastics (NPs) smaller than 1 μm, has been noted in a variety of forms, compositions, and textures across atmospheric, terrestrial, and aquatic environments. These particles can become a part of the food chain through ingestion or inhalation (Susanti et al., 2020).

Due to inefficient waste management, little amount of plastic waste is recycled (9%) and almost half (50%) of plastic waste is dumped in landfills, out of the total generated plastic waste 22% is released to land and aquatic regions, from this plastic trash around 10% of the plastic waste enters the marine habitat (Watts et al., 2021). This fast increase in the release of plastic waste into the marine ecosystem may lead to an increase of plastic more than fish in the future (Anik et al., 2021). The range of Plastic trash is not limited to terrestrial and aquatic regions but they have also been found in isolated regions such as the Mariana Trench (Jamieson et al., 2019), polar regions (Bessa et al. 2019), and Mount Everest (Napper et al., 2020). As there is a continuous increase in the quantity of plastic, it has also led to the accumulation of plastic in every corner of the world and has entered into the food chain of all the species. This chapter highlights the current status of micro/nano plastics in the marine ecosystem and their physiological and toxicological effects on marine birds. This will further help in creating awareness towards at-risk species and their declining population due to plastic exposure and improve waste management policies for the conservation of the marine environment.

2. MARINE SEA BIRDS

Marine/sea birds are one of the most threatened groups of megafauna which have been negatively impacted due to anthropogenic intervention out of which the seabirds like Suliformes, Sphenisciformes, Stercorariidae, Laridae, Procellariiformes, Alcidae, and Phaethontiformes are severely affected (Jackson et al., 2008; Croxall P, 2012). These seabirds form approximately 3% of the bird species but their biomass outweighs that of the land birds (Brooke et al., 2004). Their relative abundance and distribution throughout the oceans provide insights related to marine ecosystems and structure as well as food chain dynamics (Burger & Gochfeld., 2004). Any change in the marine ecosystem directly impacts the sea birds as they are top consumers in the food chain and help assess any changes in biological and physical parameters of a marine ecosystem like salinity of the ocean, sea-surface temperature, and depth of the thermocline, etc (Diamond et al., 2003). Monitoring their community regarding forage ecology, migration patterns, breeding biology, and season distribution opens up new arenas for the conservation of endangered species as well as the status of the marine environment (Raymond et al., 2010). These seabirds nest in the coastal areas or cliffs and feed in the marine environment whereas marine birds like Stercorariidae, Gaviidae, phalaropes, and Anatidae depend on the marine areas during migration and prefer inland as their feeding ground and habitat, especially during breeding season. During this period, these birds routinely move from their foraging habitats in oceans to their breeding grounds located on islands or coastal areas which in turn makes them essential in nutrient recycling also known as biogeochemical cycling and transferring nutrients. The ecological importance of the nutrient flux due to seabirds has fuelled the nutrient-poor ecosystems with phosphorus and nitrogen which has led to enhanced diversity and productivity in plant communities (Hentati, 2020; Anderson,1999). Sea/marine birds depend on fishes and planktonic invertebrates as potential prey (sardines, sand eels, anchovies, herring, squids, mollusks, and crustaceans) (Montevecchi et al., 1993). The foraging behavior is based on physiological and morphological adaptations like body and bill shape, type of feed, ability to catch the prey, and the amount of energy consumed, which widens their food source as well as their foraging range in various ways (Furness et al., 2000).

The seabirds have adapted themselves in varied environments from Antarctica to the North Pole but are dependent on marine ecosystems like coastal areas foraging, roosting, perching, and breeding for their existence (Rajpar et al., 2018). These marine seabirds have evolved numerous adaptations that allow them to thrive in the marine ecosystem like preen glands also known as uropygial glands that secrete waxy coating for waterproof feathers and allow them to remain buoyant and maintain homeostasis while flying over the oceans. Others like webbed feet, beak shape, and distensible pouch for foraging along with wing modifications including long and narrow shapes for gliding and soaring to cover immense areas without energy consumption (Mullen., 2019). As these birds are colonial species and migrate to great distances these adaptations have helped them connect geographically diverse marine ecosystems.

2.1 Current Status of Marine Areas and Sea Birds

The marine seabirds are a globally threatened cluster of vertebrates within which almost half of the species are recorded as federally threatened with extinction or endangerment by the International Union for Conservation of Nature (IUCN) (Dias et al., 2019). Identification and management of marine areas that are indispensable to the seabirds for foraging, migration, and staging have become crucial for their conservation (Louzao et al., 2009; Lascelles et al., 2012). These important areas for the seabirds have been identified through various programs such as environmental assessment, and broad-scale integrative marine spatial planning and are designated as Important Bird Areas (IBAs). These areas contain a significant amount of population of 1 or more species as reflected in the conservation policies of the Regional Sea Conventions (State of the Baltic Sea, 2018; OSPAR, 2010) and European Union (EU)'s Marine Strategy Framework Directive and are documented, based on standardized criteria including essential habitats and breeding niche for these marine birds (National Audubon Society, 2012b; BirdLife International, 2012). Human intervention is not restricted in these IBAs, but there are binding restrictions in areas like Marine protected areas (MPAs), Particularly Sensitive Sea Areas (PSSAs), Ecologically and Biologically Significant Areas (EBSAs) on commercial, developmental, or recreational activities (Nur et al., 2011; Ronconi et al., 2012).

Presently, less than 1% of the marine environment comes under the designated protected area in comparison to the land in which the covered protected areas are approximately 11% of the total land surface (Toropova et al., 2010). The MPAs have become valuable to combat the swift depletion of marine seabirds and the whole ecosystem (McCauley et al., 2015). Currently, these encompass 18 very large MPAs > 100,000 km2 and are surrounded by 60 terrestrial Conservation of these marine seabirds presents unique challenges for their documentation due to asynchronous, nocturnal, secretive, and aseasonal breeding as well as inaccessible breeding areas (Nelson 1978; Newman et al. 2009; VanderWerf and Young 2017). Recognition of these priority sites based on spatially and temporally limited data has become crucial for the development of IBA and MPAs (Smith et al., 2014). The distribution of seabirds is influenced according to their foraging and breeding habitats which is displayed by biogeographic patterns at different spatial and temporal scales in response to the oceanographic characteristics (Smith et al, 2014; Serratosa et al., 2020).

The current status of these marine seabirds has become concerning as they are one of the most threatened groups of vertebrates, having approximately 362 species classified under the Critically Endangered (19), Endangered (36), Near Threatened (42), and Vulnerable (58) category with a continued decline in the population. According to the ICUN Red List species of seabirds like petrels, albatrosses, penguins, and storm petrels are at a higher risk of extinction shortly than any other order (Phillips., 2022). This alarming decline in their population can be a result of climate change, alien species introduction, overfishing, and mortality from bycatch in fisheries (Dias et al. 2019; Anderson, 2011).

2.2 Sea Birds as Bioindicators of the Marine Environment

Seabirds serve as sentinel organisms for monitoring changes in the marine ecosystem and this evaluation is based on changes in the population of these marine birds. This purpose is served as seabirds have a wide range of habitats and occupy the top level on the food web and are affected by anthropogenic pressures like pollution from industrial discharge, plastic pollution, oil spills, and overexploitation of food resources (Bost and le Maho, 1993; Croxall et al. 2012). Research related to seabirds provides insights into ecosystem variation, the impact of climate change, and the health of fish stocks as a breeding failure of seabirds is directly linked to a decline in fish stocks (Einoder, 2009). Habitat degradation, introduction of invasive species, global warming, and overfishing are the major threats to marine conservation (Kingford, 2009). A long-term study based in the North Sea shows the breeding success, population size and survival of *Rissa tridactyla* are dependent upon the marine ecosystem i.e., the number of fish, planktons, and the sea-surface temperature (Aebischer et al., 1990).

These seabirds are selected as ecological bioindicators of the marine environment as they are easily identifiable, have a consistent diet, and feed on prey available in the food chain, they can accumulate high concentrations of contaminants and are resistant to some toxic effects. They occur in large colonies with consistent breeding grounds, physiology, and ecology and are less affected by human intervention (ICES, 2018). In current years marine seabirds have been used to sample the biological and physical parameters of the marine ecosystem in real time (Wilson et al., 2002a). They have been largely used to examine the presence, concentration, and effects of contamination due to pollutants and also provide insights into the consequences of environmental variability (Furness & Camphuysen, 1997). These are also preferred as ecological indicators due to their behavioral (increase in effort at the time of foraging) and demographic (breeding success) parameters (Cairns, 1987). Behavioral patterns vary as there are limitations in foraging as they are highly mobile and wide-ranging predators but their predictability decreases in marine environments as they are consistent with the same on the continental area (Bost & Le Maho 1993; Weimerskirch, 2007). As seabirds can integrate contaminant levels across the marine ecosystem, they can be used for pollution assessment (Furness & Camphuysen, 1997). One such example is Penguins as they have emerged as an effective and reliable bioindicator of mercury contamination in the southern hemisphere (Cusset et al.2023). These birds have large biomass and remain concentrated in the Southern Hemisphere all through the annual cycle which helps in the accumulation of mercury (Brasso et al., 2015; Carravieri et al., 2016).

The pelagic foraging distribution, body size, surface feeding habits, and olfactory sense of Procellariforms belonging to vulnerable order are quite significant as they have the highest ingestion rate of anthropogenic debris (Provencher et al., 2017; Tavares et al., 2017). A study conducted by Matos et al explored the effect of anthropogenic debris on three species Procellariiformes, suliformes, and phaethontiformes which come in vulnerable categories and play a pivotal role in understanding habitat as well as conservation status (Matos et al, 2023). Human activities have turned the marine environment into their final destination for xenobiotics out of which plastic debris has become a major pollutant in oceans (Faggio et al., 2014; Fazio et al., 2012). Arctic seabirds have become bioindicators of plastic pollution in marine environments especially in the North Sea region as approximately 50 species have been evaluated for plastic ingestion out of which the northern fulmar (Fulmarus glacialis) tend to ingest a large number of microplastics due to their foraging habits and gut morphology (Furness, 1985; Provencher et al., 2009). These features make them a good candidate for monitoring plastic pollution and have become official indicators of marine plastic pollution in the North Sea since 2002 (OSPAR Commission, 2008).

2.3 Threat to the Sea Birds

The key source of disturbance that displaces the seabird species from their habitats used for nesting, feeding, and roosting is human intervention. These coastal areas are also in high demand for human activities and may not directly implicate mortality but indirectly impact survival and reproduction as it also promotes the involvement of predators like skunks' gulls and raccoons that prey upon eggs and young ones of these birds (Shepherd and Boates 1999; Peters and Otis 2007). It has become variable due to activities like bulkheading, dredging, beach replenishment, and jetty building which can alter the extent and quality of these habitats (Shepherd and Boates 1999).

Anthropogenic pollution is the leading cause of a decline in the population of these marine seabirds (BirdLife International, 2022; Croxall et al., 2012). Anthropogenic factors have become a serious threat which includes oil, plastic pollution, human disturbance or introduction of invasive species, degradation of habitat, and fisheries. Among these, bycatch in fisheries is one of the most severe threats especially petrels, sea ducks, albatross, and penguins as they are attracted to these vessels carrying out fishing operations to catch easy prey or bait and other unwanted catch (Dias et al. 2019). A bycatch in seabirds occurs if they swallow the hook or get dragged, entangled, and drown (Da Rocha et al. 2021). Overfishing is another leading threat as depletion of fish stocks can have a direct effect on seabirds, especially during breeding season as the foraging area is restricted and has led to a decline of 24 species including cormorants, sea ducks, gulls, frigatebirds, and penguins (Barbraud et al. 2018; Dias et al. 2019). The impact of invasive alien species affects the host endemic seabirds, those growing up in the absence of mammalian ground-predator birds like petrels, albatrosses, and shearwaters are particularly impacted (Dias et al. 2019). Predators like *Felis cattus, Rattus* spp, and *Mus musculus* (Barbraud et al. 2021b). Approximately 100 seabird species (27%) of which albatrosses and penguins are considered most vulnerable are most affected by changes in climate. Warming temperatures have a direct physiological effect on these seabirds as they are endothermic and heat stress causes distress in reproduction and survival (Dias et al. 2019). Environmental pollution, chemical pollution, and loss of plastic in the environment magnify along the food web and accumulate in the tissues. (Borrelle et al. 2020; Cherel et al. 2018).

3. ROUTE OF ENTRY FOR PLASTICS AND THEIR DISTRIBUTION INTO THE MARINE ENVIRONMENT

Different human activities (like household, coastal, and industrial) in the terrestrial and aquatic ecosystem release harmful micro and nano plastic in the environment. These plastics are composed of a variety of polymers with a hydrophobic nature with diverse molecular conformations having the ability to adsorb metals and organic pollutants. Microplastics are present in a variety of daily household products, including sea salts, toothpaste, cleansers, and facial scrubs (Chang., 2015). The source of microplastics and nanoplastics in the aquatic ecosystem is mostly domestic runoff containing microbeads, microplastic fragments, even household laundering, and synthetic fibers released from textiles (Periyasamy et al., 2022). According to investigations, laundry contributes to approximately 500,000 tons of microparticles dissipating into the sea annually (Chatziparaskeva et al., 2022).

Figure 1. Schematic drawing showing the route of entry of various forms of plastics/ debris in the oceans

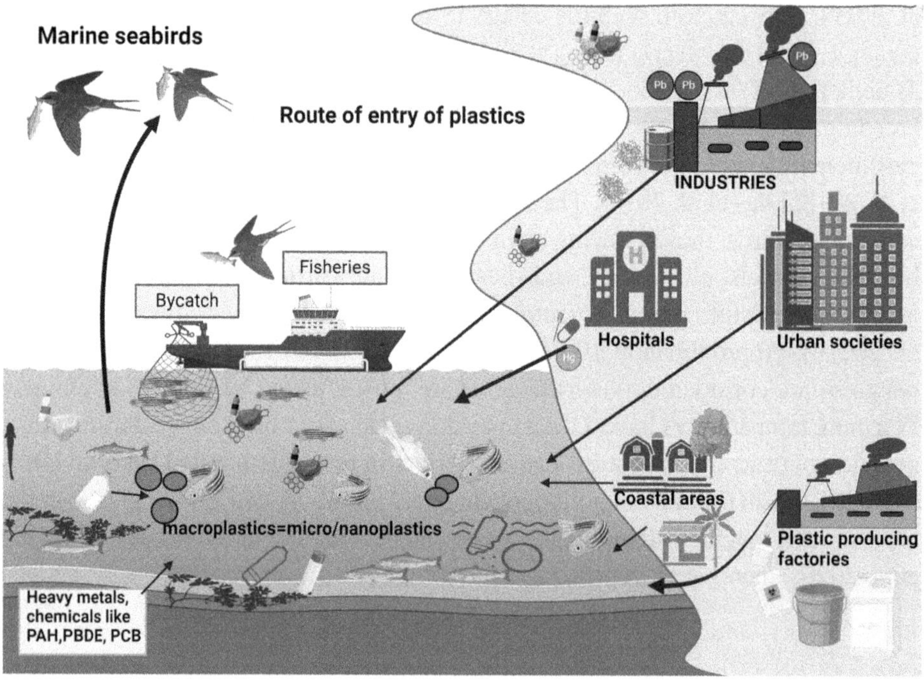

Sea-based sources include coastal activities like marine industries, aqua tourism activities, fishing practices, and aquaculture structures. Aqua culture structure consists of different forms of particles, harsh environmental conditions damage the aqua culture structure and can produce significant amounts of plastic debris. Building blocks of plastics like monomers and polymers may also be released since they turn brittle and trickle down in the marine ecosystem like styrene monomers, that are detected in the ocean and coastal areas (Kwon et al., 2015). Effluents from wastewater treatment plants signify the second largest source of macroparticles released into marine ecosystems with traces of microfibers of polyamide, acrylic, polyester, and polymeric materials (Ramasamy et al., 2022). Air-based origin includes plastic manufacturing industries that release plastics from air-blasting in the form of resin powders and pellets which ultimately can contaminate the aquatic ecosystem. Multiple activities such as plastic waste incineration, plastic production, traffic emissions, urban mining operations, and deterioration of roads and streets are responsible for plastic dust dissipation (Wang et al., 2021). These anthropogenic activities have caused in significant increase in plastic pollutants hampering the marine as well as all ecosystems as seen in Figure 1.

4. PLASTIC EXPOSURE RISK

Risk assessment needs expertise in the understanding of the vulnerability of petrels to ingestion associated with foraging near human populations, shipping lanes, and fisheries with floating debris. Exposure risk can be estimated between the hazards and organisms depending on the amount of time spent by the animals in different ranges (Clark, 2023). Ecological data of seabird species consisting of distribution and vulnerability coupled with numerical modeling of plastic transport in the ocean can help in risk assessment. There has even been a comparison between plastic concentration and species distribution and their habitats (Wilcox et al., 2015). A study evaluated the prevalence of plastic entanglement for multiple seabirds and the spatial distribution of floating plastic debris to map sensitivity distribution (Høiberg et al, 2022). Another method like the plastic: zooplankton ratio was used by Fabri-Ruiz et al. (2023) to assess risk to pelagic fish and marine megafauna in the Mediterranean.

5. EFFECT OF MICRO AND NANO PLASTIC ON MARINE SEA BIRDS

There is an uneven distribution of plastic pollutants within the ocean and coastal areas, often these plastic debris float for thousands of kilometers into the ocean currents (van Sebille, 2012). Identifying at-risk species has become crucial with the significant rise in plastic production and waste generation for targeting conservation action. The range of vulnerability relates to exposure to hazards and sensitivity to damage which impacts reproduction or survival. Plastic ingestion can have sublethal and lethal impacts due to contaminants and can cause physical injury and damage to the internal organs of these seabirds as seen in Figure 2 (Roman et al, 2019). Foraging patterns of seabirds like petrels and albatrosses can lead to the accumulation of plastic in their tissues and alter gut morphology (Young,2009).

Figure 2. The physiological and toxicological impact of environmental plastic pollution on marine seabirds

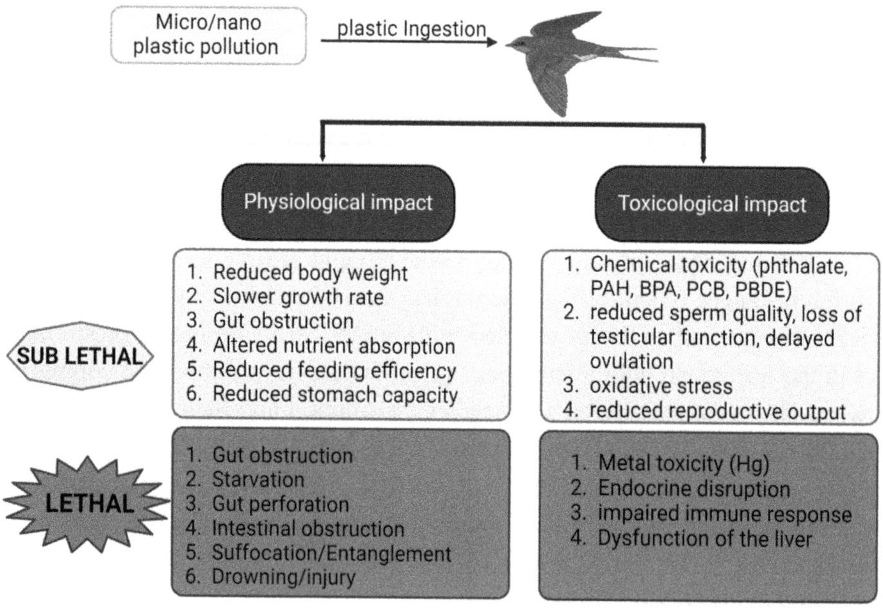

5.1 Physiological Impact

Plastic debris affects marine fauna through entanglement and ingestion (Laist, 1987). Entanglement can result in the spontaneous death of the animal due to less oxygen and drowning, or it may initiate other factors that will limit its survival by struggle in feeding, growth complications, weakness, causing injuries, and loss of limbs, (Derraik, 2002). Earlier a study on 1033 birds revealed the fact, that around 55% of the species had plastic particles in their gut which has been reported all over the food chain, from small invertebrates to zooplanktons to large predators sharks (Devriese et al., 2015). Many other indirect pathways of plastic ingestion have been reported, such as by preying on other aquatic creatures that are on plastic feed (Farrell and Nelson, 2013) or unintentionally while the aeration process particles present in the surrounding water are taken by the aquatic animals (Watts et al., 2014). Some studies suggest that various species of seabirds select explicit plastic colors and shapes, misidentifying them as probable prey (Ryan, 2016). This similar pattern was found in sea turtles (*Caretta caretta*) of the Central Mediterranean region (Gramentz, 1988). Results in the accumulation of plastic in tissues were obtained in migrating red phalaropes (*Phalaropus fulicarius*) which indirectly affect their reproductive effort on breeding and long-distance migration. Plastic ingestion gradually weakens the animal by creating a false feeling of satiety and can cause severe injuries along the gastrointestinal tract such as ulcers, and obstructions (Gregory, 1991) and these plastics can also rupture the GI tract, and cause severe damage to the proventriculus. The remaining particles can move to the gizzard, where they are again fragmented into smaller particles and move along the intestine through which they can enter the spleen and kidneys for waste disposal and production of nitrogenous waste. (Lavers et al., 2019) as seen in Table 1.

Ingestion of plastics can also cause detrimental physiological effects like reproductive failure, lowered steroid hormone levels, blockage of gastric enzyme secretion, delayed ovulation, and diminished feeding stimulus (Azzarello and Van-Vleet, 1987). Some species are more susceptible due to their inability to bring back ingested plastics to the mouth. A study on the chicks of Laysan albatrosses (*Diomedea immutabilis*) reported that microplastic accumulates in their stomachs as these birds are unable to regurgitate such materials and become a significant source of mortality in these species (Furness, 1985; Azzarello and Van-Vleet, 1987). A case study was done on two species recovered from Massachusetts, USA, *Puffinus gravis,* and *Morus bassanus,* and their cause of death was plastic ingestion, obstruction, and subsequent starvation, these were a part of a larger study on seabird mortality (SEANET) due to fisheries (Pierce et al. 2004). Data was collected on 8 species of birds including *Buteo lineatus, Coragyps atratus, Cathartes aura, Vultur gryphus* from the Audubon Center for Birds of Prey in central Florida, USA in 2018 and it was found that GI tracts contained microplastics (Carlin et al, 2020). A similar study was conducted on chicks of *Coturnix japonica* and *Pelagodroma marina* (Roman et al,2019), and *Puffinus carneipes* (Lavers et al. 2014) and showed a decreased growth rate induced by plastic ingestion as a result of reduced stomach capacity. Upon dissection of Fulmars from the Faroe Islands, plastic content in the stomach was quantified and tissues were analyzed and contained organic pollutants: polychlorinated biphenyls, polybrominates, and organophosphates (Neumann et al, 2021). Grasman,2002 investigated *Larus argentatus* and measured immunological organs including bursa of Fabricius, thymus, and spleen, and looked at peripheral white blood cell counts and observed altered results due to contaminants (Grasman,2002). The prevalence of plastic particles was assessed in dead Cory's shearwater fledglings (*Calonectris Diomedea)* and it was found that Eighty-three percent of birds were affected, containing on average 8.0 plastic pieces per bird (Rodríguez et al. 2012). A similar study done in *Passer montanus* showed that heavy metals have detrimental effects on sperm quality and testicular function (Yang et al, 2020). In a study done by Lavers and Bond, 2023 it was observed that *Ardenna carneipes* chick body mass including culmen, wing, bill, and head declined over time due to plastic ingestion (Lavers and Bond, 2023). McCarty and second examined patterns of plumage color in Tachycineta bicolor in Hudson river, USA and found that the increased concentrations of PCB contamination changed the plumage patterns consistent with disruption of the endocrine system (McCarty and Second, 2000).

Table 1. Toxicological and physiological impact on marine sea birds

Scientific name	Common name	Habitat	Toxicological effect	Physiological effect	References
Morus bassanus	Northern Gannet	Coast of North Atlantic	-	Reduced body weight, slower growth rate	Pierce et al. 2004
Buteo lineatus	Red-shouldered Hawk	Coast of California	-	Bleeding, ulcers, blockage of the digestive tract	Carlin et al. 2020
Coturnix japonica	Japanese Quail	Coast of East Asia	Dysfunction of the liver and reduced level of 17βestradiol in female	Reduced stomach capacity	Roman et al. 2019
Fulmarus glacialis)	fulmars	Northern sea	Chemical toxicity	Mechanical obstruction, inflammation	Neumann, 2021
Phalaropus fulicarius	red phalaropes	Arctic regions of North America and Eurasia.	Reproduction affected	Unable to fly for long distance	Gregory, 1991

continued on following page

Table 1. Continued

Scientific name	Common name	Habitat	Toxicological effect	Physiological effect	References
Ardenna gravis	great shearwaters	rocky islands in the South Atlantic	-	tissue damage, oxidative stress, and inflammation,	van Franeker et al., 2011
Calonectris diomedea	Cory's shearwaters	rocky islands in the eastern Atlantic	-	inflammation, oxidative stress, and tissue damage	Rodríguez et al. 2012
Larus argentatus	European herring gulls	northern and western coasts of Europe	Suppressed immune response		Grasman,2002
Sturnus vulgaris	European starlings	Great Britain and Ireland,	fertility impairments, endocrine disruption, and reduced reproductive success	Weight gain	O'Shea and Stafford, 1980
Passer montanus	Eurasian Tree Sparrows		Detrimental effects like reduced sperm quality and loss of testicular function	Impaired reproductive function	Yang et al. 2020
Ardenna carneipies	Fledgling Flesh-footed Shearwaters	Lord Howe Island, Australia	Low foraging capacity	reduced organ function, in the kidney and spleen, fibrosis	Lavers and Bond, 2023
Puffinus *carneipes*	Flesh-footed Shearwaters	Lord Howe Island, Australia	Low breeding capacity	reduced organ function, in the kidney and spleen, fibrosis	Lavers et al. 2014
Diomedea immutabilis	Laysan albatrosses	Hawaiian Islands	-	Death of birds due to accumulation of plastic in their stomach	Furness, 1985; Azzarello and Van-Vleet, 1987
Pelagodroma marina	white-faced storm-petrel	Chatham Islands (New Zealand)	-	No food content in its stomach while its gizzard was packed with plastic pellets caused death	Roman, 2019
Puffinus gravis	Great Shearwater	Rocky Islands in the South Atlantic	reduced body weight slower growth rate, obstruction of the gut,	tissue damage, oxidative stress, and inflammation	Susanti et al. 2020, Pierce 2004
Vultur gryphus	Andean Condor	Ecuador, Colombia, and Venezuela	-	intestinal obstructions, nutritional problem	Carlin, 2020

continued on following page

Table 1. Continued

Scientific name	Common name	Habitat	Toxicological effect	Physiological effect	References
Coragyps atratus	Black Vulture	Central Chile, Uruguay in South America, southeastern United States to Perú,	oxidative stress, redox unbalance, and cholinesterase effects	intestinal obstructions, infections, and metabolic alterations	Carlin et al 2020
Cathartes aura	Turkey Vulture	Canada, Tierra del Fuego	-	intestinal obstructions, nutritional problems, infections, and metabolic alterations	Carlin et al. 2020
Pachyptila belcheri	Slender-billed Prion	Oceans, the Kerguelen Islands, the Islands, and Noir Island off the coast of southern Chile. Crozet Islands	Endocrine disruption, issues in reproduction	Malnutrition	Susanti et al. 2020
Procellaria aequinoctialis	White-chinned Petrel	Crozet Islands, South Georgia, Campbell Islands, Kerguelen Islands, Antipodes Islands,Prince Edward Islands, Auckland Islands	Endocrine disruption, issues in reproduction	Malnutrition	Susanti et al. 2020
Tachycineta bicolor	Tree Swallow	US and Canada, the Gulf Coast, Panama, the northwestern coast of South America, the Indies	Reduce reproduction	Negative effects on growth and development	McCarty and Second 2000

5.2 Toxicological Impact

Plastic debris collected from oceans, and remote and urban beaches contains a wide range of these toxic contaminants that are adsorbed from the ecosystem (Hirai et al. 2011), which may lead to a multitude of dire conditions for wildlife in marine ecology (Tanaka et al. 2020). Debris contains resin pellets, microscopic plastic fragments, and organic contaminants like dichloro-diphenyl-trichloroethane and its metabolites (DDTs), polybrominated diphenyl ethers (PBDEs), polychlorinated biphenyls (PCBs), and polycyclic aromatic hydrocarbons (PAHs), concentrations of these contaminants varied in different regions (ranges from 1 to 10,000 ng/g).

European Chemicals Agency(ECA), analyzed almost 400 plastic additives, which include organotin(s), triclosan, phthalates, brominated flame retardants(BFR), diethyl hexyl phthalate and bisphenols and found at different locations (Du et al.

2017). The accumulation of plastic debris has been seen in several seabirds, like Streaked Shearwater (*Calonectris leu comelas*) (Teuten et al. 2009), Short-tailed Shearwaters(Yamashita et al. 2011), and Flesh-footed Shearwaters (Lavers et al. 2014), Kelp Gull (*Larus dominicanus*), Short-tailed Shearwater (Tanaka et al. 2013), White-chinned Petrel (*Procellaria aequinoctialis*), Slender-billed Prion (Pachyptila belcheri), Great Shearwater, Black-browed Albatross (*Thalassarche melanophrys*), and Southern Giant Petrel (*Macronectes giganteus*) (Susanti et al. 2020) suggesting that plastics are a direct carrier of chemicals to seabirds. Experimental annotation studies done on streaked shearwater chicks through mathematical models using equilibrium partitioning, and feeding experiments constantly illustrate that the amount of polyethylene accumulated is much more than organic contaminants like polyvinyl chloride and polypropylene.

Earlier studies confirmed that among these organic contaminants, diethyl hexyl phthalate may lead to weight increase in *Sturnus vulgaris* (European Starling) (O'Shea and Stafford, 1980) and further various studies revealed that diethyl hexyl phthalates are related to toxicity in liver (Zhang et al. 2018), kidneys (Li et al. 2018), and cerebellum (Du et al. 2017) in Japanese Quail. Researchers designed a study of growing Japanese quail on the Norwegian coast (to evaluate the toxicity of polypropylene and polyethylene particles collected from there. Birds were fed 600 mg of macroparticles over 5 weeks, containing small macroparticles of size 25 μm and large macroparticles size of 3 mm, both separately and in a mixture. Various sublethal endpoints in quail are evaluated, including, blood-biochemical parameters, oxidative stress, reproductive hormones in the blood, cytokine levels, and body mass. Different sizes of macroparticles have different impacts on birds, microparticles of size 25 μm induced the function of antioxidant enzymes like glutathione peroxidase, glutathione-S-transferase, and catalase (Monclus, 2022). The large size of macroplastics size of 3 mm amplified the levels of aspartate aminotransferase (liver enzyme), increased level of AST in blood shows the dysfunction of the liver and reduced the 17β-estradiol (females major sex hormone) levels in females (Reddy, 2019).

Microplastics and macroplastics are hydrophobic and thus a large surface can adsorb various harmful contaminants of different ecosystems like endocrine disrupting compounds (EDCs), polychlorinated biphenyls (PCBs), polycyclic aromatic hydrocarbons (PAHs), heavy metals, antibiotics (Reddy et al. 2019). Experimental observation on Japanese Quail demonstrated that these toxic substances on macroparticles can lead to endocrine disruption, malnutrition, and reproductive toxicity (Fossi et al. 2018; Roman et al. 2019). Japanese Quail chicks were studied for plastic ingestion and their symptoms included a significant number of cases of cysts (epididymal-intra-epithelial) in males as well as delay in sexual maturity (Roman et al. 2019). Studies also revealed the existence and effect of microplastic in various tissues of Flesh-footed Shearwaters *Ardenna carneipes*, a well-known species of marine birds that consume substantial amounts of macroparticles. Histopathological techniques of this study inspected all the visceral organs and measured their physiological responses as well as the inflammatory responses due to plastic ingestion. They found that organs like the spleen, kidney, and proventriculus had embedded microplastic particles and extensive tissue impairment which also led to tubular glands impairment and changes in the folds of rugae in the proventriculus. Alterations in the structures of organs like the spleen and kidney along with signs of inflammation and fibrosis show the severity of degradation. This study suggested that microplastics can be mobilized throughout the body and cause injury directly at the place of contact. Studies also suggest that these chemicals like phthalates and bisphenol A (BPA), can also accumulate in plastics as a result of fragmentation and drifting in the ocean and be passed to marine birds. These plastic additives produce genetic malformations, that can also affect the reproduction of fish and crustaceans, by shifting their hormonal systems (Oehlmann et al., 2009).

Different additives, among which UV stabilizers containing benzotriazole groups such as UV-236, UV-328 and UV-237 and a benzophenone group, BP-12 containing microplastic by-products, have been reported in seabirds which can disturb the endocrine system, and flame retardants like decabromodiphenyl and hexabromocyclododecane, are registered as of persistent organic pollutants (Tanaka et al., 2019). Organic UV filters found in microplastics have cocontaminants like, 4-methylbenzylidene camphor, octocrylene, ethyl hexyl dimethyl p-aminobenzoic acid, benzophenone3, and octyl-methoxycinnamate (Cadena-Aizaga et al., 2020). These types of organic pollutants are transported via plastics and become a direct pathway that can introduce risky compounds to marine birds and other marine biota. Collectively, these results show the severity of the health impact on the seabirds due to plastic pollution.

6. CASE STUDIES RELATED TO THE DETRIMENTAL EFFECTS OF PLASTICS ON MARINE BIRDS

Seabirds have been known as a model group for the avian environment as they get exposed to significant levels of contaminants from the atmosphere, land, and sea that get transferred over long distances and accumulate along the food web. These seabirds play a crucial role as nutrient transporters and ecosystem engineers. *Diomedea epomophora* (southern royal albatross) has a long lifespan as well as a low capability to adapt to environmental changes which can cause reduced reproductive success and increased mortality rate (Wooller et al., 1992). There have been cases of severe biological defects due to the accumulation of chemicals in the aquatic ecosystem including offspring developmental defects (impaired growth, abnormalities) (Letcher et al., 2010; Jenssen et al., 2010) loss of thickness in eggshells (Cortinovis et al., 2008) as well as altered vitamin homeostasis and endocrine disruption (Letcher et al., 2010). These physiological disturbances can be caused due to oxidative stress induced by the contaminants (Valavanidis et al., 2006). The Gulf of Gabes, along the Mediterranean coast is considered a hotspot for pollution as high concentrations of heavy metals and polycyclic aromatic hydrocarbons (PAH) have been detected in the ocean (Zaghden et al., 2014). Common birds like *Sterna hirundo* are known bioindicators of pollution and establish their breeding grounds in these areas. These species are spread over a wide geographical area consuming fish and present with adverse effects of pollution including oxidative stress which can cause low reproductive yield due to the accumulation of xenobiotics (Rowe, 2008).

A study was conducted in Washington, USA, and coastal waters of British Columbia, Canada, using bird species like Common Murre (*Uria aalge*), Ancient Murrelets (*Synthliboramphus antiquus*), Pigeon Guillemot (*Cepphus columba*), great shearwaters (*Ardenna gravis),* Rhinoceros Auklet (*Cerohinca mono cerata*), and Marbled Murrelets (*Brachyramphus marmoratus*). This study surveyed the insides of the stomachs of 20 bird species revealing the amount of plastic ingestion in the area. A total of 115 samples over 10 years (2001-2011) were collected from various sources like beached birds, fisheries bycatch, dietary studies and birds salvaged at sea. The ingested plastics were removed through necropsy techniques and quantified and separated based on user plastic (foamed, sheet-like, fragmented, or threadlike) and industrial pellets (van Franeker, 2004; van Franeker et al., 2011; Avery-Gomm et al, 2013). Overall, there has been a significant rise in plastic accumulation in tissues of these marine birds which has caused severe physiological and toxicological issues leading to a decline in their population and failure of water treatment plans which leach out these plastic contaminants into the oceans need to be addressed for the survival of the wildlife.

7. CONCLUSION

In conclusion, there is a dire need to work on policies for better management of environmental pollution and to address the declining population of seabirds. The physiological and toxicological impact of plastic ingestion occurs across various tissues as well as organs and alters spatial and temporal scales. These microplastics can cause substantial alterations in the pathology of the gastrointestinal tract causing inflammation and fibrosis. They are also responsible for physical impairments like entanglement and suffocation for many marine species. Accumulation of heavy metals as well as chemical contaminants poses a serious threat to these endangered species as well as the transport of these micro/nanoplastics to various areas also threatens the marine biodiversity and food web. Investigations are required to determine the hotspots of marine pollution and will help in prioritizing conservation measures for the currently endangered and vulnerable species.

ACKNOWLEDGMENT

Figures were prepared using Biorender application software.

REFERENCES

Aebischer, N. J., Coulson, J. C., & Colehsook, J. M. (1990). Parallel long-term trends across four marine trophic levels and weather. *Nature*, 147(6295), 753–755. DOI: 10.1038/347753a0

Anderson, O. R., Small, C. J., Croxall, J. P., Dunn, E. K., Sullivan, B. J., Yates, O., & Black, A. (2011). Global seabird bycatch in longline fisheries. *Endangered Species Research*, 14(2), 91–106. DOI: 10.3354/esr00347

Anderson, W., & Polis, G. (1999). Nutrient fluxes from water to land: Seabirds affect plant nutrient status on Gulf of California islands. *Oecologia*, 118(3), 324–332. DOI: 10.1007/s004420050733 PMID: 28307276

Andrady, A. L. (2015). Persistence of plastic litter in the oceans. In Bergmann, M., Gutow, L., & Klages, M. (Eds.), *Marine Anthropogenic Litter* (pp. 57–72). Springer. DOI: 10.1007/978-3-319-16510-3_3

Anik, A. H., Hossain, S., Alam, M., & Sultan, M. B. (2021). Microplastics pollution: A comprehensive review on the sources, fates, effects, and potential remediation. *Environmental Nanotechnology, Monitoring & Management*, 16, 100530. DOI: 10.1016/j.enmm.2021.100530

Arthur, C., Baker, J., & Bamford, H. (2009). Proceedings of the International Research Workshop on the Occurrence, Effects, and Fate of Microplastic Marine Debris. Group.

Avery-Gomm, S., Provencher, J. F., Morgan, K. H., & Bertram, D. F. (2013, July 15). Plastic ingestion in marine-associated bird species from the eastern North Pacific. *Marine Pollution Bulletin*, 72(1), 257–259. DOI: 10.1016/j.marpolbul.2013.04.021 PMID: 23683586

Azzarello, M. Y., & Van Vleet, E. S. (1987). Marine Birds and Plastic Pollution. *Marine Ecology Progress Series*, 37(2/3), 295–303. DOI: 10.3354/meps037295

Barbraud, C., Bertrand, A., Bouchón, M., Chaigneau, A., Delord, K., Demarcq, H., Gimenez, O., Torero, M. G., Gutiérrez, D., Oliveros-Ramos, R., Passuni, G., Tremblay, Y., & Bertrand, S. (2018). Density dependence, prey accessibility, and prey depletion by fisheries drive Peruvian seabird population dynamics. *Ecography*, 41(7), 1092–1102. DOI: 10.1111/ecog.02485

Barbraud, C., Delord, K., Le Bouard, F., Harivel, R., Demay, J., Chaigne, A., & Micol, T. (2021). Seabird population changes following mammal eradication at oceanic Saint-Paul Island, Indian Ocean. *Journal for Nature Conservation*, 63, 126049. DOI: 10.1016/j.jnc.2021.126049

Bessa, F., Ratcliffe, N., Otero, V., Sobral, P., Marques, J. C., Waluda, C. M., Trathan, P. N., & Xavier, J. C. (2019). Microplastics in gentoo penguins from the Antaretic region. *Scientific Reports*, 9(1), 14191. DOI: 10.1038/s41598-019-50621-2 PMID: 31578393

BirdLife International. (2012). *Important Bird Areas (IBAs)*. BirdLife International.

BirdLife International. (2022). Retrieved from https://www.birdlife.org/worldwide/news/single-use-plastic

Borrelle, S. B., Ringma, J., Law, K. L., Monnahan, C. C., Lebreton, L., McGivern, A., Murphy, E., Jambeck, J., Leonard, G. H., Hilleary, M. A., Eriksen, M., Possingham, H. P., De Frond, H., Gerber, L. R., Polidoro, B., Tahir, A., Bernard, M., Mallos, N., Barnes, M., & Rochman, C. M. (2020). Predicted growth in plastic waste exceeds efforts to mitigate plastic pollution. *Science*, 369(6510), 1515–1518. DOI: 10.1126/science.aba3656 PMID: 32943526

Bost, C. A., & le Maho, Y. (1993). Seabirds as bio-indicators of changing marine ecosystems: New perspectives. *Acta Oecologica*, 14, 463–470.

Brasso, R. L., Chiaradia, A., Polito, M. J., Raya Rey, A., & Emslie, S. D. (2015). A comprehensive assessment of mercury exposure in penguin populations throughout the Southern Hemisphere, using trophic calculations to identify sources of population-level variation. *Marine Pollution Bulletin*, 97(1-2), 97. DOI: 10.1016/j.marpolbul.2015.05.059 PMID: 26072048

Brooke, M. D. (2004). The food consumption of the world's seabirds. *Proceedings of the Royal Society of London. Series B, Biological Sciences*, 271, S246–S248. PMID: 15252997

Burger, J., & Gochfeld, M. (2004). Marine birds as sentinels of environmental pollution. *EcoHealth*, 1(3), 263–274. DOI: 10.1007/s10393-004-0096-4

Cadena-Aizaga, M. I., Montesdeoca-Esponda, S., Torres-Padrón, M. E., Sosa-Ferrera, Z., & Santana-Rodríguez, J. J. (2020). Organic UV filters in marine environments: An update of analytical methodologies, occurrence and distribution. *Trends in Environmental Analytical Chemistry*, 25, 25. DOI: 10.1016/j.teac.2019.e00079

Cairns, D. K. (1987). Seabirds as indicators of marine food supplies. *Biological Oceanography*, 5, 261–271.

Carlin, J., Craig, C., Little, S., Donnelly, M., Fox, D., Zhai, L., & Walters, L. (2020). Microplastic accumulation in the gastrointestinal tracts in birds of prey in central Florida, USA. *Environmental Pollution*, 264, 114633. DOI: 10.1016/j.envpol.2020.114633 PMID: 32388295

Carravieri, A., Cherel, Y., Jaeger, A., Churlaud, C., & Bustamante, P. (2016). Penguins as bioindicators of mercury contamination in the southern Indian Ocean: Geographical and temporal trends. *Environmental Pollution*, 213, 195–205. DOI: 10.1016/j.envpol.2016.02.010 PMID: 26896669

Chang, M. (2015). Reducing microplastics from facial exfoliating cleansers in wastewater through treatment versus consumer product decisions. *Marine Pollution Bulletin*, 101(1), 330–333. DOI: 10.1016/j.marpolbul.2015.10.074 PMID: 26563542

Chatziparaskeva, G., Papamichael, I., & Zorpas, A. A. (2022). Microplastics in the coastal en- vironment of Mediterranean and the impact on sustainability level. *Sustainable Chemistry and Pharmacy*, 29, 100768. DOI: 10.1016/j.scp.2022.100768

Cherel, Y., Barbraud, C., Lahournat, M., Jaeger, A., Jaquemet, S., Wanless, R. M., Phillips, R. A., Thompson, D. R., & Bustamante, P. (2018). Accumulate or eliminate? Seasonal mercury dynamics in albatrosses, the most contaminated family of birds. *Environmental Pollution*, 241, 124–135. DOI: 10.1016/j.envpol.2018.05.048 PMID: 29803026

Clark, B. L., Carneiro, A. P. B., Pearmain, E. J., Rouyer, M.-M., Clay, T. A., Cowger, W., Phillips, R. A., Manica, A., Hazin, C., Eriksen, M., González-Solís, J., Adams, J., Albores-Barajas, Y. V., Alfaro-Shigueto, J., Alho, M. S., Araujo, D. T., Arcos, J. M., Arnould, J. P. Y., Barbosa, N. J. P., & Dias, M. P. (2023). Global assessment of marine plastic exposure risk for oceanic birds. *Nature Communications*, 14(1), 3665. DOI: 10.1038/s41467-023-38900-z PMID: 37402727

Cole, M., Lindeque, P., Halsband, C., & Galloway, T. S. (2011). Microplastics as contaminants in the marine environment: A review. *Marine Pollution Bulletin*, 62(12), 2588–2597. DOI: 10.1016/j.marpolbul.2011.09.025 PMID: 22001295

Cortinovis, S., Galassi, S., Melone, G., Saino, N., Porte, C., & Bettinetti, R. (2008). Organochlorine contamination in the Great Crested Grebe (Podiceps cristatus): effects on eggshell thickness and egg steroid levels. Chemosphere, 73(3), 320–325.

Croxall, J. P., Butchart, S. H. M., Lascelles, B., Stattersfield, A. J., Sullivan, B., Symes, A., & Taylor, P. (2012). Seabird conservation status, threats and priority actions: A global assessment. *Bird Conservation International*, 22(1), 1–34. DOI: 10.1017/S0959270912000020

Cusset, F., Bustamante, P., Carravieri, A., Bertin, C., Brasso, R., Corsi, I., Dunn, M., Emmerson, L., Guillou, G., Hart, T., Juáres, M., Kato, A., Machado-Gaye, A. L., Michelot, C., Olmastroni, S., Polito, M., Raclot, T., Santos, M., Schmidt, A., & Cherel, Y. (2023). Circumpolar assessment of mercury contamination: The Adélie penguin as a bioindicator of Antarctic marine ecosystems. *Ecotoxicology (London, England)*, 32(8), 1024–1048. DOI: 10.1007/s10646-023-02709-9 PMID: 37878111

Da Rocha, N., Oppel, S., Prince, S., Matjila, S., Shaanika, T. M., Naomab, C., Yates, O., Paterson, J. R. B., Shimooshili, K., Frans, E., Kashava, S., & Crawford, R. (2021). Reduction in seabird mortality in Namibian fisheries following the introduction of bycatch regulation. *Biological Conservation*, 253, 108915. DOI: 10.1016/j.biocon.2020.108915

Derraik, J. G. (2002). The pollution of the marine environment by plastic debris: A review. *Marine Pollution Bulletin*, 44(9), 842–852. DOI: 10.1016/S0025-326X(02)00220-5 PMID: 12405208

Devriese, L. I., van der Meulen, M. D., Maes, T., Bekaert, K., Paul-Pont, I., Frère, L., Robbens, J., & Vethaak, A. D. (2015). Microplastic contamination in brown shrimp (Crangon crangon, Linnaeus 1758) from coastal waters of the Southern North Sea and Channel area. *Marine Pollution Bulletin*, 98(1-2), 179–187. DOI: 10.1016/j.marpolbul.2015.06.051 PMID: 26456303

Diamond, A., & Devlin, C. (2003). Seabirds as indicators of changes in marine ecosystems: Ecological monitoring on Machias Seal Island. *Environmental Monitoring and Assessment*, 88(1/3), 153–181. DOI: 10.1023/A:1025560805788 PMID: 14570414

Dias, M. P., Martin, R., Pearmain, E. J., Burfield, I. J., Small, C., Phillips, R. A., Yates, O., Lascelles, B., Borboroglu, P. G., & Croxall, J. P. (2019). Threats to seabirds: A global assessment. *Biological Conservation*, 237, 525–537. DOI: 10.1016/j.biocon.2019.06.033

Du, Z. H., Xia, J., Sun, X. C., Li, X. N., Zhang, C., Zhao, H.-S., Zhu, S.-Y., & Li, J.-L. (2017). A novel nuclear xenobiotic receptor (AhR/ PXR/CAR)-mediated mechanism of DEHP-induced cerebellar toxicity in quails (Coturnix Japonica) via disrupting CYP enzyme system homeostasis. *Environmental Pollution*, 226, 435–443. DOI: 10.1016/j.envpol.2017.04.015 PMID: 28413083

Einoder, L. D. (2009). A review of the use of seabirds as indicators in fisheries and ecosystem management. *Fisheries Research*, 95(1), 6–12. DOI: 10.1016/j.fishres.2008.09.024

Fabri-Ruiz, S., Baudena, A., Moullec, F., Lombard, F., Irisson, J.-O., & Pedrotti, M. L. (2023). Mistaking plastic for zooplankton: Risk assessment of plastic ingestion in the Mediterranean sea. *The Science of the Total Environment*, 856(2), 159011. DOI: 10.1016/j.scitotenv.2022.159011 PMID: 36170920

Faggio, C., Tsarpali, V., & Dailianis, S. (2018). Mussel digestive gland as a model for assessing xenobiotics: An overview. *The Science of the Total Environment*, 613, 220–229. DOI: 10.1016/j.scitotenv.2018.04.264 PMID: 29704717

Farrell, P., & Nelson, K. (2013). Trophic level transfer of microplastic: Mytilus edulis (L.) to Carcinus maenas (L.). *Environmental Pollution*, 177, 1–3. DOI: 10.1016/j.envpol.2013.01.046 PMID: 23434827

Fazio, F., Faggio, C., Marafioti, S., Torre, A., Sanfilippo, M., & Piccione, G. (2012). Comparative study of haematological profiles on Gobius niger in two different habitat sites: Faro Lake and Tyrrhenian Sea. *Cahiers de Biologie Marine*, 53, 213–219.

Fossi, M. C., Panti, C., Baini, M., & Lavers, J. L. (2018). A review of plastic-associated pressures: Cetaceans of the Mediterranean Sea and Eastern Australian shearwaters as case studies. *Frontiers in Marine Science*, 5, 1. DOI: 10.3389/fmars.2018.00173

Furness, R. W. (1985). Plastic particle pollution: Accumulation by procellariiform seabirds at scottish colonies. *Marine Pollution Bulletin*, 16(3), 103–106. DOI: 10.1016/0025-326X(85)90531-4

Furness, R. W., & Camphuysen, K. (1997). Seabirds as monitors of the marine environment. *ICES Journal of Marine Science*, 54(4), 726–737. DOI: 10.1006/jmsc.1997.0243

Furness, R. W., & Tasker, M. L. (2000). Seabird-fishery interactions: Quantifying the sensitivity of seabirds to reductions in sand eel abundance and identification of key areas of sensitive seabirds in the North Sea. *Marine Ecology Progress Series*, 202, 253–264. DOI: 10.3354/meps202253

Geyer, R., Jambeck, J. R., & Law, K. L. (2017). Production, use, and fate of all plastics ever made. *Science Advances*, 3(7), e1700782. DOI: 10.1126/sciadv.1700782 PMID: 28776036

Gramentz, D. (1988). Involvement of loggerhead turtle with the plastic, metal, and hydrocarbon pollution in the central Mediterranean. *Marine Pollution Bulletin*, 19(1), 11–13. DOI: 10.1016/0025-326X(88)90746-1

Gregory, M. R. (1991). The hazards of persistent marine pollution: Drift plastics and conservation islands. *Journal of the Royal Society of New Zealand*, 21(2), 83–100. DOI: 10.1080/03036758.1991.10431398

Hentati-Sundberg, J., Raymond, C., Sköld, M., Svensson, O., Gustafsson, B., & Bonaglia, S. (2020). Fueling of a marine-terrestrial ecosystem by a major seabird colony. *Scientific Reports*, 10(1), 15455. DOI: 10.1038/s41598-020-72238-6 PMID: 32963305

Hirai, H., Takada, H., Ogata, Y., Yamashita, R., Mizukawa, K., Saha, M., Kwan, C., Moore, C., Gray, H., Laursen, D., Zettler, E. R., Farrington, J. W., Reddy, C. M., Peacock, E. E., & Ward, M. W. (2011). Organic micropollutants in marine plastics debris from the open ocean and remote and urban beaches. *Marine Pollution Bulletin*, 62(8), 1683–1692. DOI: 10.1016/j.marpolbul.2011.06.004 PMID: 21719036

Høiberg, M. A., Woods, J. S., & Verones, F. (2022). Global distribution of potential impact hotspots for marine plastic debris entanglement. *Ecological Indicators*, 135, 108509. DOI: 10.1016/j.ecolind.2021.108509

ICES2018. Report of the Joint OSPAR/HELCOM/ICES Working Group on Marine Birds (JWGBIRD). 1–5 October 2018, Ostende, Belgium. icescm2017/acom:24, 75 pp. (link) 2018

Jackson, J. B. C. (2008). Ecological extinction and evolution in the brave new ocean. *Proceedings of the National Academy of Sciences of the United States of America*, 105(Supplement 1), 11458–11465. DOI: 10.1073/pnas.0802812105 PMID: 18695220

Jamieson, A. J., Brooks, L. S. R., Reid, W. D. K., Piertney, S. B., Narayanaswamy, B. E., & Linley, T. D. (2019). Microplastics and synthetic particles ingested by deep-sea amphipods in six of the deepest marine ecosystems on Earth. *Royal Society Open Science*, 6(2), 180667. DOI: 10.1098/rsos.180667 PMID: 30891254

Jenssen, B. M., Aarnes, J. B., Murvoll, K. M., Herzke, D., & Nygård, T. (2010). Fluctuating wing asymmetry and hepatic concentrations of persistent organic pollutants are associated in European shag (Phalacrocorax aristotelis) chicks. *The Science of the Total Environment*, 408(3), 578–585. DOI: 10.1016/j.scitotenv.2009.10.036 PMID: 19896702

Keith, A. (2002, February). Grasman, Assessing Immunological Function in Toxicological Studies of Avian Wildlife. *Integrative and Comparative Biology*, 42(1), 34–42. DOI: 10.1093/icb/42.1.34 PMID: 21708692

Kingsford, R. T., Watson, J. E. M., Lundquist, C. J., Venter, O., Hughes, L., Johnston, E. L., Atherton, J., Gawel, M., Keith, D. A., Mackey, B. G., Morley, C., Possingham, H. P., Raynor, B., Recher, H. F., & Wilson, K. A. (2009). Major conservation policy issues for biodiversity in Oceania. *Conservation Biology*, 23(4), 834–840. DOI: 10.1111/j.1523-1739.2009.01287.x PMID: 19627315

Kwon, B. G., Saido, K., Koizumi, K., Sato, H., Ogawa, N., Chung, S.-Y., Kusui, T., Kodera, Y., & Kogure, K. (2014). Regional distribution of styrene analogues generated from polystyrene degradation along the coastlines of the North-East Pacific Ocean and Hawaii. *Environmental Pollution*, 188, 45–49. DOI: 10.1016/j.envpol.2014.01.019 PMID: 24553245

Laist, D. W. (1987). Overview of the biological effects of lost and discarded plastic debris in the marine environment. *Marine Pollution Bulletin*, 18(6, Suppl. B), 319–326. DOI: 10.1016/S0025-326X(87)80019-X

Laist, D. W. (1997). Impacts of marine debris: Entanglement of marine life in marine debris including a comprehensive list of species with entanglement and ingestion records. *Marine Pollution Bulletin*, 18(6, Suppl. B), 99–139.

Lascelles, B. G., Langham, G. M., Ronconi, R. A., & Reid, J. B. (2012). From hotspots to site protection: Identifying Marine Protected Areas for seabirds around the globe. *Biological Conservation*, 156, 5–14. DOI: 10.1016/j.biocon.2011.12.008

Lavers, J. L., & Bond, A. L. (2023). Long-term decline in fledging body condition of Flesh-footed Shearwaters (Ardenna carneipes). *ICES Journal of Marine Science*, 0(4), 1–7. DOI: 10.1093/icesjms/fsad048

Lavers, J. L., Bond, A. L., & Huton, I. (2014). Plastic ingestion by flesh-footed shearwaters (Puffinus carneipes): Implications for fledgling body condition and the accumulation of plastic-derived chemicals. *Environmental Pollution*, 187, 124–129. DOI: 10.1016/j.envpol.2013.12.020 PMID: 24480381

Lavers, J. L., Hutton, I., & Bond, A. L. (2019). Clinical pathology of plastic ingestion in marine birds and relationships with blood chemistry. *Environmental Science & Technology*, 53(15), 9224–9231. DOI: 10.1021/acs.est.9b02098 PMID: 31304735

Letcher, R. J., Bustnes, J. O., Dietz, R., Jenssen, B. M., Jørgensen, E. H., Sonne, C., Verreault, J., Vijayan, M. M., & Gabrielsen, G. W. (2010). Exposure and effects assessment of persistent organohalogen contaminants in arctic wildlife and fish. *The Science of the Total Environment*, 408(15), 2995–3043. DOI: 10.1016/j.scitotenv.2009.10.038 PMID: 19910021

Li, P. C., Li, X. N., Du, Z. H., Wang, H., Yu, Z.-R., & Li, J.-L. (2018). Di(2-ethylhexyl) phthalate (DEHP) induced kidney injury in quail (Coturnix Japonica) via inhibiting HSF1/ HSF3-dependent heat shock response. *Chemosphere*, 209, 981–988. DOI: 10.1016/j.chemosphere.2018.06.158 PMID: 30114749

Louzao, M., Becares, J., Rodriguez, B., Hyrenbach, K. D., Ruiz, A., & Arcos, J. M. (2009). Combining vessel-based surveys and tracking data to identify key marine areas for seabirds. *Marine Ecology Progress Series*, 391, 183–197. DOI: 10.3354/meps08124

MacLeod, M., Arp, H. P. H., Tekman, M. B., & Jahnke, A. (2021). The global threat from plastic pollution. *Science*, 373(6550), 61–65. DOI: 10.1126/science.abg5433 PMID: 34210878

Matos, D. M., Ramos, J. A., Bessa, F., Silva, V., Rodrigues, I., Antunes, S., dos Santos, I., Coentro, J., Brandão, A. L. C., Batista de Carvalho, L. A. E., Marques, M. P. M., Santos, S., & Paiva, V. H. (2023). Anthropogenic debris ingestion in a tropical seabird community: Insights from taxonomy and foraging distribution. *The Science of the Total Environment*, 898, 165437. DOI: 10.1016/j.scitotenv.2023.165437 PMID: 37437636

McCarty, J. P., & Second, A. L. (2000). Possible effects of PCB contamination on female plum age color and reproductive success in Hudson River Tree Swallows. *The Auk*, 117(4), 987–995. DOI: 10.1093/auk/117.4.987

McCauley, D. J., Pinsky, M. L., Palumbi, S. R., Estes, J. A., Joyce, F. H., & Warner, R. R. (2015). Marine defaunation: Animal loss in the global ocean. *Science*, 347(6219), 1255641–1255647. DOI: 10.1126/science.1255641 PMID: 25593191

Monclús L, McCann Smith E, Ciesielski TM, Wagner M, Jaspers VLB. (2022). Microplastic Ingestion Induces Size-Specific Effects in Japanese Quail. Environ Sci Technol., 56(22), 15902-15911.

Montevecchi, W. A. (1993). Birds as indicators of change in marine prey stocks. In Furness, W. R., & Greenwood, J. J. D. (Eds.), *Birds as Monitors of Environmental Change* (pp. 217–266). Chapman and Hall. DOI: 10.1007/978-94-015-1322-7_6

Mullen, G. R., & Durden, L. A. (2019). Medical and Veterinary Entomology (Third Edition). Academin Press.

Napper, I. E., Davies, B. F. R., Clifford, H., Elvin, S., Koldewey, H. J., Mayewski, P. A., Miner, K. R., Potocki, M., Elmore, A. C., Gajurel, A. P., & Thompson, R. C. (2020). Reaching new heights in plastic pollution-preliminary findings of microplasties on mount everest. *One Earth*, 3(5), 621–630. DOI: 10.1016/j.oneear.2020.10.020

National Audubon Society. (2012b). *Important Bird Areas Program: A Global Currency for Bird Conservation*. National Audubon Society.

National Oceanic and Atmospheric Administration (NOAA). (2014). NOAA Marine Debris Program. 2014 Report on the Entanglement of Marine Species in Marine Debris With an Emphasis on Species in the United States. (Silver Spring, MD. 28).

Nelson, J. B. (1978). *The Sulidae: Gannets and boobies*. Oxford University Press.

Neumanna, S., Harju, M., & Herzke, D. (2021). Ingested plastics in northern fulmars (Fulmarus glacialis): A pathway for polybrominated diphenyl ether (PBDE) exposure? *The Science of the Total Environment*, 778, 146313. DOI: 10.1016/j.scitotenv.2021.146313 PMID: 33721646

Newman, J., Fletcher, D., Moller, H., Bragg, C., Scott, D., & McKechnie, S. (2009). Estimates of productivity and detection probabilities of breeding attempts in the sooty shearwater (Puffinus griseus). *Wildlife Research*, 36(2), 159–168. DOI: 10.1071/WR06074

Nur, N., Jahncke, J., Herzog, M. P., Howar, J., Hyrenbach, K. D., Zamon, J. E., Ainley, D. G., Wiens, J. A., Morgan, K., Ballance, L. T., & Stralberg, D. (2011). Where the wild things are: Predicting hotspots of seabird aggregations in the California Current System. *Ecological Applications*, 21(6), 2241–2257. DOI: 10.1890/10-1460.1 PMID: 21939058

O'Shea, T. J., & Stafford, C. J. (1980). Phthalate plasticizers: Accumulation and effects on weight and food consumption in captive starlings. *Bulletin of Environmental Contamination and Toxicology*, 25(1), 345–352. DOI: 10.1007/BF01985536 PMID: 7426782

Oehlmann, J., Schulte-Oehlmann, U., Kloas, W., Jagnytsch, O., Lutz, I., Kusk, K. O., Wollenberger, L., Santos, E. M., Paull, G. C., Van Look, K. J. W., & Tyler, C. R. (2009). A critical analysis of the biological impacts of plasticizers on wildlife. *Philosophical Transactions of the Royal Society of London. Series B, Biological Sciences*, 364(1526), 2047–2062. DOI: 10.1098/rstb.2008.0242 PMID: 19528055

OSPAR Commission. (2008). Background Document for the EcoQO on Plastic Particles in Stomachs of Seabirds.

Periyasamy, A. P., & Tehrani-Bagha, A. (2022). A review on microplastic emission from textile materials and its reduction techniques. *Polymer Degradation & Stability*, 199, 109901. DOI: 10.1016/j.polymdegradstab.2022.109901

Peters, K. A., & Otis, D. L. (2007). Shorebird roost-site selection at two temporal scales: Is human disturbance a factor? *Journal of Applied Ecology*, 44(1), 196–209. DOI: 10.1111/j.1365-2664.2006.01248.x

Phillips, R., Fort, J., & Dias, M. (2022). Conservation status and overview of threats to seabirds. In *Conservation of Marine Birds* (pp. 33–56). Elsevier.

Pierce, K. E., Harris, R. J., Larned, L. S., & Pokras, M. A. (2004). Obstruction and starvation associ ated with plastic ingestion in a northern gannet morus bassanus and a greater shearwater Puffinus Gravis. *Marine Ornithology*, 32, 187–189.

Provencher, J. F., Bond, A. L., Avery-Gomm, S., Borrelle, S. B., Bravo Rebolledo, E. L., Hammer, S., Kühn, S., Lavers, J. L., Mallory, M. L., Trevail, A., & van Franeker, J. A. (2017). Quantifying ingested debris in marine megafauna: A review and recommendations for standardization. *Analytical Methods*, 9(9), 1454–1469. DOI: 10.1039/C6AY02419J

Provencher, J. F., Gaston, A. J., & Mallory, M. L. (2009). Evidence for increased ingestion of plastics by northern fulmars (Fulmarus glacialis) in the Canadian Arctic. *Marine Pollution Bulletin*, 58(7), 1092–1095. DOI: 10.1016/j.marpolbul.2009.04.002 PMID: 19403145

Rajpar, M. N., Ozdemir, I., Zakaria, M., Sheryar, S., & Rab, A. (2018). *Seabirds as bioindicators of marine ecosystems*. InTech. DOI: 10.5772/intechopen.75458

Ramasamy, R., Aragaw, T. A., & Balasaraswathi Subramanian, R. (2022). Wastewater treat- ment plant effluent and microfiber pollution: Focus on industry-specific wastewater. *Environmental Science and Pollution Research International*, 29(34), 51211–51233. DOI: 10.1007/s11356-022-20930-7 PMID: 35606585

Raymond, B., Shaffer, S., Sokolov, S., Woehler, E., Costa, D., Einoder, L., Hindell, M., Hosie, G., Pinkerton, M., Sagar, P. M., Scott, D., Smith, A., Thompson, D. R., Vertigan, C., & Weimerskirch, H. (2010). Shearwater foraging in the Southern Ocean: The roles of prey availability and winds. *PLoS One*, 5(6, e10960), e10960. DOI: 10.1371/journal.pone.0010960 PMID: 20532034

Reddy, A. V. B., Moniruzzaman, M., & Aminabhavi, T. M. (2019). Polychlorinated biphenyls (PCBs) in the environment: Recent updates on sampling, pretreatment, cleanup technologies and their analysis. *Chemical Engineering Journal*, 358, 1186–1207. DOI: 10.1016/j.cej.2018.09.205

Robards, M. D., Piatt, J. F., & Wohl, K. D. (1995). Increasing frequency of plastic particles ingested by seabirds in the subarctic north Pacific. *Marine Pollution Bulletin*, 30(2), 151–157. DOI: 10.1016/0025-326X(94)00121-O

Rodríguez, A., Rodríguez, B., & Nazaret Carrasco, M. (2012, October). High prevalence of parental delivery of plastic debris in Cory's shearwaters (Calonectris diomedea). *Marine Pollution Bulletin*, 64(10), 2219–2223. DOI: 10.1016/j.marpolbul.2012.06.011 PMID: 22784377

Roman, L., Hardesty, B. D., Hindell, M. A., & Wilcox, C. (2019). A quantitative analysis linking seabird mortality and marine debris ingestion. *Scientific Reports*, 9(1), 3202. DOI: 10.1038/s41598-018-36585-9 PMID: 30824751

Roman, L., Lowenstine, L., Parsley, L. M., Wilcox, C., Hardesty, B. D., Gilardi, K., & Hindell, M. (2019). Is plastic ingestion in birds as toxic as we think? Insights from a plastic feeding experiment. *The Science of the Total Environment*, 665, 660–667. DOI: 10.1016/j.scitotenv.2019.02.184 PMID: 30776638

Ronconi, R. A., Lascelles, B. G., Langham, G. M., Reid, J. B., & Oro, D. (2012). The role of seabirds in marine protected area identification, delineation, and monitoring: Introduction and synthesis. *Biological Conservation*, 156, 1–4. DOI: 10.1016/j.biocon.2012.02.016

Rowe, C. L. (2008). "The calamity of so long life": Life histories, contaminants, and potential emerging threats to long-lived vertebrates. *A.I.B.S. Bulletin*, 58(7), 623–631.

Ryan, P. G. (2016). Ingestion of plastics by marine organisms. Hazardous Chemicals Associated With Plastics in the Marine Environment 235-238.

Serratosa, J., Hyrenbach, K. D., Miranda-Urbina, D., Portflitt-Toro, M., Luna, N., & Luna-Jorquera, G. (2020). Environmental drivers of seabird at-sea distribution in the Eastern South Pacific Ocean: Assemblage composition across a longitudinal productivity gradient. *Frontiers in Marine Science*, 6, 838. DOI: 10.3389/fmars.2019.00838

Shepherd, P. C. F., & Boates, J. S. (1999). Effects of a commercial baitworm harvest on Semipalmated Sandpipers and their prey in the Bay of Fundy hemispheric shorebird reserve. *Conservation Biology*, 13(2), 347–356. DOI: 10.1046/j.1523-1739.1999.013002347.x

Smith, M. A., Walker, N. J., Free, C. M., Kirchhoff, M. J., Drew, G. S., Warnock, N., & Stenhouse, I. J. (2014). Identifying marine Important Bird Areas using at-sea survey data. *Biological Conservation*, 172, 180–189. DOI: 10.1016/j.biocon.2014.02.039

Susanti, N. K., Mardiastuti, A., & Wardiatno, Y. (2020). Microplastics and the impact of plastic on wildlife: A literature review. *Environmental Earth Sciences*, 528, 012013.

Tanaka, K., van Franeker, J. A., Deguchi, T., & Takada, H. (2019). Piece-by-piece analysis of additives and manufacturing byproducts in plastics ingested by seabirds: Implication for risk of exposure to seabirds. *Marine Pollution Bulletin*, 145, 36–41. DOI: 10.1016/j.marpolbul.2019.05.028 PMID: 31590798

Tanaka, K., Watanuki, Y., Takada, H., Ishizuka, M., Yamashita, R., Kazama, M., Hiki, N., Kashiwada, F., Mizukawa, K., Mizukawa, H., Hyrenbach, D., Hester, M., Ikenaka, Y., & Nakayama, S. M. M. (2020). In vivo accumulation of plastic-derived chemicals into seabird tissues. *Current Biology*, 30(4), 723–728. DOI: 10.1016/j.cub.2019.12.037 PMID: 32008901

Tavares, D. C., de Moura, J. F., Merico, A., & Siciliano, S. (2017). Incidence of marine debris in seabirds feeding at different water depths. *Marine Pollution Bulletin*, 119(2), 68–73. DOI: 10.1016/j.marpolbul.2017.04.012 PMID: 28431744

Teuten, E. L., Saquing, J. M., Knappe, D. R. U., Barlaz, M. A., Jonsson, S., Björn, A., Rowland, S. J., Thompson, R. C., Galloway, T. S., Yamashita, R., Ochi, D., Watanuki, Y., Moore, C., Viet, P. H., Tana, T. S., Prudente, M., Boonyatumanond, R., Zakaria, M. P., Akkhavong, K., & Takada, H. (2009). Transport and release of chemicals from plastics to the environment and to wildlife. *Philosophical Transactions of the Royal Society of London. Series B, Biological Sciences*, 364(1526), 2027–2045. DOI: 10.1098/rstb.2008.0284 PMID: 19528054

Toropova, C., Meliane, I., Laffoley, D., Matthews, E., & Spalding, M. (Eds.). (2010). *Global Ocean Protection: Present Status and Future Possibilities. Brest, France: Agence des aires marines protégées; Gland, Switzerland: IUCN WCPA; Cambridge, UK: UNEP-WCMC; Arlington, USA: TNC; Tokyo, Japan: UNU.* WCS.

Valavanidis, A., Vlahogianni, T., Dassenakis, M., & Scoullos, M. (2006). Molecular biomarkers of oxidative stress in aquatic organisms in relation to toxic environmental pollutants. *Ecotoxicology and Environmental Safety*, 64(2), 178–189. DOI: 10.1016/j.ecoenv.2005.03.013 PMID: 16406578

Van Franeker, J. A. (2004). *Save the North Sea e Fulmar Study Manual 1: Collection and Dissection Procedures. Alterra Rapport 672.* Alterra.

van Franeker, J. A., Blaize, C., Danielsen, J., Fairclough, K., Gollan, J., Guse, N., Hansen, P.-L., Heubeck, M., Jensen, J.-K., Gilles, L. G., Olsen, B., Olsen, K.-O., Pedersen, J., Stienen, E. W. M., & Turner, D. M. (2011). Monitoring plastic ingestion by the northern fulmar (Fulmarus glacialis) in the North Sea. *Environmental Pollution*, 159(10), 2609–2615. DOI: 10.1016/j.envpol.2011.06.008 PMID: 21737191

Van Sebille, E., Wilcox, C., Lebreton, L., Maximenko, N., Hardesty, B. D., van Franeker, J. A., Eriksen, M., Siegel, D., Galgani, F., & Law, K. L. (2015). A global inventory of small floating plastic debris. *Environmental Research Letters*, 10(12), 124006. DOI: 10.1088/1748-9326/10/12/124006

VanderWerf, E. A., & Young, L. C. (2017). *A summary and gap analysis of seabird monitoring in the US Tropical Pacific. Report prepared for the US Fish and Wildlife Service, Region 1.* Pacific Rim Conservation.

Wang, Y., Huang, J., Zhu, F., & Zhou, S. (2021). Airborne microplastics: A review on the occurrence, migration and risks to humans. *Bulletin of Environmental Contamination and Toxicology*, 107(4), 657–664. DOI: 10.1007/s00128-021-03180-0 PMID: 33742221

Watts, A. J. R., Lewis, C., Goodhead, R. M., Beckett, S. J., Moger, J., Tyler, C. R., & Galloway, T. S. (2014). Uptake and retention of microplastics by the shore crab Carcinus maenas. *Environmental Science & Technology*, 48(15), 8823–8830. DOI: 10.1021/es501090e PMID: 24972075

Weimerskirch, H., Pinaud, D., Pawlowski, F., & Bost, C. A. (2007). Does prey capture induce area-restricted search? A fine-scale study using GPS in a marine predator, the wandering albatross. *American Naturalist*, 170(5), 734–743. DOI: 10.1086/522059 PMID: 17926295

Wilcox, C., van Sebille, E., & Hardesty, B. D. (2015). Threat of plastic pollution to seabirds is global, pervasive, and increasing. *Proceedings of the National Academy of Sciences of the United States of America*, 112(38), 11899–11904. DOI: 10.1073/pnas.1502108112 PMID: 26324886

Wilson, R. P., Grémillet, D., Syder, J., Kierspel, M. A. M., Garthe, S., Weimerskirch, H., Schäfer-Neth, C., Scolaro, J. A., Bost, C. A., Plötz, J., & Nel, D. (2002). Remote-sensing systems and seabirds: Their use, abuse and potential for measuring marine environmental variables. *Marine Ecology Progress Series*, 228, 241–261. DOI: 10.3354/meps228241

Wooller, R. D., Bradley, J. S., & Croxall, J. P. (1992). Long-term population studies of seabirds. *Trends in Ecology & Evolution*, 7(4), 111–114. DOI: 10.1016/0169-5347(92)90143-Y PMID: 21235974

Yamashita, R., Takada, H., Fukuwaka, M. A., & Watanuki, Y. (2011). Physical and chemical effects of ingested plastic debris on short-tailed shearwaters, Puffinus tenuirostris, in the North Pacific Ocean. *Marine Pollution Bulletin*, 62(12), 2845–2849. DOI: 10.1016/j.marpolbul.2011.10.008 PMID: 22047741

Yang, Y., Zhang, W., Wang, S., Zhang, H., & Zhang, Y. (2020). Response of male reproductive function to environmental heavy metal pollution in a free-living passerine bird, *Passer montanus.The Science of the Total Environment*, 747, 141402. DOI: 10.1016/j.scitotenv.2020.141402 PMID: 32771794

Young, L. C., Vanderlip, C., Duffy, D. C., Afanasyev, V., & Shaffer, S. A. (2009). Bringing home the trash: Do colony-based differences in foraging distribution lead to increased plastic ingestion in Laysan Albatrosses? *PLoS One*, 4(10), e7623. DOI: 10.1371/journal.pone.0007623 PMID: 19862322

Zaghden, H., Kallel, M., Elleuch, B., Oudot, J., Saliot, A., & Sayadi, S. (2014). Evaluation of hydrocarbon pollution in marine sediments of Sfax coastal areas from the Gabes Gulf of Tunisia, Mediterranean Sea. *Environmental Earth Sciences*, 72(4), 1073–1082. DOI: 10.1007/s12665-013-3023-6

Chapter 12
Microplastic Menace: Ecological Ramifications and Solutions for a Sustainable Future

Manjunatha Badiger
https://orcid.org/0000-0001-8073-0270
NMAM Institute of Technology, Nitte University (Deemed), India

Jose Alex Mathew
A.J. Institute of Engineering and Technology, India

Savidhan Shetty C. S.
https://orcid.org/0009-0007-3032-1736
National Institute of Technology, Karnataka, India

Varuna Kumara
Moodlakatte Institute of Technology, India

Ganesh Srinivasa Shetty
Shri Madhwa Vadiraja Institute of Technology, Bantakal, India

Pratheksha Rai N.
A.J. Institute of Engineering and Technology, India

Mehnaz Fathima C.
Sahyadri College of Engineering and Management, India

ABSTRACT

Microplastics and nanoplastics, particles smaller than 5 millimeters, have induced profound ecological imbalances in aquatic environments, posing threats to habitats, food chains, and organisms through pollution and bioaccumulation. Their capacity to adsorb harmful chemicals raises concerns for aquatic life and human health via contaminated seafood consumption. Furthermore, terrestrial ecosystems are not spared, with soil quality and nutrient cycling impacted by these pollutants. Given their global dispersion through wind and water currents, even remote areas are affected. Addressing these challenges mandates significant actions, including reducing plastic production, improving waste management, and implementing strategies for environmental remediation. Public awareness and education are pivotal

DOI: 10.4018/979-8-3693-3447-8.ch012

for fostering sustainable practices and mitigating the pervasive impact of plastic contamination on ecosystems worldwide.

1. INTRODUCTION

Microplastics are plastic fragments that typically measure one-fifth of a centimeter or less, making it impossible for the naked eye to distinguish them from larger plastics. We can classify them into two groups: the principal microplastics, designed to be as small as possible, such as microbeads in care products, and the secondary microplastics, originating from the disintegration of larger plastic items like bottles, bags, and packaging, among others. We can form these tiny and persistent particles into fibers, bits, pellets, circles, or spheres (Enyoh et al. 2019). Because they are small and almost exist in all places, microplastics can invade ecosystems, which also has dangerous implications for the environment and the lives of species. Every biome, including marine, freshwater, soil, and the atmosphere, most likely contains them. Microplastics pose a significant risk to marine life due to their potential for ingestion, which can lead to discomfort, suffocation, or even death, as well as their ability to trap animals, resulting in entanglement and damage. Furthermore, microplastics have the potential to alter habitats, posing a threat not only to individual species but also to the stability of entire ecosystems. Microplastics can pump dangerous compounds and adsorb them, thus delivering these substances to the environment. Researchers have also discovered that microplastics can infiltrate the human body, settling in human organs and tissues, potentially posing more long-term risks to human health. Thus, it is only possible to provide a deep and comprehensive approach to solving the issue of microplastics by addressing all of the mentioned factors. That is why you should focus on minimizing plastic waste at the source, improving waste management, and designing effective plastic waste handling mechanisms. These should be the key priorities of the efforts undertaken. Thus, it is necessary to recognize that microplastics are a global problem and implement measures that will minimize the impact of these particles and enhance the protection of the environment and human health (Gasperi et al. 2018).

2. ECOLOGICAL RAMIFICATIONS OF MICROPLASTICS

Microplastics have huge implications for freshwater ecosystems because they pollute drinking water and have negative impacts on wildlife and ecosystems. Microplastics can reach these freshwater habitats through direct or indirect pathways, such as water runoff from urban areas, treated sewage, and sedimentation from the atmosphere. Once in freshwater systems, sediments, the water column, and the tissues of some waterborne organisms trap microplastics. This raises issues in the identification of microplastics in tap water, bottled water, and even pristine sources of water at remote locations raising concerns about the quality of drinking water sources. Water pollutants, microplastics, in particular are known to be damaging to organisms living in water. For instance, zooplankton, fish larvae, and bottom-dwelling organisms, including benthic invertebrates, can accidentally swallow the MPs, thinking that they are food particles (Anon n.d.-a). It can also result in physical harm, intestinal stenosis, inferior feeding conversion efficiency, and poor growth and maturation. Also, microplastics may act as vectors of toxic substances, including persistent organic pollutants and heavy metals, which may elevate the bioavailability and toxic effects of pollutants on organisms inhabiting aquatic ecosystems. The constant presence of microplastics also poses a threat to the ecological processes and food chains within the freshwater systems. Microplastic pollution alters the behavioral distribution and abundance of freshwater organisms, as well as the interactions and dynamics of the affected ecosystem and biodiversity. Moreover, we have yet to fully realize the impacts of microplastics on communities and ecosystems inhabiting freshwater ecosystems, which could include changes in nitrogen and phosphorus cycling, a loss of habitat quality, and a reduction in overall ecosystem health. On balance, the experiences of microplastics in freshwater systems indicate that there is a dire need for integrated management measures that would address problems of plastic pollution to conserve the physiochemical qualities of water and the services of aquatic ecosystems to human beings. The effects of microplastics are expected to be damaging to human health from the consumption of affected seafood and drinking water. Various organisms consume these microplastics in marine and freshwater ecosystems, forming a chain of food sources. In that way, humans may experience the direct intake of microplastics via drinking water and consuming fish.

The possible effects of microplastics on human health filtered through the consumption of contaminated seafood and drinking water are relatively anticipated (da Costa 2018). Huge pieces of plastic can break down finally into tiny particles that marine animals can consume from the water and as such become part of the human diet system through taking marine products. Scientific studies have detected microplastics in various kinds of seafood, like fish, molluscs, and crustaceans, suggesting the potential for humans to be in contact with these contaminants. Eating contaminated seafood again presents the possibility of swallowing microplastics along with the edible organ tissues. Although the overall effect of ingesting microplastics has not yet been scientifically determined, some of the likely ways through which they pose a danger to human health are as follows. Microplastics pose a threat that may lead to physical harm of the digestive system or may even hinder it.

Moreover, microplastics possess certain characteristics, including the capacity to accumulate life-threatening substances from their environment, such as persistent organic pollutants and heavy metals. If fish in the sea absorb certain chemicals, these chemical products could potentially enter their gastrointestinal systems, leading to harmful effects. It implies that microplastics are not only present in the seafood we consume, tap water, and bottled water we drink every day. There are studies that have established that microplastics exist in several water sources, raising concern about possible human ingestion of the toxin through drinking water. Fish and other seafood, tap water, and bottled water, commonly used for drinking water, also contain microplastics. Multiple research studies have described how microplastics have been detected in various water sources, raising alarm about the potential for people's ingestion of water containing these particles. As for the second aspect of ingesting microplastics through the intake of drinking water, there is the potential of health risks that are similar to those of consuming seafood, even though this has not been fully outlined.

First and foremost, further investigation of the effects microplastics can have on human health is necessary, yet the challenges affecting research at the basic steps, starting with sample collection through analysis of data, are manifold. Sampling adequate, representative samples from various environmental compartments such as water, sediment, soil, and biota present a major challenge, compounded by the potential for bias or sample loss during fieldwork and sample processing. There is a significant challenge in identifying the extent of microplastics, as samples come in multiple shapes and sizes as well as are made of various polymers and can be sorted only by an expert or with the help of complicated sorting procedures or special analytical tools (Anon n.d.-c). Furthermore, microplastics come in various particle sizes, from micrometers to millimeters, and there be new and smaller particles, such as nanoplastics, which are even tinier, thus, would require higher precision equipment and methods for identification. Quantification of microplastics from samples is a complex process that often encounters issues such as ethical concerns, trace levels of microplastics, heterogeneity, and matrix interference, necessitating the use of standard protocols across research studies. First and foremost, further investigation of the effects microplastics can have on human health is necessary, yet the challenges affecting research at the basic steps, starting with sample collection through analysis of data, are manifold. Sampling adequate, representative samples from various environmental compartments such as water, sediment, soil, and biota present a major challenge, compounded by the potential for bias or sample loss during fieldwork and sample processing. There is a significant challenge in identifying the extent of microplastics, as samples come in multiple shapes and sizes as well as are made of various polymers and can be sorted only by an expert or with the help of complicated sorting procedures or special analytical tools (Randhawa

2023). Furthermore, microplastics come in various particle sizes, from micrometers to millimeters, and there be new and smaller particles, such as nanoplastics, which are even tinier, thus, would require higher precision equipment and methods for identification. Quantification of microplastics from samples is also a complicated exercise, encountering problems such as ethical concerns, trace levels of microplastics, heterogeneity, and matrix interference; thus, the need for standard protocols across research studies. Moreover, identification of chemical constituents of microplastics, their categorization according to the known or potential hazard they present to the environment, their capability to cause harm and analysis of possible consequences call for sophisticated analytical capabilities. A comprehensive overview of physical and biological characteristics of microplastics such as particle size, density, frequency, spatial distribution, and their behaviours and effects on other organisms and ecosystems (Kedzierski et al. 2022). There are no standard protocols or calibration tests for quantitative microplastic identification and quantification, making it difficult to compare findings from various investigations. This is why it's important to improve the quality of the research being done through validation approaches and ring trials. Furthermore, as the field of microplastics develops, other problematic areas like identifying and describing nanoplastics, the impacts of deterioration and their subsequent effects on weathering and assessing the exposure and toxicity of chemicals associated with microplastics remain to be noted due to the dynamism of the research. All of these concerns call for a multi-disciplinary approach, methodological innovations, and continued endeavours to further enhance knowledge about microplastics, their ubiquity, and their impacts. These limitations can be addressed, and through such efforts, researchers can produce sound data that will benefit policymakers and identify ways through which the occurrence of microplastic pollution could be prevented. New technologies sound like science fiction, but they are also providing new solutions for the identification and quantification of microplastic particles. Nanoparticle Tracking Analysis (NTA) effectiveness in quantifying the particles that are present in liquid samples in real time will help to gain knowledge in the movement of micro plastics in water systems(Peters et al. 2022). Due to the typing of the microplastics, RAMAN confirmed the type of polymer present in the sample, with minimal interference forming the basis for classification(Araujo et al. 2018). Using the DNA barcoding technique, researchers map microplastics back to biological entities which reveals information on ecological roles. Hyperspectral imaging enables the use of microplastics imaging without damaging the samples and specimens involved, and machine learning and artificial intelligence improve the data analysis and computational processes. These concepts are used in addition to conventional strategies, deepening the knowledge of microplastic contamination and their elimination tactics.

Microplastics management necessitates regulation and legislation to address the complex issues caused by anthropogenic activities. With regard to the challenges, opportunities, and best practices, it is possible to outline several crucial factors(Khan et al. 2022).

Barriers to successful implementation of effective systems:

1. Complex microplastic sources: Complex microplastic sources include debris, beads from washed-off cosmetics, fine fabrics from washed garments, and microfibers. Extracting these microplastics requires a multi-step strategy to address these challenges (Schymanski et al. 2018).
2. Inadequate understanding of the consequences: Nevertheless, the primary data about the microplastic presence in the environment and its influence on human health and the environment in the living space remain sparse, though awareness of the issue increases day by day. The absence of certain information can impact the formation and enactment of policies.
3. Cross-border nature: The absence of clear national laws regarding microplastics pollution complicates the implementation of any autonomous standards. To achieve their goals, international organizations require good cooperation and partnership in executing their mandates.
4. Lack of Standardized Measurement and Monitoring: At present, there is a minimal standard, let alone specific protocols, for identifying and quantifying microplastics, which raises questions about the credibility of the assessment of levels of pollution and the effectiveness of measures implemented from this standpoint.
5. Technological and Financial Constraints: Modern approaches to dealing with microplastic detection and washing are significant because they fundamentally require the investment of millions of dollars, particularly in regions where funds are insufficient.

(Tuyen n.d.)Opportunities to improve legal frameworks and foster international cooperation:

1. Standardization: Developing standardized procedures for measuring and monitoring microplastics will help enterprises compare and plan their data more efficiently. International organizations, such as the United Nations Environment Programme (UNEP), may play a significant role in creating standards.
2. Integrated Approaches: Some of the strategies that can be implemented to reduce the level of microplastic pollution include implementation of good waste management practices from the time of usage of plastics in industries to recycling and disposal.

3. Product Design and Regulation: When there are manufacturing and utilizing plastic merchandise, including beauty products with microbeads, the right actions that can be taken include the prohibition of the merchandise or encouraging recyclable merchandise that will in one way or another decrease the emission of microplastics.
4. Public Awareness and Education: Researching cases, increasing public awareness of how microplastics affect them, and changing behaviors including banning plastics and encouraging purchasers to buy plastic products, which are made up of recycled plastics, might help in enforcing laws.
5. Research and Innovation: Further research and development suggestions aimed at the management of the issue are the enhancement of the technological resources for microplastics identification and eradication
6. International Agreements and Partnerships: Enhancing existing global treaties, such as the Basel Convention on the Control of Transboundary Movement of Hazardous Wastes and their Disposal and forming new associations may encourage international cooperation in combating microplastic pollution.
7. Capacity Development and Technical Aid: By offering research support and development interventions to the least developed countries, they can be better equipped to regulate microplastic pollution.

Addressing these concerns and their relationship with new actions can help officials establish a strong legislative base and enhance global cooperation in preventing the use of microplastics.

3. TECHNOLOGICAL AND ANALYTICAL METHODS FOR MICROPLASTIC DETECTION

Scientific methods for the identification and estimation of microplastics have recently emerged with noticeable improvements in synthesis investigation in the last few years due to the rising consciousness of microplastics in the environment (Ziani et al., 2023). Here's an overview of various typical approaches, including upcoming technologies: Here's an overview of various typical approaches, including upcoming technologies:

Microscopy Techniques

Optical Microscopy: Stereomicroscopy and fluorescence microscopy aid in the imaging and counting of microplastics in samples.

Scanning Electron Microscopy (SEM): SEM provides photographs of microplastics with relatively high resolutions, which enable the morphological description of the plastic particles.

Spectroscopy Techniques

Fourier transform infrared spectroscopy (FTIR) is the most commonly used method of identifying microplastics based on the spectra of the collected sample's IR transmission.

Raman spectroscopy captures the vibrations of basic molecules, which is useful for studying different types of microplastics.

Chromatography Techniques

The identification of additives and degradation products of microplastics within their chemical structure is a common application of Gas Chromatography-Mass Spectrometry (GC-MS).

Emerging Technologies

Nanoparticle Tracking Analysis (NTA): NTA can follow the motion of nanoparticles, and it allows to measure and sort microplastics in water samples.

DNA barcoding: This method entails isolating and applying DNA sequencing to microorganisms attached to microplastics to establish their origins and elaborate on interactions between microplastics and microbes.

Pyrolysis Gas Chromatography (Py-GC): Py-GC can pinpoint the pyrolysis products of microplastics and offers data on their chemical characteristics and composition of linked groups.

In terms of microplastic analysis, there are some challenges and limitation that are connected with the research.

Microplastic analysis presents challenges and limitations, including:

Sample Preparation: Sample preparation is another problem where microplastics interlink with both organic and inorganic materials, hence making a process lengthy and being at risk of contamination.

Size Range: This might fail to identify all elements in the microplastics spectrum, particularly nanoparticles less than 100 nm in size.

Identification: Discriminating microplastics from other natural and anthropogenic particles is even more challenging with regard to individual microplastics and poly-fragmented samples.

Quantification: Apart from method validation, we are still developing the calibration processes for these programs, as they require well-established protocols to assess the concentration of microplastics in environmental samples.

Cost and accessibility: The complexities of some processes necessitate expensive strategies, specialized instruments, and skills, so they are used sparingly.

In this regard, the emerging technologies are considered to present solutions to the existing difficulties in microplastic unveil and identification through enhanced features in sensitivity, selectivity, and efficiency. Of course, more investment and exploration of this research topic needs to be carried out as a way of establishing the exact measures and effects of microplastic pollution to ecosystems and human beings.

4. POLICY AND REGULATORY FRAMEWORKS FOR MICROPLASTIC MANAGEMENT

Policy and regulations at the global and country levels have addressed microplastic pollution since its recognition as an environmental issue. International organisations such as the United Nations Environmental Programme (UNEP) and international conventions like the Basel Convention are slowly starting to address the issue of microplastics(Usman et al. 2022). The UNEP has continuously raised the alarm on marine plastic debris, including microplastics, through projects like the Global Partnership on Marine Litter and the Clean Seas campaign. Besides, the Basel Convention, which focuses on the management of hazardous waste and transboundary waste movements, has begun to address plastic waste, including microplastics. In 2019, the Basel Convention underwent amendments to establish proscriptive regulations for the transfer of plastic waste between countries. At the national and regional levels, different governments have put in place various laws and policies that deal with microplastics in different forms, such as manufacturing, application, and discharge(Karasik et al. n.d.). These regulations include, for example, the banning or limitation of some microplastics in products such as cosmetics and personal care products, measures to reduce plastic waste generation, and the enhancement of proper waste disposal mechanisms. However, there are several barriers that hinder civil society from implementing and enforcing policies on the management of microplastics.

1. Complex supply chains: Numerous commodities and substances incorporate microplastics, making the production, utilisation, and elimination of these particles challenging.
2. Lack of Standardised Methods: Probably due to the relatively recent discovery of microplastics, there are no universally acceptable methodologies to quantify and assess the impact of microplastics.
3. Global Trade: Because of the international nature of the plastics supply chain and waste exportation as a form of trade, enacting laws presents challenges.
4. Enforcement: A number of factors, such as inadequate funds, low public compliance, and reluctance from industries that the regulation is likely to negatively impact, often make the implementation of regulations a challenging process.
5. Scientific Uncertainty: Even in the case of exposed microplastics, there are scientifically grey areas as to where they originate and where they end up, and the effect they have on the ecosystems, and this hampers effort towards policy-making.

Despite these challenges, there are opportunities for strengthening regulatory frameworks and promoting international cooperation: Despite these challenges, there are opportunities for strengthening regulatory frameworks and promoting international cooperation:

1. Standardisation of criteria: Implementing uniform rules and protocols for the identification and evaluation of microplastics may improve the capacity to compare data with other monitored regions and increase the effectiveness of regulatory measures.
2. Capacity Building: Assisting governments mainly those in the developing nations in improving on the kind of abilities they have in implementing and enforcing such required rules.
3. Public Awareness and Engagement: Improving the public's understanding of micro-plastics in the environment is crucial for garnering support for legislative action towards positive change.
4. Collaborating with industry: In order to build comprehensive corporate social responsibility (CSR) programmes and making commitments to reduce emissions may bolster the efforts of authorities and laws in combating microplastic pollution.
5. International Cooperation: It is suggested that the Global Partnership for Phasing Out Microplastics, UNEP, and international treaties pooling system-based approaches to exchange information, build capacity, and cooperate on managing microplastics.

By addressing and managing these challenges and perceiving opportunities for cooperation, the policymakers might try to achieve the goal of developing an integral set of legislative measures in order to deal with the problem of microplastics and initiate the processes of expectable and effective utilization of a sustainable future (Onyena et al. 2022)

5. MITIGATION AND REMEDIATION STRATEGIES

Measures adopted in the fight against microplastic pollution entail prevention measures, measures of managing the already existing plastic waste and measures that seek to rehabilitate affected ecosystems. Here's a breakdown of these strategies:

1. Prevention Measures: Bans on Single-Use Plastics: Evaluating and subsequently imposing the outright ban or limitation on the use of single use plastics or certain product categories such as the bags, straws or cutlery for instance will greatly assist in minimizing on generation of micro plastics right from their source.
    1. Product Design Innovations: The usage of eco-friendly substitutes and the advocacy for green product development help curb the employments of plastics and lessen their invasion to the environment.
    2. Plastic Waste Management and Recycling: Promoting Recycling: Recycling programs, proper disposal, sensitisation, and creation of proper disposal points can create a recycling channel and hence reduce plastic waste which otherwise can end up in the environment.

Circular Economy Principles: An aspect of the circular economy that has revolutionised many economies is policies like extended producer responsibility policies, which make manufacturers use their skills to produce products that are either reusable or recyclable.

3. Cutting-edge Technologies for the Capture and Elimination of Microplastics: Water Filtration Systems: The viability of microplastic elimination from waste-water and surface water may be achieved through microfiltration and ultrafiltration membrane technology which are far efficient than the membranes in water treatment plants.

Sediment Dredging Techniques: Sediment dredging and remediation methods that can be applied in water bodies include; these help in eradicating loosely compacted microplastics which are often released in water bodies.

4. Restoration Approaches for Contaminated Ecosystems: Habitat Restoration: Replanting the ecosystems like the coastal wetlands and mangroves can make the ecosystem stronger to capture and retain the microplastics in the natural processes.

Bioremediation: Microbial degradation of plastics including the use of microorganisms to break down plastics in the affected environments is deemed as a viable option in the long-term elimination of microplastic pollution.

Thus, to address microplastic pollution it is necessary to apply the package approach implying regulatory impacts, technological solutions, and environmentally friendly approaches. Through measures aimed at stopping the emission of microplastics, raising awareness of plastic waste disposal and recycling, using new technologies of microplastics removal, and supporting the restoration of ecosystems, stakeholders can help minimize the negative effects of microplastics on the environment and build the earth's sustainable future(Kibria et al. 2023).

6. PUBLIC AWARENESS AND EDUCATION INITIATIVES

Increasing people's knowledge about microplastics in their environment is vital to encouraging consciousness, changing behaviour, and advancing community participation towards combating this enemy of the environment (Baechler et al., 2024). Public awareness serves several important purposes: Public awareness serves several important purposes:

1. Educating the public: The general public is often unaware of just how severe the problem of microplastic pollution really is in terms of the environment and the effects it has on people (Henderson & Green, 2020). Awareness-making engages the individuals and gives them a rough view of the problem and its impacts.
2. Empowering Individuals: This enlightens the public, hence embracing change in their diets, methods of disposing waste, and policies pertaining to the control of microplastic pollution.
3. Changing People's Behaviour: Raising knowledge of the people to change their behaviour for the better and healthy society by using less plastic and throwing away trash in the right way. This helps to clean up the mess that plastic leaves.
4. Building Support for Policy Action: Enhanced public awareness may engage citizens' support for policy factors and regulation of microplastic pollution, which may prompt policymakers to commit necessary actions.

The outreach programmes and education section also have a critical role to play in the promotion of change in people's behaviour and moderation of their tendency to consume.

1. School curricula and educational programs: Education that encompasses information about issues with plastics and how to avoid them is an effective way to end the problem since it will create a culture among the young people who will effectively practice sustainability.
2. Organising workshops, seminars, and community activities concerning microplastic pollution affects the residents and prompts them to participate in combating the problem at the household level.
3. Online Resources and Campaigns: Using websites, social networking sites, and other online tools ensures that the information gets to a larger audience, and there is always a way of keeping in touch with the public.

Cooperating with people-influenced governments, non-governmental and international organisations, industries, and academia is crucial for the establishment of sound communication strategies (Rajabi et al., 2021).

1.Multi-Stakeholder Partnerships: Assembling the members from different fields is helpful in creating unified and elaborate strategies of communication to use with the success of various audiences.

2. Utilising Diverse Expertise: Engaging scientists, communication specialists, educators, and community leaders in the communication process guarantees that the information being delivered is grounded in research, respects the culture of the target group, and is culturally sensitive in reaching out to everyone.
3. Tailoring Messages: Europeanization of messages makes messages more applicable to the specific targeted demography as well as the cultural setting, as this increases their capacity to bring about desired practices and paideia.

Engagement successes and other initiatives are helpful when planning more effective future engagement. The Ocean Cleanup by Boyan Slat have made people from across the world considerate of this vice through inventions and the massive clearing of plastic in the seas. Such popular initiatives as Plastic Free July offer guides, recommendations, and people's support for minimising the usage of plastic and the production of waste. Removal of plastic litter from the beaches and riversides by volunteers or through local groups does not only remove plastic pollution but also helps create awareness among the public about the menace at a local level(van Emmerik and Schwarz 2020).

All in all, it is imperative to increase environmental consciousness and engage in public education to solve the problem of microplastic pollution. Developing communities, enlightening the masses, and empowering the populace for change will ensure that there are efforts put in place to ensure that there is limited use of plastics in the environment, hence improving the quality of the ecosystem. Action coordination with other actors and appealing to best practices can also increase the efficiency of these activities.

7. RESEARCH DIRECTIONS AND FUTURE OUTLOOK

Understanding the consequences and further destiny of microplastics in the environment becomes critical for designing measures that will help solve problems and create a future that will be safe for the Earth. Areas for further research include:

1. Ecological Impacts: Researching the effects of microplastics on different levels of biological organization in aquatic environments, as well as the consequences of microplastic pollution on species composition, trophic structure, and ecosystem processes.
2. Human Health Effects: Utilizing the environmental-matrix methods it intends to examine the possible routes of microplastics entering a human body and accumulating there; the possible consequences for human health.
3. Environmental Fate and Transport: Understanding the movement, transformation, eventual encounter, and storage of microplastics in water, soil, and air.
4. Sources and Pathways: Detailing and qualifying the origins and channels of microplastics (e.g., agricultural, industrial, or municipal), and so on, estimating their relative contribution.
5. Cumulative Impacts: Identifying links between the complexity of microplastics pollution and other stressors like climate change, chemical pollution, and habitat alterations. In this context, the inter-disciplinary approach plays a crucial role in solving the real problem of microplastic pollution. The interdisciplinary approach combines the knowledge of ecology, chemistry, engineering, and social sciences in order to study microplastic pollution and its impact on society as a whole (Belontz et al., 2019). Professional involvement in their respective fields ensures the synthesis of new technologies, policies, and behavioral changes that will help to address the issue of microplastic pollution in the most efficient way possible. Microplastics research presents opportunities for innovation and technological development, including:

1. Advanced Detection Methods: Fine and selective analysis of microplastics in water, sediment, soil, and biota.
2. Mitigation Technologies: We are developing new efficient methods for collecting and eradicating microplastics from the environment. These methods could include new filters, biodegradable polymers, and microorganisms that consume microplastics.
3. Remediation Strategies: Applying methods of eliminating microplastic pollution in the source areas, including coastal regions, rivers, and some urban locations, through practices like sediment removal, bioremediation, and restoration of the ecological systems.
4. Risk Assessment Tools: Additional formulating and creating the approaches for the evaluation of the impacts of microplastics on ecosystems and human health, taking into account the tendencies of exposure, accumulation, and toxicity.

Major and lasting solutions to addressing the microlastics menace and the proposed improvements to populations' long-term quality of life involve a systemic approach that requires global, regional, and local collaboration. Key components of such strategies include:

1. Policy and Regulation: This can be done through the effective implementation of specific and strict rules that may include restrictions on the use, reuse, and recycling of plastics. The activation of principles of sustainable consumption and production and enhancing waste management.
2. Education and Outreach: To increase public knowledge of the effects of microplastics and enhance the community's environmentally friendly attitude through increased participation.
3. Innovation and Research: focusing on the funding for conducting research and launching innovations aimed at creating efficient technologies, measures, and other solutions concerning the problem of microplastic pollution and the impact it has on the environment and people's health.
4. International Collaboration: coordinating with different countries, groups, and institutions around the world to share information, experiences, powerful methods, and plans for solving the problem in the most complex way possible.

Consequently, a world without microplastics can only be obtained when scientists, administration, representatives of producing industries, and citizens all combine their forces and initiatives. With further development of scientific investigations, extension of inter- and multidisciplinary cooperation, encouragement of new ideas and perspectives, and integration of long-term approaches, it is possible to start equalising the effects of microplastic pollution and maintaining the earth's health and unity for generations to come. The research into microplastics has covered a plethora of information about what microplastics are, what they are made of, where they emanate from, and the global prevalence of microplastics. The smallest type of plastic pollutants is in the form of microplastics, which have entered many ecosystems, including seas, freshwater bodies, soil, and air, with severe effects on the ecosystems, making them one of the most important global environmental issues. It has, for instance, investigated the impacts of the substances on the marine, terrestrial, and freshwater ecosystems in a systematic manner. These concerns relate to the uptake of these substances by marine organisms, pollution of soils, and concern about their impact on human health through their influence on seafood and water resources. This report conducts a comprehensive review of the various methods of microplastic sampling and analysis currently in use, reflecting the challenges and upcoming issues facing this sector. Furthermore, the analysis of policy and regulation trends at the international, national, and regional levels has identified the challenges of cross-border implementation and opportunities for improving the collective effort. Contemporary environmental challenges include microplastics pollution, prompting the employment of a variety of strategies such as prevention, futuristic technology, and restoration solutions to systematically and effectively address the problem. As a broader area of public health, we have emphasized the significance of traits such as public awareness and education programmes for the development of behaviours and increasing societal participation. It is worth pointing out that we highlight fully-fledged campaigns and initiatives in order to discuss them and potentially replicate them. Understanding the core areas to focus on, methods involving multiple disciplines, and opportunities for creative collaboration are critical for gaining a perspective on the microplastics-free future(Lusher et al. 2021). Such steps will assist in a more effective appreciation and subsequent control of the problem.

8. CONCLUSION

The presence of microplastics surrounds the environment and infiltrates human lives with significant negative consequences for both the biosphere and people's health, which requires urgent and concerted effort at the international, continental, national, and local levels. As a result, this research has highlighted how microplastics are present in virtually all ecosystems and the harm they cause to marine, terrestrial, and freshwater environments. Moreover, it has sparked a growing concern about the detrimental impacts of food and water pollution on human health, particularly when individuals consume contaminated fish and water. Despite significant advancements in various methods of microplastic identification and analysis, which hold significant potential for microplastic distribution and its environmental impacts, fundamental challenges persist. Meticulous routines and protocols, economical processes, and collaborative effort among professionals are mandatory to solve the challenges of microplastic pollution. Government policies and laws play a central role in controlling pollution by providing measures for sustainable production and consumption systems that reduce the use of microplastics. Although there are international treaties and agreements to work together, improvements in enforcement mechanisms and cooperation are vital to dealing with this type of problem. Measures that focus on prevention and compensation related to microplastics offer potential solutions for preventing microplastic pollution and restoring contaminated areas. Thus, the measures in question imply primary prevention, waste reduction management, and scientific approaches for the elimination of microplastics. Raising the public's awareness of the sustainable consumption topic and educating communities and industries are critical for boosting sales of more sustainable products and discouraging the purchase of unsustainable ones. Subsequent research should focus on further studies employing interdisciplinary methodologies to enhance the availability of knowledge on microplastics' processes and identify effective ways of reducing their detrimental effects on the environment. This is due to the fact that only cooperation, along with appropriate and harmonised policies, can enhance the ability of global society to ensure a future free from microplastic pollution and to safeguard the health of ecosystems and humans.

In conclusion, the growing issue of microplastics necessitates the critical collaboration of governments, industries, CSOs, and universities. To enact the solution, there will be extensive synergy and collaboration with various governmental units, continuous advancement in technologies, scientific findings, and regular international education. Thus, we can prevent the problem of microplastic pollution and retrieve ecosystems for numerous living organisms' support.

REFERENCES

Araujo, C. F., Nolasco, M. M., Ribeiro, A. M. P., & Ribeiro-Claro, P. J. A. (2018). Identification of microplastics using Raman spectroscopy: Latest developments and future prospects. *Water Research*, 142, 426–440. DOI: 10.1016/j.watres.2018.05.060 PMID: 29909221

Baechler, B. R., De Frond, H., Dropkin, L., Leonard, G. H., Proano, L., & Mallos, N. J. (2024). Public awareness and perceptions of ocean plastic pollution and support for solutions in the United States. *Frontiers in Marine Science*, 10, 1323477. DOI: 10.3389/fmars.2023.1323477

Belontz, S. L., Corcoran, P. L., Davis, H., Hill, K. A., Jazvac, K., Robertson, K., & Wood, K. (2019). Embracing an interdisciplinary approach to plastics pollution awareness and action. *Ambio*, 48(8), 855–866. DOI: 10.1007/s13280-018-1126-8 PMID: 30448996

da Costa, J. P. (2018). Micro- and nanoplastics in the environment: Research and policymaking. *Current Opinion in Environmental Science & Health*, 1, 12–16. DOI: 10.1016/j.coesh.2017.11.002

Enyoh, C. E., Verla, A. W., Verla, E. N., Ibe, F. C., & Amaobi, C. E. (2019). Airborne microplastics: A review study on method for analysis, occurrence, movement and risks. *Environmental Monitoring and Assessment*, 191(11), 668. DOI: 10.1007/s10661-019-7842-0 PMID: 31650348

Gasperi, J., Wright, S. L., Dris, R., Collard, F., Mandin, C., Guerrouache, M., Langlois, V., Kelly, F. J., & Tassin, B. (2018). Microplastics in air: Are we breathing it in? *Current Opinion in Environmental Science & Health*, 1, 1–5. DOI: 10.1016/j.coesh.2017.10.002

Henderson, L., & Green, C. (2020). Making sense of microplastics? Public understandings of plastic pollution. *Marine Pollution Bulletin*, 152, 110908. DOI: 10.1016/j.marpolbul.2020.110908 PMID: 32479284

Karasik, R., Vegh, T., Diana, Z., Bering, J., Caldas, J., Pickle, A., Rittschof, D., & Virdin, J. (n.d.). *20 Years of Government Responses to the Global Plastic Pollution Problem.*

Kedzierski, M., Palazot, M., Soccalingame, L., Falcou-Préfol, M., Gorsky, G., Galgani, F., Bruzaud, S., & Pedrotti, M. L. (2022). Chemical composition of microplastics floating on the surface of the Mediterranean Sea. *Marine Pollution Bulletin*, 174, 113284. DOI: 10.1016/j.marpolbul.2021.113284 PMID: 34995887

Khan, N. A., Khan, A. H., López-Maldonado, E. A., Alam, S. S., López López, J. R., Méndez Herrera, P. F., Mohamed, B. A., Mahmoud, A. E. D., Abutaleb, A., & Singh, L. (2022). Microplastics: Occurrences, treatment methods, regulations and foreseen environmental impacts. *Environmental Research*, 215, 114224. DOI: 10.1016/j.envres.2022.114224 PMID: 36058276

Kibria, M., Masuk, N. I., Safayet, R., Nguyen, H. Q., & Mourshed, M. (2023). Plastic Waste: Challenges and Opportunities to Mitigate Pollution and Effective Management. *International Journal of Environmental Research*, 17(1), 20. DOI: 10.1007/s41742-023-00507-z PMID: 36711426

Lusher, A. L., Hurley, R., Arp, H. P. H., Booth, A. M., Bråte, I. L. N., Gabrielsen, G. W., Gomiero, A., Gomes, T., Grøsvik, B. E., Green, N., Haave, M., Hallanger, I. G., Halsband, C., Herzke, D., Joner, E. J., Kögel, T., Rakkestad, K., Ranneklev, S. B., Wagner, M., & Olsen, M. (2021). Moving forward in microplastic research: A Norwegian perspective. *Environment International*, 157, 106794. DOI: 10.1016/j.envint.2021.106794 PMID: 34358913

Microplastics in environment: Global concern, challenges, and controlling measures—PMC. (n.d.). Retrieved July 1, 2024, from https://www.ncbi.nlm.nih.gov/pmc/articles/PMC9135010/

Onyena, A. P., Aniche, D. C., Ogbolu, B. O., Rakib, M. R. J., Uddin, J., & Walker, T. R. (2022). Governance Strategies for Mitigating Microplastic Pollution in the Marine Environment: A Review. *Microplastics*, 1(1), 1. Advance online publication. DOI: 10.3390/microplastics1010003

Peters, R. J. B., Relou, E., Sijtsma, E. L. E., & Undas, A. K. (2022). *Evaluation of Nanoparticle Tracking Analysis (NTA) for the Measurement of Nanoplastics in Drinking Water*. DOI: 10.21203/rs.3.rs-1809144/v1

Rajabi, M., Ebrahimi, P., & Aryankhesal, A. (2021). Collaboration between the government and nongovernmental organizations in providing health-care services: A systematic review of challenges. *Journal of Education and Health Promotion*, 10(1), 242. DOI: 10.4103/jehp.jehp_1312_20 PMID: 34395679

Randhawa, J. S. (2023). Advanced analytical techniques for microplastics in the environment: A review. *Bulletin of the National Research Center*, 47(1), 174. DOI: 10.1186/s42269-023-01148-0

Schymanski, D., Goldbeck, C., Humpf, H.-U., & Fürst, P. (2018). Analysis of microplastics in water by micro-Raman spectroscopy: Release of plastic particles from different packaging into mineral water. *Water Research*, 129, 154–162. DOI: 10.1016/j.watres.2017.11.011 PMID: 29145085

Tuyen, P. T. T. (n.d.). *International legal framework for ocean plastic pollution prevention and enforcement in Vietnam.*

Usman, S., Abdull Razis, A. F., Shaari, K., Azmai, M. N. A., Saad, M. Z., Mat Isa, N., & Nazarudin, M. F. (2022). The Burden of Microplastics Pollution and Contending Policies and Regulations. *International Journal of Environmental Research and Public Health*, 19(11), 6773. DOI: 10.3390/ijerph19116773 PMID: 35682361

van Emmerik, T., & Schwarz, A. (2020). Plastic debris in rivers. *WIREs. Water*, 7(1), e1398. DOI: 10.1002/wat2.1398

Ziani, K., Ioni ă-Mîndrican, C.-B., Mititelu, M., Neac u, S. M., Negrei, C., Moro an, E., Drăgănescu, D., & Preda, O.-T. (2023). Microplastics: A Real Global Threat for Environment and Food Safety: A State of the Art Review. *Nutrients*, 15(3), 617. DOI: 10.3390/nu15030617 PMID: 36771324

Chapter 13
Policy and Regulatory Approaches to Mitigating Micro- and Nano Plastic Pollution

Mohit Yadav
https://orcid.org/0000-0002-9341-2527
O.P. Jindal Global University, India

Ashutosh Pandey
https://orcid.org/0000-0002-8255-8459
FORE School of Management, New Delhi, India

Xuan-Hoa Nghiem
https://orcid.org/0000-0003-2292-0257
International School, Vietnam National University, Hanoi, Vietnam

ABSTRACT

Microplastic and nano-plastic pollution poses a significant global challenge. This chapter examines the sources, pathways, and impacts of these contaminants. It explores existing policies, regulations, and treatment technologies while highlighting the need for a comprehensive approach. Prevention strategies, including design modifications, consumer education, and waste management improvements, are emphasized. Building trust and cooperation among stakeholders is crucial for effective governance and implementation. Continued research, focusing on nano-plastics, human health, and environmental impacts, is essential. By combining policy, technology, and stakeholder engagement, it is possible to mitigate microplastic and nano-plastic pollution and protect ecosystems.

DOI: 10.4018/979-8-3693-3447-8.ch013

INTRODUCTION

This century has experienced an unprecedented proliferation of plastic, a material that revolutionized industries while posing a formidable threat to our planet. While the detrimental impacts of macroscopic plastic waste have raised wide attention, a more insidious challenge is emerging: the omnipresence of microplastics and nano-plastics.

Microplastics—plastic particles smaller than 5 millimeters—and nano-plastics—less than 100 nanometers in size—blanket the environment. These small particles have a wide variety of sources: from the degradation of larger plastic items to industrial processes and personal care products. After their release, they start their risky journey through air, water, and soil into the ecosystem, going so far as to present a potential risk for human health.

The impact of microplastic and nano-plastic pollution in general is a broad and multifaceted issue. This is because marine ecosystems are prone to such due to the fact that particles are ingested by most organisms, starting from the simplest ones, therefore disrupting food webs and impairing reproductive capabilities (Kurniawan et al., 2023). Soil in terrestrial environments is contaminated with microplastics, and this can probably have an impact on the growth of plants and human supplies of food. Most worrying is the ability of microplastics and nano-plastics to vector hazardous chemicals and pathogens, raising concerns about their role in disease transmission.

This is a global crisis and requires a global, well-coordinated, comprehensive response. The first efforts to handle plastic wastes have focused on macroscopic wastes; this nature of microplastics and nano-plastics requires changing focus and nuance. In this, the paper endeavors to delve into the complexities of microplastic and nano-plastic pollution and look through the sources, pathways, and impacts. It will work to identify and analyze existing policy and regulatory frameworks in place to address the problem and review the effectiveness of the same. Propose mitigation and prevention strategies to put into operation. It takes cognizance of the magnitude of the challenge at hand and paves ways for robust solutions toward a goal where microplastics and nano-plastics do not pose threats to our planet and its inhabitants.

Sources and Pathways of Micro and Nano Plastic Pollution

Microplastics and nano-plastics come from a variety of direct primary sources, most of which are associated with human activities.

- **Industrial Processes:** Industrial processes are connected with large inputs of microplastic pollution. Microplastics become the by-products of production (Hettiarachchi & Meegoda, 2023). Manufacturing and washing connected with synthetic fibers—say, in the cases of textiles and fishing nets—make microplastic fibers enter into the environment. Plastic pellet production can result in inadvertent spills onto land or water, which eventually contaminate the environment.

- **Tire Wear:** Tire abrasion from vehicles is one of the largest sources of microplastics. When tires wear out, very small particles of the rubber that contain plastic additives remain and eventually find their way into waterways and the soil.
- **Personal Care Products:** Many personal care products, including facial scrubs, toothpaste, and shampoo contain microplastic beads that get washed down drains and into eventual environmental circulation. Again, their use has been banned in some countries, but their legacy continues as they persist in the environment.
- **Paints and Coatings:** Paints and coatings containing plastic additives, during their application, weathering, or removal, can release microplastics. This has the potential to introduce microplastics into the aquatic environment through stormwater runoff and atmospheric deposition.

Secondary Sources of Microplastics and Nano-Plastics

The secondary sources of microplastics and nano-plastics are derived or resulted from larger plastic items.

- **Plastic Debris:** Larger plastic items, bottles, bags, food packaging, etc., undergo physical, chemical, and biological breakdown into smaller fragments. Sunlight, winds, wave actions, etc. cause fragmentation of plastics and finally result in micro- and nano-plastics (Kumar et al., 2021).
- **Plastic Additives:** Most plastic products normally contain additives to enhance their properties, among them plasticizers, stabilizers, and flame retardants. All these additives can leach from plastic and contribute to microplastic and nano-plastic pollution.
- **Microbial Degradation:** Plastics may be degraded through the action of microorganisms, leading to the generation of microplastics and probably nano-plastics. Natural microorganism actions could act in a catalytic way in the decomposition of plastic wastes.

Pathways of Microplastic and Nano-Plastic Pollution

Microplastics and nano-plastics are moved along different pathways, reaching various environmental compartments.

- **Aquatic Environments:** Waterways, such as rivers, lakes, and oceans, present major sinks for microplastics and nano-plastics (da Costa, 2018). The particles are discharged through stormwater runoffs, wind, and wastewater discharges. After their deposition in aquatic media, those particles could be ingested by organisms, accumulated in sediments, or transported over long distances by ocean currents.
- **Terrestrial Environments:** Contamination of soils occurs by the following: agricultural practice, wastewater irrigation, and atmospheric deposition. Ingestion by soil organisms exposes soil health and affects ecosystem functioning.
- **Atmosphere:** By wind, microplastics and nano-plastics can be transported over long distances through the atmosphere into remote areas and finally deposit on land or water.

Information on the sources and pathways of microplastic and nano-plastic pollution is critically needed to inform the design of effective prevention and mitigation strategies. Unmasking primary contributors, understanding how particles move through an environment helps policy decision-makers and researchers to target interventions that reduce impacts from microplastic and nano-plastic pollution (Kumar. M et al, 2023).

Challenges in Monitoring and Assessment

One of the major challenges in assessing microplastic and nano-plastic pollution regards their accurate extent and impact evaluation. Because of their small size and diversified composition, these are set against complex environmental matrices that require very sophisticated analytical techniques and rigorous methodologies.

Analytical Challenges

- **Detection and Quantification**: Because of the minute size of microplastics and nano-plastics, technical detection and quantification have been very challenging. While microscopy techniques are suitable for big-sized microplastics, the smaller ones require specialized instruments like scanning electron microscopy and Fourier-transform infrared spectroscopy. These techniques, however, are time-consuming and most of them require expert interpretation.
- **Sample Preparation:** Extraction of microplastics and nano-plastics from environmental samples is tricky due to organic matter, salts, and other contaminants. Proper analysis, therefore, requires fast and reproducible sample preparation protocols.

- **Distinguishing Plastics from Other Particles:** Since naturally occurring particles, like fragments of minerals or remains of organic origin, can be easily mistaken for microplastics, it requires much more advanced analytical techniques with care in interpretation to avoid overestimation of microplastic abundance.
- **Standardization:** The lack of standardization in sampling, extraction methods, and analytical methods presents an incomparability problem among various studies. There is a need for a standardized protocol so that the data generated will be reliable and comparable.

Data Gaps and Inconsistencies

- **Global disparities:** In regard to microplastics and nano-plastics, countries, and even regions, have an enormous gap in monitoring (Meegoda & Hettiarachchi, 2023). High-income countries usually have infrastructures and resources that can conduct monitoring; from developing countries, data are very limited.
- **Data comparability:** Due to the use of diverse methods for sampling, analytics, and formats for reporting data from such research studies, it makes the integration and comparison of data really challenging in setting up global trends or hotspots of microplastic pollution.
- **Lack of Historical Data**: The case of microplastic and even nano-plastic pollution is established only recently. Therefore, in general, historically comparative data is lacking, which constrains understanding the long-term trends and impacts.

The Challenge of Nano-Plastics

Nano-plastics are of even greater challenges to monitoring and assessment due to their very small size. Most analytical techniques lack the sensitivity to detect and quantify nano-plastics properly. Moreover, there is a possibility that interaction of nano-plastics with other constituents in the environment may further complicate the analysis.

To overcome these challenges, we need to conduct relentless research and development, pursue cooperation between scientists and policy actors, and put in place global monitoring networks. Investment in advanced analysis techniques can enhance a great deal of extra analytical capacity and standardized protocols so that we may understand to an enhanced level the extent and impacts of microplastic and nano-plastic pollution.

Policy and Regulatory Frameworks

The global response to microplastic and nano-plastic pollution is still in its nascent stages, patching out a quilt of policies and regulations across a host of jurisdictions. While the general framework for plastic pollution is slowly developing, measures against microplastics and nano-plastics are very often scattered and varied.

International Agreements and Conventions

While concern over microplastic and nano-plastic pollution is growing, there is to date no dedicated international treaty specifically addressing such pollutants. Nevertheless, a number of existing international agreements provide at least a partial framework for the tackling of plastic pollution more generally.

- **Basel Convention on the Control of Transboundary Movements of Hazardous Wastes and Their Disposal:** The Basel Convention has essentially been concerned with hazardous wastes; however, it played a vital role in controlling transboundary movement regarding plastic wastes, including scraps of certain types of plastics.
- **London Protocol to the Convention on the Prevention of Marine Pollution by Dumping of Wastes and Other Matter:** This protocol is explicitly oriented against the dumping of all forms of plastic waste at sea, including microplastics and nano-plastics.
- **Marpol Annex V:** This is an international convention controlling the discharge of garbage by ships at sea. While this principally tries to deal with the issue of marine litter, it has a bearing on microplastic pollution.

While these agreements provide a foundation for addressing plastic pollution, their scope is very limited with respect to targeting microplastics and nano-plastics.

Regional and National Policies

Responding to increased public concern and scientific evidence, many countries and regions have proposed a number of policies and regulations aimed at reducing plastic pollution. By scope and stringency, they very significantly differ (Meegoda & Hettiarachchi, 2023).

- **Plastic bag bans:** Probably one of the most common ways this could be done is through a ban on single-use plastic bags or by adding a fee to use them. This would somehow affect the problem of macroscopic plastic waste, although in an indirect way it could also help decrease the extent of microplastic pollution by reducing the total volume of plastic in circulation.

- **Extended Producer Responsibility:** Many countries have introduced EPR schemes for plastic packaging, making producers responsible for the final disposal of plastic at its end-of-life stage. This can drive innovation in package design and encourage recycling.
- **Microbead Bans:** More and more countries have banned the use of microplastic beads in personal care products as a direct contribution to microplastic pollution in aquatic environments (Das, R. K et al., 2021).
- **Deposit-Return Systems:** Several governments have enacted deposit-return systems on beverage containers to enhance the recycling rate and contribute towards reducing plastic litter.

These developments are most welcome and almost invariably oriented to the prevention of new microplastics entering the environment, not existing contamination.

Challenges and Gaps

Notwithstanding all the negative impacts on health and the environment, policies and regulations aimed at tackling microplastic and nano-plastic pollution are hindered by several challenges.

· **Insufficient Data:** Sources, pathways, and impacts of microplastics and nano-plastics are not completely known, hence, a target-oriented policy formulation suffers.
· **Scientific Uncertainty**: The new microplastic and nano-plastic science implies that knowledge gaps remain concerning their impacts on the environment and health.
· **Economic Considerations:** Effective policies often involve trade-offs and are costly; therefore, this creates another challenge for the policymakers.

- **International Cooperation:** Since pollution is global in nature, it is very important that there be international cooperation for developing and implementing effective solutions.

The victory over these challenges requires the joint commitment of governments, industry, and the scientific community to develop and enforce comprehensive, well-coordinated policies on how to tackle microplastic and nano-plastic pollution.

Prevention and Reduction Strategies

This requires mitigation strategies that are targeted at sources and pathways of microplastic and nano-plastic pollution. Multi-faceted intervention at the stages of design modification, consumer education, and policy would be required to be effective in reducing their emission. Design and Production Phase

- **Eco-design and Material Substitution:** The promotion of ecodesign principles during the development and adoption can drastically help reduce the formation of microplastics and nano-plastics. This would include designing products that contain as minimal plastic content as possible, reusing and recycling of materials whenever possible, and increasing the life span of a particular product.
- **Material Innovation:** Heavy investment in research and development will have to be made in developing alternatives to conventional plastics (Nielsen et al., 2023). Biodegradable and compostable materials can significantly contribute to a reduction in plastic wastes, as well as subsequently fracturing into microplastics, along with recycled content.
- **Closed-Loop Systems:** plastic wastes can be minimized and microplastic release reduced through closed-loop production. This means returning plastic materials from the output end of the production cycle to the input stage for recycling, to reduce the amount of virgin plastic resources being used.

Consumer Awareness and Behavioral Change

- **Public Education and Awareness:** Public awareness of microplastic and nano-plastic pollution is a key strategy in behavior change. This information campaign shall focus on the sources of microplastics, the implications of accumulation, as well as actions that any individual can do to reduce personal plastic use (Meegoda & Hettiarachchi, 2023).
- **Consumer Choices:** Much of the microplastic pollution can be attributed to consumer decisions on product purchases. Promoting sustainable consumption patterns in terms of reducing plastic use, preferring reusable products, and responsible disposal will cut back on microplastic emissions to a great extent.
- **Extended Producer Responsibility:** The implementation of EPR schemes makes producers liable for the entire lifecycle of their products down to end-of-life management. This thus incentivizes producers toward designing products that can be recyclable and minimizes plastics used for packaging.

Waste Management and Recycling

- **Improved Waste Collection and Sorting:** The improvement in infrastructure for waste collection and sorting is the most requisite measure for preventing microplastics from getting into the environment. Funding waste management systems in which plastics are separated from other waste streams allows high recycling rates while, at the same time, reducing plastic waste directed to landfills or incinerators (Le, V. G et al., 2023).
- **Recycling and Recovery:** Enhanced recycling and recovery of plastic wastes would reduce virgin plastic production. Investment in technologies for recycling and development of markets for recycled plastics shall motivate interests in its recycling and reduce the quantum of micro-plastics generated (Ogunola et al., 2018).
- **Waste Prevention:** Source reduction, minimization of packaging, re-usable products, repair, and re-use would reduce wastes and in turn, reduce microplastic pollution.

Combination of prevention and reduction strategies put into action will significantly reduce the amount of microplastics and nano-plastics reaching the environment. An approach that covers the whole life cycle of plastic products is core to long-term sustainability.

Treatment and Remediation Technologies

Even though prevention is the best way to deal with microplastic and nano-plastic pollution, the fact remains that they already proliferate the environment. In the light of this fact, it becomes very critical to develop and apply feasible treatment and remediation technologies.

Physical Treatment Methods

The physical methods lean toward the removal mechanism for microplastics and nano-plastics through mechanical processes.

- **Filtration:** It involves filters of different pore sizes for capturing microplastics from water bodies, wastewater, and even from the air. However, with the idea of removing smaller particles, membrane filtration proves to be very effective (Nielsen et al., 2023).

- **Sedimentation:** It is a process based on the gravitational setting of microplastics in water or wastewater treatment tanks. Its effectiveness in the case of smaller particles is not high.
- **Flocculation and Coagulation:** These techniques involve the addition of chemicals to form flocs of microplastics for sedimentation or filtration.

Chemical Treatment Methods

The chemical methods aim to degrade or transform microplastics into less harmful substances.

- **Advanced Oxidation Processes:** AOPs use powerful oxidants like ozone, hydrogen peroxide, or ultraviolet radiation to break down microplastics into smaller, less harmful compounds.
- **Sorption**: Adsorbents, such as activated carbon or zeolites, are utilized to bind microplastics on their surface, thus removing them from the water.

Biological Treatment Methods

These methods make use of microorganisms that degrade or transform microplastics.

- **Biodegradation:** Some microorganisms have the capacity to degrade certain types of plastics. However, the biodegradability of microplastics is still under study and it can be very slow.
- **Bioaccumulation:** Some organisms may bioaccumulate microplastics, which would decrease the available concentration of plastic debris in the environment but may, inversely, negatively impact the organisms.

Challenges and Considerations

- **Efficiency and Scalability:** Most treatment and remediation technologies are still at different development stages and not efficient and scalable for practical applications.
- **Cost-Effectiveness**: Treatment and remediation technologies can be quite expensive; thus, the cost is always a huge entry barrier.
- **Secondary Pollution**: The second output of the treatment methodologies may consist of byproducts that are themselves pollutants. These byproducts should be properly accounted for since they are released into the environment.

- **Nano-plastic Challenges:** Nano-plastics are very small in size. They can pass through most conventional treatment systems, and treatment methods are difficult to apply.

Treatment and remediation of microplastic and nano-plastic pollution are rather challenging and would require continuous research and development in physical, chemical, and biological methods to reach efficient and sustainable solutions.

Economic Instruments

Economic instruments can be represented by an effective toolkit on microplastic and nano-plastic pollution challenges. These instruments can offer financial incentives or disincentives, stimulating innovation and encouraging friendly environmental behavior, generating revenue to manage the pollution.

Taxes and Fees

- **Plastic Taxation:** So, the consumption of the types of plastic materials like plastic bags, plastic bottles, or plastic packaging can be prevented by a tax on plastic products and bring some revenue for the purpose of pollution prevention and cleaning (Le, V. G et al., 2023).
- **Environmental User Fees:** The charges for plastic waste disposal can help bring about behavioral change towards reducing waste and increasing recycling.
- **Carbon Taxes**: Though aimed mainly at GHG, carbon taxes indirectly decrease plastic pollution by encouraging technologies that are low in carbon and high in resource efficiency.

Subsidies and Incentives

- **Research and Development Subsidies:** The subsidization or offering of tax breaks to research and develop alternative materials, new recycling technologies, and pollution prevention techniques will stimulate innovation and raise the pace of transition to a more circular economy.
- **Recycling Incentives:** Financial incentives or rebates for recycling plastic products will increase participation and hence raise recycling rates.
- **Green Procurement Policies:** Governments can set examples and buy products that have less impact on the environment, which would include reduced plastic content or have been produced with recycled material.

Market-Based Instruments

- **Emission Trading:** This creates a market in permits to emit microplastics or nano-plastics to the environment. Companies with fewer emissions can sell their excess permits to other businesses that are highly polluting (Allan et al., 2021).
- **Deposit Refund Systems:** Implement deposit-refund systems for plastic bottles and containers to take them back for recycling. Consumers receive a refund upon the return of used containers.
- **Extended Producer Responsibility:** While this is often categorized as a regulatory instrument, it also has an economic dimension (Silva et al., 2018). On the condition that producers are held liable financially for the management of plastic wastes, the scheme can encourage or incentivize them to reduce packaging, improve product design, and increase the rate of recycling. The challenges and considerations
- **Distributional Impacts:** From a policy perspective, distributional impacts are relevant for the setting of economic instruments so that the burden is not more significantly borne by low-income households. If this is not the case, then it is likely that an unpopular policy will result.
- **Leakage:** Another issue is that of "leakage" whereby pollution is shifted to another location or industry through the use of economic instruments.
- **Effectiveness and Efficiency:** Careful attention needs to be paid to how economic instruments are designed and introduced if they are to be effective in achieving environmental goals yet cause minimal economic distortion (Ogunola et al., 2018).

Economic instruments can be designed and implemented with flexibility and adaptability to address microplastic and nano-plastic pollution. In relation to the context and outcomes pursued, careful design of economic instruments as incentives will trigger behavioral change toward more sustainable practices.

Governance and Implementation

Effective governance and implementation should be in place for policy frameworks to be translated into concrete actions of microplastic and nano-plastic pollution mitigation. This paper calls for collaboration from different stakeholders involving governments, industries, civil society, and research institutions.

Governmental Role

Governments have a very vital role in the scenario of microplastic and nano-plastic pollution. This may involve the following roles of the government in that context:

- **Policy Development and Implementation:** The prime responsibilities for controlling microplastic and nano-plastic emissions should be to develop full-scale and effectively enforceable regulations through bans, restrictions, and standards.
- **Monitoring and Enforcement:** Development of strong monitoring programs regarding the levels of microplastic and nano-plastic pollution and strict enforcement of regulations to ensure compliance.
- **Research and Development:** This would entail investment in research to advance our understanding of microplastic and nano-plastic pollution, develop innovative solutions, and support new technology development.

  Public Awareness and Education: Conducting public awareness campaigns to inform citizens of the impacts from microplastic and nano-plastic pollution and promote best sustainable behaviors.
- **Collaboration Across Borders:** Sharing efforts among different countries to combat microplastic and nano-plastic pollution due to its transboundary nature.

Role of Industry

The involvement of the private sector can play a vital role in mitigating microplastic and nano-plastic pollution. Its major responsibilities include:

- **Designing Products and New Product Innovations**: Designing products with a minimal content of plastics, using recycled materials in the manufacturing process, making the products easy to recycle or reuse.
- **Supply Chain Management:** To ensure that on the whole sustainable practices covering the entire supply chain, involving suppliers and partners, towards source reduction of microplastic and nano-plastic emissions are implemented.
- **Research and Development:** Further research into novel solutions to address plastic pollution through material replacement and through new recycling technologies.
- **Corporate Social Responsibility**: Incorporating environmental sustainability, including goals and metrics on microplastics and nano-plastics in company strategy and reporting.

- **Industry Association:** Industry peers working together to identify best practices and voluntary self-regulatory actions.

Role of Civil Society

The CSOs can do much toward advocacy for better policies, raising greater public awareness, and actively engaging in monitoring and cleanup exercises. Their roles include:

- **Advocacy and Lobbying:** The actualization of policies governing microplastic and nano-plastic pollution, and holding governments and industries to account.
- **Public Awareness:** Organizing public education campaigns and outreach activities with communities on how to embrace sustainable behaviors.
- **Monitoring and Data Collection:** Contributing to citizen science initiatives dealing with data collection on microplastic and nano-plastic pollution.
- **Collaboration**: Collaboration with governments, industries, and academia in the development of effective solutions and their implementation.

Collaboration of Multiple Stakeholders

The issue of microplastic and nano-plastic pollution can be comprehensively addressed only when there is collaboration among governments, industries, civil society, and research institutions (Mitrano et al., 2021). Doing so can make knowledge sharing, resource mobilization, and comprehensive solution development possible.

- **Public-Private Partnerships:** Partnerships between public agencies and industries can be formed to pool resources and competence to address microplastic and nano-plastic pollution.
- **Stakeholder Participation:** Consultation with stakeholders will help bring aboard many perspectives available on the status, causes, and consequences of microplastic and nano-plastic pollution into the policy process for its solution.
- **Capacity Building:** Enhancement in the capability of stakeholders through some training programs in those skills and knowledge could make the response to microplastics and nano-plastics much more effective.

Effective governance and implementation are therefore needed to address the rather complex challenge of microplastic and nano-plastic pollution. By building collaboration and cooperation among various stakeholders, a set of effective solutions as regards protecting human health and the environment can be developed and thereafter implemented.

Case Studies of Successful Governance and Implementation Initiatives

While challenges abound, there are also exhilarating examples of successful governance and implementation initiatives.

- **European Union's Strategy on Plastics:** To a certain extent, the European Union is regarded as having been very proactive on issues related to plastic pollution, including microplastics. The most recent adoption of the Single-Use Plastics Directive has affected a blanket ban or restriction on single-use plastic items (Walker & Fequet, 2023). Apart from this, extended producers' responsibility and stringent recycling targets set by the EU depict a holistic approach towards the problem.
- **California Microbead Ban:** California became the first US state to prohibit the use of microbeads in personal care products. This legislation is at the forefront of reducing microplastic pollution in aquatic environments.
- **The Ocean Cleanup:** Although not a government initiative, The Ocean Cleanup is the prime example of private sector innovation in the face of plastic pollution. Their work in making technology solutions for cleaning plastic from oceans underscores the role that public-private partnerships could play.

The examples show how clear policy frameworks, great leadership, and collaboration among many stakeholders drive progress in microplastic and nano-plastic pollution management.

Future Directions and Research Needs

The field of microplastic and nano-plastic pollution is rapidly developing, and challenges, as well as opportunities, are constantly appearing. A forward-looking approach needs to be adopted that incorporates emerging issues and filling knowledge gaps; it has to be interdisciplinary in nature.

Emerging Issues and Challenges

- **Nano-plastics:** Enhanced analytical techniques now mean the presence of nano-plastics in the environment can be established. If mitigation strategies are to be developed, this needs to be underpinned by a primordial understanding of their behavior, fate, and impacts.

- **Microplastics in the Food Chain:** The potential for transfer of microplastics from primary producers through the food chain to humans is a severe health risk. This requires research into the exposure and related human health effects.
- **Microplastics and Climate Change:** The connection of microplastic pollution in climate change is another growing concern. More research is needed to understand how microplastics affect climate processes and how climate change affects microplastic degradation and distribution.
- **Microplastics in Indoor Environments:** The presence of microplastics in the indoor air, dust, and consumer products is an area of studies that has emerged recently. This means the sources, pathways, and health implications need to be understood quite massively.

Knowledge Gaps and Research Priorities

- **Standardized Analytical Methods: There** are requirements for standardized methods with regard to sampling, extraction, and analysis for microplastics and nano-plastics, so that comparability of data is ensured through appropriate assessment of pollution levels (Rai et al., 2021).
- **Fate and Transport:** Knowing the fate and transport of microplastics and nano-plastics in various environmental compartments is very important for understanding their distribution and persistence.
- **Ecological Impacts:** More studies are needed to determine the ecological impacts of microplastics and nano-plastics on various organisms and ecosystems, particularly their long-term effects.
- **Human Health Impacts:** Research on the potential health risk associated with exposures to microplastics and nano-plastics is critical to inform public health policy.
- **Economic Impacts:** Research on the economic costs of microplastic and nano-plastic pollution to fisheries, tourism, and human health may be used to justify investments in prevention and remediation.

Synergies With Other Environmental Policies

Action on microplastic and nano-plastic pollution may be aligned with other policies on environmental protection to contribute to wider sustainable development objectives. Some potential synergies are:

- **Circular Economy:** Circular economy principles can bring about the reduction in plastic waste and the generation of microplastics.
- **Improved Waste Management:** Implement ways of better waste management, such as recycling, composting, and reducing generation of waste, to prevent microplastics from reaching the natural environment.
- **Water Quality:** Taking steps to solve microplastic pollution will contribute towards improving water quality in general.
- **Climate Change Mitigation:** Plastic production and usage reduction has the potential to bring down greenhouse gases and hence would help in mitigating climate change.

This will be obtained by addressing future directions and research needs, which in such a way will enhance our knowledge concerning microplastic and nano-plastic pollution and some relevant strategies taken to reduce it. Interdisciplinary collaboration and international cooperation in continual research efforts are important if a plastic-free future is to be achieved (Mitrano et al., 2021).

CONCLUSION

Widespread occurrence of microplastics and nano-plastics in the environment presents a potential threat to ecological health and human well-being (Usman et al., 2022). Given the complexity of this problem, it requires a multi-faceted approach that has both preventive, treatment, and remediation strategies.

Summary of the Findings

This paper pointed out the fact that there is a critical concern with mitigating microplastic and nano-plastic pollution. Key findings are:

- The widespread nature of microplastics and nano-plastics across several environmental compartments.
- Complex, often unknown, impacts of those particles on ecosystems and human health;
- Challenges in monitoring, assessment, and data collection, associated problems;
- Preventive measures to be implemented, including design changes, consumer education, improvement in waste management;
- Technologies of treatment and remediation against existing pollution;
- Economic instruments to promote sustainable practices;

- Effective governance, collaboration, and implementation related to policy setting.

Research and development, in particular, should focus on the new emerging challenges.

Recommendations

The following recommendations are required to effectively address microplastic and nano-plastic pollution:

- **Better Policy and Legislation:** Full policies on microplastic and nano-plastic pollution should be established at the national and international levels, and as appropriate, implemented (Onyena et al., 2021).
  Investment in Research: Research to advance our understanding of sources, pathways, impacts, and mitigation options for microplastic and nano-plastic should be significantly and actively promoted.
- **Public Awareness and Education:** Run public education campaigns that will in turn make people more aware of the problem and lead to sustainable behaviors;
- **Industry Engagement:** Engage industries towards sustainable practices and reduce plastic consumption by encouraging innovative solutions;
- **International Cooperation:** Organize collaboration between countries to address the international transboundary nature of microplastic and nano-plastic pollution;
- **Monitoring and Assessment:** Effective monitoring programs shall be implemented to be able to measure the levels of microplastics and nano-plastics, and to effectively identify where mitigation measures are appropriate.

Outlook for the Future

Though the case for microplastic and nano-plastic contamination is enormous, there are also reasons to be optimistic. Rapidly developing technology, growing awareness among citizens, and the priming of policies by different governments provide some hope that there may yet be a future with significantly reduced contaminant loads. The collaboration of governments, industry, civil society, and the scientific community should give rise to effective solutions for the protection of our planet now and into the future.

It requires long-term commitment and sustained effort to eradicate microplastic and nano-plastic pollution. We can mitigate these impacts of the emerging pollutants with decisive action taken now for a more sustainable future.

REFERENCES

Allan, J., Belz, S., Hoeveler, A., Hugas, M., Okuda, H., Patri, A., Rauscher, H., Silva, P., Slikker, W., Sokull-Kluettgen, B., Tong, W., & Anklam, E. (2021). Regulatory landscape of nanotechnology and nano-plastics from a global perspective. *Regulatory Toxicology and Pharmacology*, 122, 104885. DOI: 10.1016/j.yrtph.2021.104885 PMID: 33617940

da Costa, J. P. (2018). Micro-and nano-plastics in the environment: Research and policymaking. *Current Opinion in Environmental Science & Health*, 1, 12–16. DOI: 10.1016/j.coesh.2017.11.002

Das, R. K., Sanyal, D., Kumar, P., Pulicharla, R., & Brar, S. K. (2021). Science-society-policy interface for microplastic and nano-plastic: Environmental and biomedical aspects. *Environmental Pollution*, 290, 117985. DOI: 10.1016/j.envpol.2021.117985 PMID: 34454195

Hettiarachchi, H., & Meegoda, J. N. (2023). Microplastic pollution prevention: The need for robust policy interventions to close the loopholes in current waste management practices. *International Journal of Environmental Research and Public Health*, 20(14), 6434. DOI: 10.3390/ijerph20146434 PMID: 37510666

Kumar, M., Chen, H., Sarsaiya, S., Qin, S., Liu, H., Awasthi, M. K., Kumar, S., Singh, L., Zhang, Z., Bolan, N. S., Pandey, A., Varjani, S., & Taherzadeh, M. J. (2021). Current research trends on micro-and nano-plastics as an emerging threat to global environment: A review. *Journal of Hazardous Materials*, 409, 124967. DOI: 10.1016/j.jhazmat.2020.124967 PMID: 33517026

Kumar, R., Verma, A., Shome, A., Sinha, R., Sinha, S., Jha, P. K., Kumar, R., Kumar, P., Shubham, , Das, S., Sharma, P., & Vara Prasad, P. V. (2021). Impacts of plastic pollution on ecosystem services, sustainable development goals, and need to focus on circular economy and policy interventions. *Sustainability (Basel)*, 13(17), 9963. DOI: 10.3390/su13179963

Kurniawan, T. A., Haider, A., Mohyuddin, A., Fatima, R., Salman, M., Shaheen, A., Ahmad, H. M., Al-Hazmi, H. E., Othman, M. H. D., Aziz, F., Anouzla, A., & Ali, I. (2023). Tackling microplastics pollution in global environment through integration of applied technology, policy instruments, and legislation. *Journal of Environmental Management*, 346, 118971. DOI: 10.1016/j.jenvman.2023.118971 PMID: 37729832

Le, V. G., Nguyen, M. K., Nguyen, H. L., Lin, C., Hadi, M., Hung, N. T. Q., Hoang, H.-G., Nguyen, K. N., Tran, H.-T., Hou, D., Zhang, T., & Bolan, N. S. (2023). A comprehensive review of micro-and nano-plastics in the atmosphere: Occurrence, fate, toxicity, and strategies for risk reduction. *The Science of the Total Environment*, 904, 166649. DOI: 10.1016/j.scitotenv.2023.166649 PMID: 37660815

Meegoda, J. N., & Hettiarachchi, M. C. (2023). A path to a reduction in micro and nano-plastics pollution. *International Journal of Environmental Research and Public Health*, 20(8), 5555. DOI: 10.3390/ijerph20085555 PMID: 37107837

Mitrano, D. M., Wick, P., & Nowack, B. (2021). Placing nano-plastics in the context of global plastic pollution. *Nature Nanotechnology*, 16(5), 491–500. DOI: 10.1038/s41565-021-00888-2 PMID: 33927363

Nielsen, M. B., Clausen, L. P. W., Cronin, R., Hansen, S. F., Oturai, N. G., & Syberg, K. (2023). Unfolding the science behind policy initiatives targeting plastic pollution. *Microplastics and Nanoplastics*, 3(1), 3. DOI: 10.1186/s43591-022-00046-y PMID: 36748026

Ogunola, O. S., Onada, O. A., & Falaye, A. E. (2018). Mitigation measures to avert the impacts of plastics and microplastics in the marine environment (a review). *Environmental Science and Pollution Research International*, 25(10), 9293–9310. DOI: 10.1007/s11356-018-1499-z PMID: 29470754

Onyena, A. P., Aniche, D. C., Ogbolu, B. O., Rakib, M. R. J., Uddin, J., & Walker, T. R. (2021). Governance strategies for mitigating microplastic pollution in the marine environment: A review. *Microplastics*, 1(1), 15–46. DOI: 10.3390/microplastics1010003

Rai, P. K., Lee, J., Brown, R. J., & Kim, K. H. (2021). Micro-and nano-plastic pollution: Behavior, microbial ecology, and remediation technologies. *Journal of Cleaner Production*, 291, 125240. DOI: 10.1016/j.jclepro.2020.125240

Silva, A. B., Costa, M. F., & Duarte, A. C. (2018). Biotechnology advances for dealing with environmental pollution by micro (nano) plastics: Lessons on theory and practices. *Current Opinion in Environmental Science & Health*, 1, 30–35. DOI: 10.1016/j.coesh.2017.10.005

Usman, S., Abdull Razis, A. F., Shaari, K., Azmai, M. N. A., Saad, M. Z., Mat Isa, N., & Nazarudin, M. F. (2022). The burden of microplastics pollution and contending policies and regulations. *International Journal of Environmental Research and Public Health*, 19(11), 6773. DOI: 10.3390/ijerph19116773 PMID: 35682361

Walker, T. R., & Fequet, L. (2023). Current trends of unsustainable plastic production and micro (nano) plastic pollution. *Trends in Analytical Chemistry*, 160, 116984. DOI: 10.1016/j.trac.2023.116984

Compilation of References

Abbasi, S. (2021). Microplastics washout from the atmosphere during a monsoon rain event. *Journal of Hazardous Materials Advances*, 4, 100035. DOI: 10.1016/j.hazadv.2021.100035

Abdurahman, A., Cui, K., Wu, J., Li, S., Gao, R., Dai, J., Liang, W., & Zeng, F. (2020). Adsorption of dissolved organic matter (DOM) on polystyrene microplastics in aquatic environments: Kinetic, isotherm and site energy distribution analysis. *Ecotoxicology and Environmental Safety*, 198, 110658. DOI: 10.1016/j.ecoenv.2020.110658

Adamczyk, S., Chojak-Koźniewska, J., Oleszczuk, S., Michalski, K., Velmala, S., Zantis, L. J., Bosker, T., Zimny, J., Adamczyk, B., & Sowa, S.L. J. (2023). Polystyrene nanoparticles induce concerted response of plant defense mechanisms in plant cells. *Scientific Reports*, 13(1), 22423. DOI: 10.1038/s41598-023-50104-5

Aebischer, N. J., Coulson, J. C., & Colehsook, J. M. (1990). Parallel long-term trends across four marine trophic levels and weather. *Nature*, 147(6295), 753–755. DOI: 10.1038/347753a0

Afrin, S., Uddin, M. K., & Rahman, M. M. (2020). Microplastics contamination in the soil from Urban Landfill site, Dhaka, Bangladesh. *Heliyon*, 6(11), e05572. DOI: 10.1016/j.heliyon.2020.e05572

Agbekpornu, P., & Kevudo, I. (2023). The Risks of Microplastic Pollution in the Aquatic Ecosystem. *The Risks of Microplastic Pollution in the Aquatic Ecosystem.*, 8(January). Advance online publication. DOI: 10.5772/intechopen.108717

Agboola, O. D., & Benson, N. U. (2021). Physisorption and chemisorption mechanisms influencing micro (nano) plastics-organic chemical contaminants interactions: A review. *Frontiers in Environmental Science*, 9, 678574. DOI: 10.3389/fenvs.2021.678574

Ahlawat, J. (2024). *Nanoparticles characterization and bioremediation-A synergy to the potential environmental benefit.*

Ahmed, M. B., Rahman, M. S., Alom, J., Hasan, M. S., Johir, M., Mondal, M. I. H., Lee, D.-Y., Park, J., Zhou, J. L., & Yoon, M.-H. (2021). Microplastic particles in the aquatic environment: A systematic review. *The Science of the Total Environment*, 775, 145793. DOI: 10.1016/j.scitotenv.2021.145793

al Ghosh, S. (2023). Microplastics as an Emerging Threat to the Global Environment and Human Health. *Sustainability (Basel)*, 15(14), 10821. DOI: 10.3390/su151410821

Alberghini, L., Truant, A., Santonicola, S., Colavita, G., & Giaccone, V. (2023). Microplastics in Fish and Fishery Products and Risks for Human Health: A Review. *International Journal of Environmental Research and Public Health*, 20(1), 789. Advance online publication. DOI: 10.3390/ijerph20010789 PMID: 36613111

AlDhaen E. (2022). Awareness of occupational health hazards and occupational stress among dental care professionals: Evidence from the GCC region. Front Public Health, 8(10), 922748. .DOI: 10.3389/fpubh.2022.922748

Aley, S., & Pranjal, A. (2023). PHA - Based Bioplastic : a Potential Alternative to Address Microplastic Pollution. In *Water, Air, & Soil Pollution*. Springer International Publishing. DOI: 10.1007/s11270-022-06029-2

Alimi, O. S., Budarz, J. F., Hernandez, L. M., Tufenkji, N., & Sinton, D. (2018). Microplastics and Nanoplastics in Aquatic Environments: Aggregation, Deposition, and Enhanced Contaminant Transport. *Environmental Science & Technology*, 52(4), 1704–1724. DOI: 10.1021/acs.est.7b05559 PMID: 29265806

Allan, J., Belz, S., Hoeveler, A., Hugas, M., Okuda, H., Patri, A., Rauscher, H., Silva, P., Slikker, W., Sokull-Kluettgen, B., Tong, W., & Anklam, E. (2021). Regulatory landscape of nanotechnology and nano-plastics from a global perspective. *Regulatory Toxicology and Pharmacology*, 122, 104885. DOI: 10.1016/j.yrtph.2021.104885 PMID: 33617940

Allen, S., Allen, D., Karbalaei, S., Maselli, V., & Walker, T. R. (2022). Micro(nano) plastics sources, fate, and effects: What we know after ten years of research. *Journal of Hazardous Materials Advances*, 6, 100057. DOI: 10.1016/j.hazadv.2022.100057

Allen, S., Allen, D., Phoenix, V. R., Le Roux, G., Durántez Jiménez, P., Simonneau, A., Binet, S., & Galop, D. (2019). Atmospheric transport and deposition of microplastics in a remote mountain catchment. *Nature Geoscience*, 12(5), 339–344. DOI: 10.1038/s41561-019-0335-5

Amelia, D., Karamah, E. F., Mahardika, M., Syafri, E., Rangappa, S. M., Siengchin, S., & Asrofi, M. (2022). Effect of advanced oxidation process for chemical structure changes of polyethylene microplastics. *Materials Today: Proceedings*, 52, 2501–2504. DOI: 10.1016/j.matpr.2021.10.438

Amélineau, F., Bonnet, D., Heitz, O., Mortreux, V., Harding, A. M., Karnovsky, N., Walkusz, W., Fort, J., & Grémillet, D. (2016). Microplastic pollution in the Greenland Sea: Background levels and selective contamination of planktivorous diving seabirds. *Environmental Pollution*, 219, 1131–1139. DOI: 10.1016/j.envpol.2016.09.017

Anderson, O. R., Small, C. J., Croxall, J. P., Dunn, E. K., Sullivan, B. J., Yates, O., & Black, A. (2011). Global seabird bycatch in longline fisheries. *Endangered Species Research*, 14(2), 91–106. DOI: 10.3354/esr00347

Anderson, W., & Polis, G. (1999). Nutrient fluxes from water to land: Seabirds affect plant nutrient status on Gulf of California islands. *Oecologia*, 118(3), 324–332. DOI: 10.1007/s004420050733 PMID: 28307276

Andrade, J., Fernández-González, V., López-Mahía, P., & Muniategui, S. (2019). A low-cost system to simulate environmental microplastic weathering. *Marine Pollution Bulletin*, 149, 110663. DOI: 10.1016/j.marpolbul.2019.110663

Andrade, M. C., Winemiller, K. O., Barbosa, P. S., Fortunati, A., Chelazzi, D., Cincinelli, A., & Giarrizzo, T. (2019). First account of plastic pollution impacting freshwater fishes in the Amazon: Ingestion of plastic debris by piranhas and other serrasalmids with diverse feeding habits. *Environmental Pollution*, 244, 766–773. DOI: 10.1016/j.envpol.2018.10.088

Andrady, A. L., & Neal, M. A. (2009). Applications and societal benefits of plastics. *Philosophical Transactions of the Royal Society of London. Series B, Biological Sciences*, 364(1526), 1977–1984. DOI: 10.1098/rstb.2008.0304 PMID: 19528050

Angnunavuri, P. N., Attiogbe, F., Dansie, A., & Mensah, B. (2020). Consideration of emerging environmental contaminants in africa: Review of occurrence, formation, fate, and toxicity of plastic particles. *Scientific African*, 9, e00546. DOI: 10.1016/j.sciaf.2020.e00546

Anik, A. H., Hossain, S., Alam, M., & Sultan, M. B. (2021). Microplastics pollution: A comprehensive review on the sources, fates, effects, and potential remediation. *Environmental Nanotechnology, Monitoring & Management*, 16, 100530. DOI: 10.1016/j.enmm.2021.100530

Anouk D'Hont. (2021) Dropping the microbead: Source and sink related microplastic distribution in the Black Sea and Caspian Sea basins. Marine Pollution Bulletin, 173(Part A), 112982. DOI: 10.1016/j.marpolbul.2021.112982

Anuli Dass. (2021) Air pollution: A review and analysis using fuzzy techniques in Indian scenario. Environmental Technology & Innovation, 22, 101441. DOI: 10.1016/j.eti.2021.101441

Anusha, J., Citarasu, T., Uma, G., Vimal, S., Kamaraj, C., Kumar, V., & Sankar, M. M. (2024). Recent advances in nanotechnology-based modifications of micro/nano PET plastics for green energy applications. *Chemosphere*, 352, 141417. DOI: 10.1016/j.chemosphere.2024.141417

Anyanwu, I. N., & Semple, K. T. (2016). Assessment of the effects of phenanthrene and its nitrogen heterocyclic analogues on microbial activity in soil. *SpringerPlus*, 5(1), 279. DOI: 10.1186/s40064-016-1918-x

Aragaw, T. A. (2020). Surgical face masks as a potential source for microplastic pollution in the COVID-19 scenario. *Marine Pollution Bulletin*, 159, 111517. DOI: 10.1016/j.marpolbul.2020.111517

Araujo, C. F., Nolasco, M. M., Ribeiro, A. M. P., & Ribeiro-Claro, P. J. A. (2018). Identification of microplastics using Raman spectroscopy: Latest developments and future prospects. *Water Research*, 142, 426–440. DOI: 10.1016/j.watres.2018.05.060 PMID: 29909221

Ariza-Tarazona, M. C., Villarreal-Chiu, J. F., Barbieri, V., Siligardi, C., & Cedillo-González, E. I. (2019). New strategy for microplastic degradation: Green photocatalysis using a protein-based porous N-TiO2 semiconductor. *Ceramics International*, 45(7), 9618–9624. DOI: 10.1016/j.ceramint.2018.10.208

Arthur, C., Baker, J., & Bamford, H. (2009). Proceedings of the International Research Workshop on the Occurrence, Effects, and Fate of Microplastic Marine Debris. Group.

Ašmonaitė, G., & Almroth, B. C. (2019). *Effects of microplastics on organisms and impacts on the environment: Balancing the known and unknown. Department of Biological and Environmental Sciences*. University of Gothenburg.

Ásmundsdóttir, Á. M., & Schulz, B. (2020). Effects of Microplastics in the Cryosphere. In Rocha-Santos, T., Costa, M. F., & Mouneyrac, C. (Eds.), *Handbook of Microplastics in the Environment*. Springer. DOI: 10.1007/978-3-030-39041-9_47

Athey, S. N., Albotra, S. D., Gordon, C. A., Monteleone, B., Seaton, P., Andrady, A. L., Taylor, A. R., & Brander, S. M. (2020). Trophic transfer of microplastics in an estuarine food chain and the effects of a sorbed legacy pollutant. *Limnology and Oceanography Letters*, 5(1), 154–162. DOI: 10.1002/lol2.10130

Avery-Gomm, S., Provencher, J. F., Morgan, K. H., & Bertram, D. F. (2013, July 15). Plastic ingestion in marine-associated bird species from the eastern North Pacific. *Marine Pollution Bulletin*, 72(1), 257–259. DOI: 10.1016/j.marpolbul.2013.04.021 PMID: 23683586

Aves, A. R., Revell, L. E., Gaw, S., Ruffell, H., Schuddeboom, A., Wotherspoon, N. E., LaRue, M., & McDonald, A. J. (2022). First evidence of microplastics in Antarctic snow. *The Cryosphere*, 16(6), 2127–2145. DOI: 10.5194/tc-16-2127-2022

Awasthi, J. P., Saha, B., Chowardhara, B., Devi, S. S., Borgohain, P., & Panda, S. K. (2018). Qualitative Analysis of Lipid Peroxidation in Plants under Multiple Stress Through Schiff's Reagent: A Histochemical Approach. *Bio-Protocol*, 8(8). Advance online publication. DOI: 10.21769/BioProtoc.2807

Ayeleru, O. O., Dlova, S., Akinribide, O. J., Ntuli, F., Kupolati, W. K., Marina, P. F., Blencowe, A., & Olubambi, P. A. (2020). Challenges of plastic waste generation and management in Sub-Saharan Africa: A review. *Waste Management (New York, N.Y.)*, 110, 24–42. DOI: 10.1016/j.wasman.2020.04.017

Azeem, I., Adeel, M., Ahmad, M. A., Shakoor, N., Jiangcuo, G. D., Azeem, K., Ishfaq, M., Shakoor, A., Ayaz, M., Xu, M., & Rui, Y. (2021). Uptake and Accumulation of Nano/Microplastics in Plants: A Critical Review. *Nanomaterials (Basel, Switzerland)*, 11(11), 2935. DOI: 10.3390/nano11112935

Azoulay, D., Villa, P., Arellano, Y., Gordon, M. F., Moon, D., Miller, K. A., & Thompson, K. (2019). Plastic & health: the hidden costs of a plastic planet. *Center for International Environmental Law (CIEL).* Available at https://www.ciel.org/plasticandhealth

Azzarello, M. Y., & Van Vleet, E. S. (1987). Marine Birds and Plastic Pollution. *Marine Ecology Progress Series*, 37(2/3), 295–303. DOI: 10.3354/meps037295

Babaei, A. A., Kakavandi, B., Rafiee, M., Kalantarhormizi, F., Purkaram, I., Ahmadi, E., & Esmaeili, S. (2017). Comparative treatment of textile wastewater by adsorption, Fenton, UV-Fenton and US-Fenton using magnetic nanoparticles-functionalized carbon (MNPs@ C). *Journal of Industrial and Engineering Chemistry*, 56, 163–174. DOI: 10.1016/j.jiec.2017.07.009

Bacha, A.-U.-R., Nabi, I., & Zhang, L. (2021). Mechanisms and the engineering approaches for the degradation of microplastics. *ACS ES&T Engineering*, 1(11), 1481–1501. DOI: 10.1021/acsestengg.1c00216

Back, H. de M., Vargas Junior, E. C., Alarcon, O. E., & Pottmaier, D. (2022). Training and evaluating machine learning algorithms for ocean microplastics classification through vibrational spectroscopy. *Chemosphere*, 287, 131903. DOI: 10.1016/j.chemosphere.2021.131903

Baechler, B. R., De Frond, H., Dropkin, L., Leonard, G. H., Proano, L., & Mallos, N. J. (2024). Public awareness and perceptions of ocean plastic pollution and support for solutions in the United States. *Frontiers in Marine Science*, 10, 1323477. DOI: 10.3389/fmars.2023.1323477

Baldwin, A. K., Corsi, S. R., & Mason, S. A. (2016). Plastic debris in 29 great lakes tributaries: Relations to watershed attributes and hydrology. *Environmental Science & Technology*, 50(19), 10377–10385. DOI: 10.1021/acs.est.6b02917

Barbraud, C., Bertrand, A., Bouchón, M., Chaigneau, A., Delord, K., Demarcq, H., Gimenez, O., Torero, M. G., Gutiérrez, D., Oliveros-Ramos, R., Passuni, G., Tremblay, Y., & Bertrand, S. (2018). Density dependence, prey accessibility, and prey depletion by fisheries drive Peruvian seabird population dynamics. *Ecography*, 41(7), 1092–1102. DOI: 10.1111/ecog.02485

Barbraud, C., Delord, K., Le Bouard, F., Harivel, R., Demay, J., Chaigne, A., & Micol, T. (2021). Seabird population changes following mammal eradication at oceanic Saint-Paul Island, Indian Ocean. *Journal for Nature Conservation*, 63, 126049. DOI: 10.1016/j.jnc.2021.126049

Barker, M., Willans, M., Pham, D.-S., Krishna, A., & Hackett, M. (2022). Explainable Detection of Microplastics Using Transformer Neural Networks. In Aziz, H., Corrêa, D., & French, T. (Eds.), *AI 2022: Advances in Artificial Intelligence* (pp. 102–115). Springer International Publishing. DOI: 10.1007/978-3-031-22695-3_8

Bashir, S. M., Kimiko, S., Mak, C. W., Fang, J. K. H., & Gonçalves, D. (2021). Personal Care and Cosmetic Products as a Potential Source of Environmental Contamination by MPs in a Densely Populated Asian City. *Frontiers in Marine Science*, 8, 683482. DOI: 10.3389/fmars.2021.683482

Bastyans, S., Jackson, S., & Fejer, G. (2022). Micro and nano-plastics, a threat to human health? *Emerging Topics in Life Sciences*, 6(4), 411–422. DOI: 10.1042/ETLS20220024

Beladi-Mousavi, S. M., Hermanova, S., Ying, Y., Plutnar, J., & Pumera, M. (2021). A maze in plastic wastes: Autonomous motile photocatalytic microrobots against microplastics. *ACS Applied Materials & Interfaces*, 13(21), 25102–25110. DOI: 10.1021/acsami.1c04559

Bellasi, A., Binda, G., Pozzi, A., Boldrocchi, G., & Bettinetti, R. (2021). Chemosphere The extraction of microplastics from sediments : An overview of existing methods and the proposal of a new and green alternative. *Chemosphere*, 278, 130357. DOI: 10.1016/j.chemosphere.2021.130357 PMID: 33823347

Belontz, S. L., Corcoran, P. L., Davis, H., Hill, K. A., Jazvac, K., Robertson, K., & Wood, K. (2019). Embracing an interdisciplinary approach to plastics pollution awareness and action. *Ambio*, 48(8), 855–866. DOI: 10.1007/s13280-018-1126-8 PMID: 30448996

Benavides, P. T., Lee, U., & Zarè-Mehrjerdi, O. (2020). Life cycle greenhouse gas emissions and energy use of polylactic acid, bio-derived polyethylene, and fossil-derived polyethylene. *Journal of Cleaner Production*, 277, 124010. DOI: 10.1016/j.jclepro.2020.124010

Benson, N. U., Fred-Ahmadu, O. H., Ekett, S. I., Basil, M. O., Adebowale, A. D., Adewale, A. G., & Ayejuyo, O. O. (2020). Occurrence, depth distribution and risk assessment of PAHs and PCBs in sediment cores of Lagos lagoon, Nigeria. *Regional Studies in Marine Science*, 37, 101335. DOI: 10.1016/j.rsma.2020.101335

Bergmann, M., Mützel, S., Primpke, S., Tekman, M. B., Trachsel, J., & Gerdts, G. (2019). White and wonderful? Microplastics prevail in snow from the Alps to the Arctic. *Science Advances*, 5(8), eaax1157. DOI: 10.1126/sciadv.aax1157

Bernhard, G. H., Neale, R. E., Barnes, P. W., Neale, P. J., Zepp, R. G., Wilson, S. R., Andrady, A. L., Bais, A. F., McKenzie, R. L., Aucamp, P. J., Young, P. J., Liley, J. B., Lucas, R. M., Yazar, S., Rhodes, L. E., Byrne, S. N., Hollestein, L. M., Olsen, C. M., Young, A. R., & White, C. C. (2020). Environmental effects of stratospheric ozone depletion, UV radiation and interactions with climate change: UNEP Environmental Effects Assessment Panel, update 2019. *Photochemical & Photobiological Sciences*, 19(5), 542–584. DOI: 10.1039/d0pp90011g

Bessa, F., Ratcliffe, N., Otero, V., Sobral, P., Marques, J. C., Waluda, C. M., Trathan, P. N., & Xavier, J. C. (2019). Microplastics in gentoo penguins from the Antaretic region. *Scientific Reports*, 9(1), 14191. DOI: 10.1038/s41598-019-50621-2 PMID: 31578393

Bhatla, R., Verma, S., Ghosh, S., & Gupta, A. (2020). Abrupt changes in mean temperature over India during 1901–2010. *Journal of Earth System Science*, 129(1), 1–11. DOI: 10.1007/s12040-020-01421-0

Bhatnagar, N., & Asija, N. (2016). *Durability of high-performance ballistic composites Lightweight ballistic composites*. Elsevier.

Bianco, V., Memmolo, P., Carcagnì, P., Merola, F., Paturzo, M., Distante, C., & Ferraro, P. (2020). Microplastic Identification via Holographic Imaging and Machine Learning. *Advanced Intelligent Systems*, 2(2), 1900153. DOI: 10.1002/aisy.201900153

Billings, A., Carter, H., Cross, R. K., Jones, K. C., Pereira, M. G., & Spurgeon, D. J. (2023). Co-occurrence of macroplastics, microplastics, and legacy and emerging plasticisers in UK soils. *The Science of the Total Environment*, 880(April), 163258. DOI: 10.1016/j.scitotenv.2023.163258 PMID: 37019241

Bindoff, N. L., Cheung, W. W., Kairo, J. G., Arístegui, J., Guinder, V. A., Hallberg, R., ... & Williamson, P. (2019). Changing ocean, marine ecosystems, and dependent communities. *IPCC Special Report on the Ocean and Cryosphere in a Changing Climate,* 477-587.

BirdLife International. (2012). *Important Bird Areas (IBAs)*. BirdLife International.

BirdLife International. (2022). Retrieved from https://www.birdlife.org/worldwide/news/single-use-plastic

Boots, B., Russell, C. W., & Green, D. S. (2019). Effects of microplastics in soil ecosystems: Above and below ground. *Environmental Science & Technology*, 53(19), 11496–11506. DOI: 10.1021/acs.est.9b03304

Borrelle, S. B., Ringma, J., Law, K. L., Monnahan, C. C., Lebreton, L., McGivern, A., Murphy, E., Jambeck, J., Leonard, G. H., Hilleary, M. A., Eriksen, M., Possingham, H. P., De Frond, H., Gerber, L. R., Polidoro, B., Tahir, A., Bernard, M., Mallos, N., Barnes, M., & Rochman, C. M. (2020). Predicted growth in plastic waste exceeds efforts to mitigate plastic pollution. *Science*, 369(6510), 1515–1518. DOI: 10.1126/science.aba3656 PMID: 32943526

Bosker, T., Bouwman, L. J., Brun, N. R., Behrens, P., & Vijver, M. G. (2019, July). Microplastics Accumulate on Pores in Seed Capsule and Delay Germination and Root Growth of the Terrestrial Vascular Plant Lepidium Sativum. *Chemosphere*, 226, 774–781. DOI: 10.1016/j.chemosphere.2019.03.163

Bost, C. A., & le Maho, Y. (1993). Seabirds as bio-indicators of changing marine ecosystems: New perspectives. *Acta Oecologica*, 14, 463–470.

Boucher, J., & Friot, D. (2017). Primary MPs in the Oceans: A Global Evaluation of Sources; IUCN: Gland, Switzerland. Environmental Science. *Geology*. Advance online publication. DOI: 10.2305/IUCN.CH.2017.01.en

Boyle, K., & Örmeci, B. (2020). Microplastics and Nanoplastics in the Freshwater and Terrestrial Environment: A Review. *Water (Basel)*, 12(9), 2633. DOI: 10.3390/w12092633

Brahney, J. (2020). Wet and dry plastic deposition data for western US National Atmospheric Deposition Program sites (2017-2019). Constraining physical understanding of aerosol loading, biogeochemistry, and snowmelt hydrology from hillslope to watershed scale in the east river scientific focus area. *ESS-DIVE repository*. DOI: 10.15485/1773176

Brasso, R. L., Chiaradia, A., Polito, M. J., Raya Rey, A., & Emslie, S. D. (2015). A comprehensive assessment of mercury exposure in penguin populations throughout the Southern Hemisphere, using trophic calculations to identify sources of population-level variation. *Marine Pollution Bulletin*, 97(1-2), 97. DOI: 10.1016/j.marpolbul.2015.05.059 PMID: 26072048

Bratovic A. (2023). European Journal of Advanced Chemistry Research. DOI: DOI: 0.24018/ejchem.2023.4.1.124

Brennecke, D., Duarte, B., Paiva, F., Caçador, I., & Canning-Clode, J. (2016). Microplastics as vector for heavy metal contamination from the marine environment. *Estuarine, Coastal and Shelf Science*, 178, 189–195. DOI: 10.1016/j.ecss.2015.12.003

Bretas Alvim, C., Mendoza-Roca, J. A., & Bes-Piá, A. (2020). Wastewater treatment plant as microplastics release source - Quantification and identification techniques. *Journal of Environmental Management*, 255, 109739. DOI: 10.1016/j.jenvman.2019.109739

Bridson, J. H., Gaugler, E. C., Smith, D. A., Northcott, G. L., & Gaw, S. (2021). Leaching and extraction of additives from plastic pollution to inform environmental risk: A multidisciplinary review of analytical approaches. *Journal of Hazardous Materials*, 414(March), 125571. DOI: 10.1016/j.jhazmat.2021.125571 PMID: 34030416

Brooke, M. D. (2004). The food consumption of the world's seabirds. *Proceedings of the Royal Society of London. Series B, Biological Sciences*, 271, S246–S248. PMID: 15252997

Bullard, J. E., Ockelford, A., O'Brien, P., & McKenna Neuman, C. (2021). Preferential Transport of MPs by Wind. *Atmospheric Environment*, 245, 118038. DOI: 10.1016/j.atmosenv.2020.118038

Burger, J., & Gochfeld, M. (2004). Marine birds as sentinels of environmental pollution. *EcoHealth*, 1(3), 263–274. DOI: 10.1007/s10393-004-0096-4

Cabernard, L., Pfister, S., Oberschelp, C., & Hellweg, S. (2022). Growing environmental footprint of plastics driven by coal combustion. *Nature Sustainability*, 5(2), 139–148. DOI: 10.1038/s41893-021-00807-2

Cadena-Aizaga, M. I., Montesdeoca-Esponda, S., Torres-Padrón, M. E., Sosa-Ferrera, Z., & Santana-Rodríguez, J. J. (2020). Organic UV filters in marine environments: An update of analytical methodologies, occurrence and distribution. *Trends in Environmental Analytical Chemistry*, 25, 25. DOI: 10.1016/j.teac.2019.e00079

Cai, J., Niu, T., Shi, P., & Zhao, G. (2019). Boron-doped diamond for hydroxyl radical and sulfate radical anion electrogeneration, transformation, and voltage-free sustainable oxidation. *Small*, 15(48), 1900153. DOI: 10.1002/smll.201900153

Cairns, D. K. (1987). Seabirds as indicators of marine food supplies. *Biological Oceanography*, 5, 261–271.

Campanale C. (2020) A Detailed Review Study on Potential Effects of Microplastics and Additives of Concern on Human Health. Int J Environ Res Public Health, 13(17), 1212. .DOI: 10.3390/ijerph17041212

Campanale, C., Savino, I., Massarelli, C. and Uricchio, V.F., 2023. Fourier transform infrared spectroscopy to assess the degree of alteration of artificially aged and environmentally

Carbery, M., O'Connor, W., & Palanisami, T. (2018). Trophic transfer of microplastics and mixed contaminants in the marine food web and implications for human health. *Environment International*, 115, 400–409. DOI: 10.1016/j.envint.2018.03.007

Carla, A., Tasso, C., Wambier, S., Luiz, A., Junior, F., & Nardes, C. (2020). *Case Studies in Chemical and Environmental Engineering Filtration, assimilation and elimination of microplastics by freshwater bivalves.* DOI: 10.1016/j.cscee.2020.100053

Carlin, J., Craig, C., Little, S., Donnelly, M., Fox, D., Zhai, L., & Walters, L. (2020). Microplastic accumulation in the gastrointestinal tracts in birds of prey in central Florida, USA. *Environmental Pollution*, 264, 114633. DOI: 10.1016/j.envpol.2020.114633 PMID: 32388295

Carolina, N., Beljanski, A., Cole, C., Fuxa, F., Setiawan, E., Singh, H., Arbor, A., Advisor, F., & Alford, L. K. (2016). *Efficiency and Effectiveness of a Low-Cost, Self-Cleaning Microplastic Filtering System for Wastewater Treatment Plants The University of Michigan – Ann Arbor.*

Carravieri, A., Cherel, Y., Jaeger, A., Churlaud, C., & Bustamante, P. (2016). Penguins as bioindicators of mercury contamination in the southern Indian Ocean: Geographical and temporal trends. *Environmental Pollution*, 213, 195–205. DOI: 10.1016/j.envpol.2016.02.010 PMID: 26896669

Carrington, D. 2019. Microplastics 'significantly contaminating the air', scientists warn. *The Guardian.* Retrieved 05 April 2024, from https://www.theguardian.com/environment/2019/aug/14/microplastics-found-at-profuse-levels-in-snow-from-arctic-to-alps-contamination

Carr, M. H., Neigel, J. E., Estes, J. A., Andelman, S., Robert, R., Carr, M. H., Neigel, J. E., Estes, J. A., Andelman, S., Warner, R. R., & Largier, J. L. (2003). Comparing Marine and Terrestrial Ecosystems : Implications for the Design of Coastal Marine Reserves Warner and John L. Largier Source : Ecological Applications, Vol. 13, No. 1, Supplement : The Science of Marine Reserves Published by : Wiley Stable. *Ecological Applications*, 13(1), S90–S107. DOI: 10.1890/1051-0761(2003)013[0090:CMATEI]2.0.CO;2

Carr, S. A., Liu, J., & Tesoro, A. G. (2016). Transport and fate of microplastic particles in wastewater treatment plants. *Water Research*, 91, 174–182. DOI: 10.1016/j.watres.2016.01.002 PMID: 26795302

Carson, H. S. (2013). The incidence of plastic ingestion by fishes: From the prey's perspective. *Marine Pollution Bulletin*, 74(1), 170–174. DOI: 10.1016/j.marpolbul.2013.07.008 PMID: 23896402

Castro, B., Citterico, M., Kimura, S., Stevens, D. M., Wrzaczek, M., & Coaker, G. (2021). Stress-induced reactive oxygen species compartmentalization, perception and signalling. *Nature Plants*, 7(4), 403–412. DOI: 10.1038/s41477-021-00887-0

Chae, Y., & An, Y. J. (2017). Effects of micro- and nanoplastics on aquatic ecosystems: Current research trends and perspectives. *Marine Pollution Bulletin*, 124(2), 624–632. DOI: 10.1016/j.marpolbul.2017.01.070 PMID: 28222864

Chamas, A., Moon, H., Zheng, J., Qiu, Y., Tabassum, T., Jang, J. H., Abu-Omar, M., Scott, S. L., & Suh, S. (2020). Degradation Rates of Plastics in the Environment. *ACS Sustainable Chemistry & Engineering*, 8(9), 3494–3511. DOI: 10.1021/acssuschemeng.9b06635

Chang, M. (2015). Reducing microplastics from facial exfoliating cleansers in wastewater through treatment versus consumer product decisions. *Marine Pollution Bulletin*, 101(1), 330–333. DOI: 10.1016/j.marpolbul.2015.10.074 PMID: 26563542

Chatziparaskeva, G., Papamichael, I., & Zorpas, A. A. (2022). Microplastics in the coastal en- vironment of Mediterranean and the impact on sustainability level. *Sustainable Chemistry and Pharmacy*, 29, 100768. DOI: 10.1016/j.scp.2022.100768

Chen L. (2017) Inflammatory responses and inflammation-associated diseases in organs. Oncotarget, 14(9), 7204-7218. .DOI: 10.18632/oncotarget.23208

Chen Q. (2023) Factors Affecting the Adsorption of Heavy Metals by Microplastics and Their Toxic Effects on Fish. Toxics, 28(11), 490. .DOI: 10.3390/toxics11060490

Chen, R., Qi, M., Zhang, G., & Yi, C. (2018). *Comparative experiments on polymer degradation technique of produced water of polymer flooding oilfield.* Paper presented at the IOP Conference Series: Earth and Environmental Science. DOI: 10.1088/1755-1315/113/1/012208

Chen, G., Li, Y., Liu, S., Junaid, M., & Wang, J. (2022, February). Effects of Micro (Nano)Plastics on Higher Plants and the Rhizosphere Environment. *The Science of the Total Environment*, 807, 150841. DOI: 10.1016/j.scitotenv.2021.150841

Chen, J., Wu, J., Sherrell, P. C., Chen, J., Wang, H., Zhang, W., & Yang, J. (2022). How to build a microplastics-free environment: Strategies for microplastics degradation and plastics recycling. *Advanced Science (Weinheim, Baden-Wurttemberg, Germany)*, 9(6), 2103764. DOI: 10.1002/advs.202103764

Chen, M., Zhao, X., Wu, D., Peng, L., Fan, C., Zhang, W., Li, Q., & Ge, C. (2022). Addition of biodegradable microplastics alters the quantity and chemodiversity of dissolved organic matter in latosol. *The Science of the Total Environment*, 816, 151960. DOI: 10.1016/j.scitotenv.2021.151960

Chen, X., Huang, G., Gao, S., & Wu, Y. (2021). Effects of permafrost degradation on global microplastic cycling under climate change. *Journal of Environmental Chemical Engineering*, 9(5), 106000. DOI: 10.1016/j.jece.2021.106000

Chen, Y., Tang, H., Cheng, Y., Huang, T., & Xing, B. (2023). Interaction between microplastics and humic acid and its effect on their properties as revealed by molecular dynamics simulations. *Journal of Hazardous Materials*, 455, 131636. DOI: 10.1016/j.jhazmat.2023.131636

Chen, Y., Wen, D., Pei, J., Fei, Y., Ouyang, D., Zhang, H., & Luo, Y. (2020). Identification and quantification of microplastics using Fourier-transform infrared spectroscopy: Current status and future prospects. *Current Opinion in Environmental Science & Health*, 18, 14–19. DOI: 10.1016/j.coesh.2020.05.004

Chen, Z., Huang, Z., Liu, J., Wu, E., Zheng, Q., & Cui, L. (2021). Phase transition of Mg/Al-flocs to Mg/Al-layered double hydroxides during flocculation and polystyrene nanoplastics removal. *Journal of Hazardous Materials*, 406, 124697. DOI: 10.1016/j.jhazmat.2020.124697

Cherel, Y., Barbraud, C., Lahournat, M., Jaeger, A., Jaquemet, S., Wanless, R. M., Phillips, R. A., Thompson, D. R., & Bustamante, P. (2018). Accumulate or eliminate? Seasonal mercury dynamics in albatrosses, the most contaminated family of birds. *Environmental Pollution*, 241, 124–135. DOI: 10.1016/j.envpol.2018.05.048 PMID: 29803026

Cheung, C. K. H., & Not, C. (2023). Impacts of extreme weather events on microplastic distribution in coastal environments. *The Science of the Total Environment*, 904, 166723. DOI: 10.1016/j.scitotenv.2023.166723

Cheung, P. K., Hung, P. L., & Fok, L. (2019). River microplastic contamination and dynamics upon a rainfall event in Hong Kong, China. *Environmental Processes*, 6(1), 253–264. DOI: 10.1007/s40710-018-0345-0

Choi, S., Kim, J., & Kwon, M. (2022). The Effect of the Physical and Chemical Properties of Synthetic Fabrics on the Release of Microplastics during Washing and Drying. *Polymers*, 14(16), 3384. Advance online publication. DOI: 10.3390/polym14163384 PMID: 36015640

Choi, Y. R., Kim, Y. N., Yoon, J. H., Dickinson, N., & Kim, K. H. (2021). Plastic contamination of forest, urban, and agricultural soils: A case study of Yeoju City in the Republic of Korea. *Journal of Soils and Sediments*, 21(5), 1962–1973. DOI: 10.1007/s11368-020-02759-0

Chomiak, K. M., Owens-rios, W. A., Bangkong, C. M., Day, S. W., Eddingsaas, N. C., Hoffman, M. J., Hudson, A. O., & Tyler, A. C. (2024). *Impact of Microplastic on Freshwater Sediment Biogeochemistry and Microbial Communities Is Polymer Specific*.

Claessens, M., Van Cauwenberghe, L., Vandegehuchte, M. B., & Janssen, C. R. (2013). New techniques for the detection of microplastics in sediments and field collected organisms. *Marine Pollution Bulletin*, 70(1–2), 227–233. DOI: 10.1016/j.marpolbul.2013.03.009 PMID: 23601693

Clark, B. L., Carneiro, A. P. B., Pearmain, E. J., Rouyer, M.-M., Clay, T. A., Cowger, W., Phillips, R. A., Manica, A., Hazin, C., Eriksen, M., González-Solís, J., Adams, J., Albores-Barajas, Y. V., Alfaro-Shigueto, J., Alho, M. S., Araujo, D. T., Arcos, J. M., Arnould, J. P. Y., Barbosa, N. J. P., & Dias, M. P. (2023). Global assessment of marine plastic exposure risk for oceanic birds. *Nature Communications*, 14(1), 3665. DOI: 10.1038/s41467-023-38900-z PMID: 37402727

Cole, D. W., Cole, R., Gaydos, S. J., Gray, J., Hyland, G., Jacques, M. L., Powell-Dunford, N., Sawhney, C., & Au, W. W. (2009). Aquaculture: Environmental, toxicological, and health issues. *International Journal of Hygiene and Environmental Health*, 212(4), 369–377. DOI: 10.1016/j.ijheh.2008.08.003 PMID: 18790671

Cole, M., & Galloway, T. S. (2015). Ingestion of nanoplastics and microplastics by Pacific oyster larvae. *Environmental Science & Technology*, 49(24), 14625–14632. DOI: 10.1021/acs.est.5b04099

Cole, M., Lindeque, P., Fileman, E., Halsband, C., Goodhead, R., Moger, J., & Galloway, T. S. (2013). Microplastic ingestion by zooplankton. *Environmental Science & Technology*, 47(12), 6646–6655. DOI: 10.1021/es400663f

Cole, M., Lindeque, P., Halsband, C., & Galloway, T. S. (2011). Microplastics as contaminants in the marine environment: A review. *Marine Pollution Bulletin*, 62(12), 2588–2597. DOI: 10.1016/j.marpolbul.2011.09.025 PMID: 22001295

Cornick S. (2015) Roles and regulation of the mucus barrier in the gut. Tissue Barriers, 3(1-2), e982426. DOI: 10.4161/21688370.2014.982426

Corradini, F., Bartholomeus, H., Lwanga, E. H., Gertsen, H., & Geissen, V. (2019). Predicting soil microplastic concentration using vis-NIR spectroscopy. *The Science of the Total Environment*, 650, 922–932. DOI: 10.1016/j.scitotenv.2018.09.101

Corte Pause, F., Urli, S., Crociati, M., Stradaioli, G., & Baufeld, A. (2024). Connecting the Dots: Livestock Animals as Missing Links in the Chain of Microplastic Contamination and Human Health. *Animals (Basel)*, 14(2), 350. DOI: 10.3390/ani14020350 PMID: 38275809

Cortinovis, S., Galassi, S., Melone, G., Saino, N., Porte, C., & Bettinetti, R. (2008). Organochlorine contamination in the Great Crested Grebe (Podiceps cristatus): effects on eggshell thickness and egg steroid levels. Chemosphere, 73(3), 320–325.

Costa, P., Duarte, A. C., Rocha-santos, T., & Prata, J. C. (2019). *Trends in Analytical Chemistry Methods for sampling and detection of microplastics in water and sediment : A critical review Density separation.* DOI: 10.1016/j.trac.2018.10.029

Cowger, W., Steinmetz, Z., Gray, A., Munno, K., Lynch, J., Hapich, H., Primpke, S., De Frond, H., Rochman, C., & Herodotou, O. (2021). Microplastic Spectral Classification Needs an Open Source Community: Open Specy to the Rescue! *Analytical Chemistry*, 93(21), 7543–7548. DOI: 10.1021/acs.analchem.1c00123 PMID: 34009953

Cózar, A., Echevarría, F., González-Gordillo, J. I., Irigoien, X., Úbeda, B., Hernández-León, S., Palma, Á. T., Navarro, S., García-de-Lomas, J., Ruiz, A., Fernández-de-Puelles, M. L., & Duarte, C. M. (2014). Plastic debris in the open ocean. *Proceedings of the National Academy of Sciences of the United States of America*, 111(28), 10239–10244. DOI: 10.1073/pnas.1314705111 PMID: 24982135

Croxall, J. P., Butchart, S. H. M., Lascelles, B., Stattersfield, A. J., Sullivan, B., Symes, A., & Taylor, P. (2012). Seabird conservation status, threats and priority actions: A global assessment. *Bird Conservation International*, 22(1), 1–34. DOI: 10.1017/S0959270912000020

Cui, Y., Zhang, Q., Liu, P., & Zhang, Y. (2022). Effects of Polyethylene and Heavy Metal Cadmium on the Growth and Development of Brassica chinensis var. Chinensis. *Water, Air, and Soil Pollution*, 233(10), 426. DOI: 10.1007/s11270-022-05888-z

Curcuruto, M., Williams, S., Brondino, M., & Bazzoli, A. (2023). Investigating the Impact of Occupational Technostress and Psychological Restorativeness of Natural Spaces on Work Engagement and Work–Life Balance Satisfaction. *International Journal of Environmental Research and Public Health*, 20(3), 2249. Advance online publication. DOI: 10.3390/ijerph20032249 PMID: 36767614

Cusset, F., Bustamante, P., Carravieri, A., Bertin, C., Brasso, R., Corsi, I., Dunn, M., Emmerson, L., Guillou, G., Hart, T., Juáres, M., Kato, A., Machado-Gaye, A. L., Michelot, C., Olmastroni, S., Polito, M., Raclot, T., Santos, M., Schmidt, A., & Cherel, Y. (2023). Circumpolar assessment of mercury contamination: The Adélie penguin as a bioindicator of Antarctic marine ecosystems. *Ecotoxicology (London, England)*, 32(8), 1024–1048. DOI: 10.1007/s10646-023-02709-9 PMID: 37878111

da Costa, J. P. (2018). Micro- and nanoplastics in the environment: Research and policymaking. *Current Opinion in Environmental Science & Health*, 1, 12–16. DOI: 10.1016/j.coesh.2017.11.002

da Costa, J. P., Santos, P. S. M., Duarte, A. C., & Rocha-Santos, T. (2016). (Nano) plastics in the environment – Sources, fates and effects. *The Science of the Total Environment*, 566–567, 15–26. DOI: 10.1016/j.scitotenv.2016.05.041

Da Rocha, N., Oppel, S., Prince, S., Matjila, S., Shaanika, T. M., Naomab, C., Yates, O., Paterson, J. R. B., Shimooshili, K., Frans, E., Kashava, S., & Crawford, R. (2021). Reduction in seabird mortality in Namibian fisheries following the introduction of bycatch regulation. *Biological Conservation*, 253, 108915. DOI: 10.1016/j.biocon.2020.108915

Dainelli, M., Castellani, M. B., Pignattelli, S., Falsini, S., Ristori, S., Papini, A., Colzi, I., Coppi, A., & Gonnelli, C. (2024). Growth, physiological parameters and DNA methylation in Spirodela polyrhiza (L.) Schleid exposed to PET micro-nanoplastic contaminated waters. *Plant Physiology and Biochemistry*, 207, 108403. DOI: 10.1016/j.plaphy.2024.108403

Dalela, M., Shrivastav, T. G., Kharbanda, S., & Singh, H. (2015). pH-Sensitive Biocompatible Nanoparticles of Paclitaxel-Conjugated Poly(styrene-co-maleic acid) for Anticancer Drug Delivery in Solid Tumors of Syngeneic Mice. *ACS Applied Materials & Interfaces*, 7(48), 26530–26548. DOI: 10.1021/acsami.5b07764

Dalu, M. T. B., Cuthbert, R. N., Muhali, H., Chari, L. D., Manyani, A., Masunungure, C., & Dalu, T. (2020). Is Awareness on Plastic Pollution Being Raised in Schools? Understanding Perceptions of Primary and Secondary School Educators. *Sustainability (Basel)*, 12(17), 6775. DOI: 10.3390/su12176775

Das, R. K., Sanyal, D., Kumar, P., Pulicharla, R., & Brar, S. K. (2021). Science-society-policy interface for microplastic and nano-plastic: Environmental and biomedical aspects. *Environmental Pollution*, 290, 117985. DOI: 10.1016/j.envpol.2021.117985 PMID: 34454195

Davaasuren, N., Marino, A., Boardman, C., Alparone, M., Nunziata, F., Ackermann, N., & Hajnsek, I. (2018). Detecting Microplastics Pollution in World Oceans Using Sar Remote Sensing. *IGARSS 2018 - 2018 IEEE International Geoscience and Remote Sensing Symposium*, 938–941. DOI: 10.1109/IGARSS.2018.8517281

de Souza Machado, A. A., Kloas, W., Zarfl, C., Hempel, S., & Rillig, M. C. (2018). Microplastics as an emerging threat to terrestrial ecosystems. *Global Change Biology*, 24(4), 1405–1416. DOI: 10.1111/gcb.14020

De Souza Machado, A. A., Lau, C. W., Kloas, W., Bergmann, J., Bachelier, J. B., Faltin, E., Becker, R., Görlich, A. S., & Rillig, M. C. (2019). Microplastics Can Change Soil Properties and Affect Plant Performance. *Environmental Science & Technology*, 53(10), 6044–6052. DOI: 10.1021/acs.est.9b01339

De-la-Torre, G. E. (2020). Microplastics: An emerging threat to food security and human health. *Journal of Food Science and Technology*, 57(5), 1601–1608. DOI: 10.1007/s13197-019-04138-1 PMID: 32327770

Derraik, J. G. (2002). The pollution of the marine environment by plastic debris: A review. *Marine Pollution Bulletin*, 44(9), 842–852. DOI: 10.1016/S0025-326X(02)00220-5 PMID: 12405208

Devriese, L. I., Van der Meulen, M. D., Maes, T., Bekaert, K., Paul-Pont, I., Frère, L., Robbens, J., & Vethaak, A. D. (2015). Microplastic contamination in brown shrimp (*Crangon crangon*, Linnaeus 1758) from coastal waters of the Southern North Sea and Channel area. *Marine Pollution Bulletin*, 98(1-2), 179–187. DOI: 10.1016/j.marpolbul.2015.06.051

Dhiman, S., Sharma, C., Kumar, A., & Pathak, P. (2023). *Microplastics in Aquatic and Food Ecosystems : Remediation Coupled with Circular Economy Solutions to Create Resource from Waste*.

Di Bella, G., Corsino, S. F., De Marines, F., Lopresti, F., La Carrubba, V., Torregrossa, M., & Viviani, G. (2022). Occurrence of Microplastics in Waste Sludge of Wastewater Treatment Plants: Comparison between Membrane Bioreactor (MBR) and Conventional Activated Sludge (CAS) Technologies. *Membranes (Basel)*, 12(4), 371. DOI: 10.3390/membranes12040371

Diamond, A., & Devlin, C. (2003). Seabirds as indicators of changes in marine ecosystems: Ecological monitoring on Machias Seal Island. *Environmental Monitoring and Assessment*, 88(1/3), 153–181. DOI: 10.1023/A:1025560805788 PMID: 14570414

Diao, T., Liu, R., Meng, Q., & Sun, Y. (2023). Microplastics derived from polymer-coated fertilizer altered soil properties and bacterial community in a Cd-contaminated soil. *Applied Soil Ecology*, 183, 104694. DOI: 10.1016/j.apsoil.2022.104694

Dias, M. P., Martin, R., Pearmain, E. J., Burfield, I. J., Small, C., Phillips, R. A., Yates, O., Lascelles, B., Borboroglu, P. G., & Croxall, J. P. (2019). Threats to seabirds: A global assessment. *Biological Conservation*, 237, 525–537. DOI: 10.1016/j.biocon.2019.06.033

Ding, L., Yu, X., Guo, X., Zhang, Y., Ouyang, Z., Liu, P., Zhang, C., Wang, T., Jia, H., & Zhu, L. (2022). The photodegradation processes and mechanisms of polyvinyl chloride and polyethylene terephthalate microplastic in aquatic environments: Important role of clay minerals. *Water Research*, 208, 117879. DOI: 10.1016/j.watres.2021.117879

Domínguez-Jaimes, L. P., Cedillo-González, E. I., Luévano-Hipólito, E., Acuña-Bedoya, J. D., & Hernández-López, J. M. (2021). Degradation of primary nanoplastics by photocatalysis using different anodized TiO2 structures. *Journal of Hazardous Materials*, 413, 125452. DOI: 10.1016/j.jhazmat.2021.125452

Dong, H., Wang, L., Wang, X., Xu, L., Chen, M., Gong, P., & Wang, C. (2021). Microplastics in a remote lake basin of the Tibetan Plateau: Impacts of atmospheric transport and glacial melting. *Environmental Science & Technology*, 55(19), 12951–12960. DOI: 10.1021/acs.est.1c03227

Dong, Y., Gao, M., Qiu, W., & Song, Z. (2021). Uptake of microplastics by carrots in presence of as (III): Combined toxic effects. *Journal of Hazardous Materials*, 411, 125055. DOI: 10.1016/j.jhazmat.2021.125055

Dore, M. H. (2005). Climate change and changes in global precipitation patterns: What do we know? *Environment International*, 31(8), 1167–1181. DOI: 10.1016/j.envint.2005.03.004

Dovidat, L. C., Brinkmann, B. W., Vijver, M. G., & Bosker, T. (2020). Plastic particles adsorb to the roots of freshwater vascular plant *Spirodela polyrhiza* but do not impair growth. *Limnology and Oceanography Letters*, 5(1), 37–45. DOI: 10.1002/lol2.10118

Duan, J., Bolan, N., Li, Y., Ding, S., Atugoda, T., Vithanage, M., Sarkar, B., Tsang, D. C. W., & Kirkham, M. (2021). Weathering of microplastics and interaction with other coexisting constituents in terrestrial and aquatic environments. *Water Research*, 196, 117011. DOI: 10.1016/j.watres.2021.117011

Dube E. (2023) Plastics and Micro/Nano-Plastics (MNPs) in the Environment: Occurrence, Impact, and Toxicity. Int J Environ Res Public Health, 28(20), 6667. DOI: 10.3390/ijerph20176667

Du, H., Xie, Y., & Wang, J. (2021). Microplastic degradation methods and corresponding degradation mechanism: Research status and future perspectives. *Journal of Hazardous Materials*, 418, 126377. DOI: 10.1016/j.jhazmat.2021.126377

Du, Z. H., Xia, J., Sun, X. C., Li, X. N., Zhang, C., Zhao, H.-S., Zhu, S.-Y., & Li, J.-L. (2017). A novel nuclear xenobiotic receptor (AhR/ PXR/CAR)-mediated mechanism of DEHP-induced cerebellar toxicity in quails (Coturnix Japonica) via disrupting CYP enzyme system homeostasis. *Environmental Pollution*, 226, 435–443. DOI: 10.1016/j.envpol.2017.04.015 PMID: 28413083

Eerkes-Medrano, D., Thompson, R. C., & Aldridge, D. C. (2015). Microplastics in freshwater systems: A review of the emerging threats, identification of knowledge gaps and prioritisation of research needs. *Water Research*, 75, 63–82. DOI: 10.1016/j.watres.2015.02.012

Egger, M., Sulu-Gambari, F., & Lebreton, L. (2020). First evidence of plastic fallout from the North Pacific Garbage Patch. *Scientific Reports*, 10(1), 7495. DOI: 10.1038/s41598-020-64465-8

Einoder, L. D. (2009). A review of the use of seabirds as indicators in fisheries and ecosystem management. *Fisheries Research*, 95(1), 6–12. DOI: 10.1016/j.fishres.2008.09.024

Ekner-Grzyb, A., Duka, A., Grzyb, T., Lopes, I., & Chmielowska-Bąk, J. (2022). Plants oxidative response to nanoplastic. *Frontiers in Plant Science*, 13, 1027608. DOI: 10.3389/fpls.2022.1027608

Ekvall, M. T., Lundqvist, M., Kelpsiene, E., Šileikis, E., Gunnarsson, S. B., & Cedervall, T. (2019). Nano-plastics formed during the mechanical breakdown of daily-use polystyrene products. *Nanoscale Advances*, 1(3), 1055–1061. DOI: 10.1039/C8NA00210J

El Baraka, N., Laknifli, A., Saffaj, N., Addich, M., Taleb, A. A., Mamouni, R., . . . Baih, M. A. (2020). *Study of coupling photocatalysis and membrane separation using tubular ceramic membrane made from natural Moroccan clay and phosphate.* Paper presented at the E3S Web of Conferences. DOI: 10.1051/e3sconf/202015001007

Emenike E.C. (2023) From oceans to dinner plates: The impact of microplastics on human health. Heliyon, 26(9), e20440. .DOI: 10.1016/j.heliyon.2023.e20440

Enfrin, M., Hachemi, C., Hodgson, P. D., Jegatheesan, V., Vrouwenvelder, J., Callahan, D. L., Lee, J., & Dumée, L. F. (2021). Nano/micro plastics–Challenges on quantification and remediation: A review. *Journal of Water Process Engineering*, 42, 102128. DOI: 10.1016/j.jwpe.2021.102128

Enyoh, C. E., Verla, A. W., Verla, E. N., Ibe, F. C., & Amaobi, C. E. (2019). Airborne microplastics: A review study on method for analysis, occurrence, movement and risks. *Environmental Monitoring and Assessment*, 191(11), 1–17. DOI: 10.1007/s10661-019-7842-0

Eriksen, M., Lebreton, L. C. M., Carson, H. S., Thiel, M., Moore, C. J., Borerro, J. C., Galgani, F., Ryan, P. G., & Reisser, J. (2014). Plastic pollution in the world's oceans: More than 5 trillion plastic pieces weighing over 250,000 tons afloat at sea. *PLoS One*, 9(12), e111913. DOI: 10.1371/journal.pone.0111913

Escalona-Segura, G., Borges-Ramírez, M. M., Estrella-Canul, V., & Rendón-von Osten, J. (2022). A methodology for the sampling and identification of microplastics in bird nests. *Green Analytical Chemistry*, 3, 100045. DOI: 10.1016/j.greeac.2022.100045

Evangeliou, N., Grythe, H., Klimont, Z., Heyes, C., Eckhardt, S., Lopez-Aparicio, S., & Stohl, A. (2020). Atmospheric transport is a major pathway of microplastics to remote regions. *Nature Communications*, 11(1), 3381. DOI: 10.1038/s41467-020-17201-9

Fabri-Ruiz, S., Baudena, A., Moullec, F., Lombard, F., Irisson, J.-O., & Pedrotti, M. L. (2023). Mistaking plastic for zooplankton: Risk assessment of plastic ingestion in the Mediterranean sea. *The Science of the Total Environment*, 856(2), 159011. DOI: 10.1016/j.scitotenv.2022.159011 PMID: 36170920

Fadda, G. (2023). CO_2 Emissions and underground storage analysis: towards achieving net-zero targets by 2030 and 2050 (*Masters' Dissertation, Politecnico di Torino*) https://webthesis.biblio.polito.it/id/eprint/29166

Fadli, M. H., Ibadurrohman, M., & Slamet, S. (2021). *Microplastic pollutant degradation in water using modified TiO2 photocatalyst under UV-irradiation.* Paper presented at the IOP Conference Series: Materials Science and Engineering. DOI: 10.1088/1757-899X/1011/1/012055

Faggio, C., Tsarpali, V., & Dailianis, S. (2018). Mussel digestive gland as a model for assessing xenobiotics: An overview. *The Science of the Total Environment*, 613, 220–229. DOI: 10.1016/j.scitotenv.2018.04.264 PMID: 29704717

Fahrenkamp-Uppenbrink, J. (2016). Earthworms on a microplastics diet. *Science*, 351(6277), 1039–1039. DOI: 10.1126/science.351.6277.1039-a

Fakolade, O. A., & Atanda, A. I. (2015). Literature review Literature review. *Literature Review*, (November), 33–37.

Fan, C., Huang, Y. Z., Lin, J. N., & Li, J. (2022). Microplastic quantification of nylon and polyethylene terephthalate by chromic acid wet oxidation and ultraviolet spectrometry. *Environmental Technology & Innovation*, 28, 102683. DOI: 10.1016/j.eti.2022.102683

Farrell, P., & Nelson, K. (2013). Trophic level transfer of microplastic: Mytilus edulis (L.) to Carcinus maenas (L.). *Environmental Pollution*, 177, 1–3. DOI: 10.1016/j.envpol.2013.01.046 PMID: 23434827

Fazio, F., Faggio, C., Marafioti, S., Torre, A., Sanfilippo, M., & Piccione, G. (2012). Comparative study of haematological profiles on Gobius niger in two different habitat sites: Faro Lake and Tyrrhenian Sea. *Cahiers de Biologie Marine*, 53, 213–219.

Fei, Y., Huang, S., Zhang, H., Tong, Y., Wen, D., Xia, X., Wang, H., Luo, Y., & Barceló, D. (2020). Response of soil enzyme activities and bacterial communities to the accumulation of microplastics in an acid cropped soil. *The Science of the Total Environment*, 707, 135634. DOI: 10.1016/j.scitotenv.2019.135634

Feng, H.-M., Zheng, J.-C., Lei, N.-Y., Yu, L., Kong, K. H.-K., Yu, H.-Q., Lau, T.-C., & Lam, M. H. (2011). Photoassisted Fenton degradation of polystyrene. *Environmental Science & Technology*, 45(2), 744–750. DOI: 10.1021/es102182g

Feng, S., Lu, H., Tian, P., Xue, Y., Lu, J., Tang, M., & Feng, W. (2020). Analysis of microplastics in a remote region of the Tibetan Plateau: Implications for natural environmental response to human activities. *The Science of the Total Environment*, 739, 140087. DOI: 10.1016/j.scitotenv.2020.140087

Fernandes, E. M. S., de Souza, A. G., Barbosa, R. F. da S., & Rosa, D. dos S. (2022). Municipal Park Grounds and MPs Contamination. *Journal of Polymers and the Environment*, 30(12), 5202–5210. DOI: 10.1007/s10924-022-02580-5

Ferrante, M. C., Monnolo, A., Del Piano, F., Mattace Raso, G., & Meli, R. (2022). The pressing issue of micro-and nanoplastic contamination: Profiling the reproductive alterations mediated by oxidative stress. *Antioxidants*, 11(2), 193. DOI: 10.3390/antiox11020193

Ferreira, P., Fonte, E., Soares, M. E., Carvalho, F., & Guilhermino, L. (2016). Effects of multi-stressors on juveniles of the marine fish *Pomatoschistus microps*: Gold nanoparticles, microplastics and temperature. *Aquatic Toxicology (Amsterdam, Netherlands)*, 170, 89–103. DOI: 10.1016/j.aquatox.2015.11.011

Foltz, G. R., Schmid, C., & Lumpkin, R. (2015). Transport of surface freshwater from the equatorial to the subtropical North Atlantic Ocean. *Journal of Physical Oceanography*, 45(4), 1086–1102. DOI: 10.1175/JPO-D-14-0189.1

Ford, H. V., Jones, N. H., Davies, A. J., Godley, B. J., Jambeck, J. R., Napper, I. E., Suckling, C. C., Williams, G. J., Woodall, L. C., & Koldewey, H. J. (2022). The fundamental links between climate change and marine plastic pollution. *The Science of the Total Environment*, 806, 150392. DOI: 10.1016/j.scitotenv.2021.150392

Forschungsverbund, B. (2018). An underestimated threat: Land-based pollution with microplastics. Retrieved 5 April 2024, from https://www.sciencedaily.com/releases/2018/02/180205125728.htm

Fossi, M. C., Panti, C., Baini, M., & Lavers, J. L. (2018). A review of plastic-associated pressures: Cetaceans of the Mediterranean Sea and Eastern Australian shearwaters as case studies. *Frontiers in Marine Science*, 5, 1. DOI: 10.3389/fmars.2018.00173

Fraissinet, S., De Benedetto, G. E., Malitesta, C., Holzinger, R., & Materić, D. (2024). Microplastics and nanoplastics size distribution in farmed mussel tissues. *Communications Earth & Environment*, 5(1), 128. DOI: 10.1038/s43247-024-01300-2

Franco, A. A., Martín-García, A. P., Egea-Corbacho, A., Arellano, J. M., Albendín, G., Rodríguez-Barroso, R., Quiroga, J. M., & Coello, M. D. (2023). Assessment and accumulation of microplastics in sewage sludge at wastewater treatment plants located in Cádiz, Spain. *Environmental pollution, 317*, 120689. DOI: 10.1016/j.envpol.2022.120689

Fraser, M. A., Chen, L., Ashar, M., Huang, W., Zeng, J., Zhang, C., & Zhang, D. (2020). Occurrence and distribution of microplastics and polychlorinated biphenyls in sediments from the Qiantang River and Hangzhou Bay, China. *Ecotoxicology and Environmental Safety*, 196, 110536. DOI: 10.1016/j.ecoenv.2020.110536

Fred-Ahmadu, O. H., Bhagwat, G., Oluyoye, I., Benson, N. U., Ayejuyo, O. O., & Palanisami, T. (2020). Interaction of chemical contaminants with microplastics: Principles and perspectives. *The Science of the Total Environment*, 706, 135978. DOI: 10.1016/j.scitotenv.2019.135978

Frias, J. P. G. L., Gago, J., Otero, V., & Sobral, P. (2016). Microplastics in coastal sediments from Southern Portuguese shelf waters. *Marine Environmental Research*, 114, 24–30. DOI: 10.1016/j.marenvres.2015.12.006 PMID: 26748246

Frias, J. P. G. L., & Nash, R. (2019). Microplastics: Finding a consensus on the definition. *Marine Pollution Bulletin*, 138, 145–147. DOI: 10.1016/j.marpolbul.2018.11.022 PMID: 30660255

Fu, L., Li, J., Wang, G., Luan, Y., & Dai, W. (2021). Adsorption behavior of organic pollutants on MPs. *Ecotoxicology and Environmental Safety*, 217, 112207. DOI: 10.1016/j.ecoenv.2021.112207

Fuller, S., & Gautam, A. (2016). A procedure for measuring microplastics using pressurized fluid extraction. *Environmental Science & Technology*, 50(11), 5774–5780. DOI: 10.1021/acs.est.6b00816

Furness, R. W. (1985). Plastic particle pollution: Accumulation by procellariiform seabirds at scottish colonies. *Marine Pollution Bulletin*, 16(3), 103–106. DOI: 10.1016/0025-326X(85)90531-4

Furness, R. W., & Camphuysen, K. (1997). Seabirds as monitors of the marine environment. *ICES Journal of Marine Science*, 54(4), 726–737. DOI: 10.1006/jmsc.1997.0243

Furness, R. W., & Tasker, M. L. (2000). Seabird-fishery interactions: Quantifying the sensitivity of seabirds to reductions in sand eel abundance and identification of key areas of sensitive seabirds in the North Sea. *Marine Ecology Progress Series*, 202, 253–264. DOI: 10.3354/meps202253

Galloway, T. S., Cole, M., & Lewis, C. (2017). Interactions of microplastic debris throughout the marine ecosystem. *Nature Ecology & Evolution, 1*(5), 0116.

Gallowaya, T. S., & Lewisa, C. N. (2016). Marine microplastics spell big problems for future generations. *Proceedings of the National Academy of Sciences of the United States of America*, 113(9), 2331–2333. DOI: 10.1073/pnas.1600715113 PMID: 26903632

Galloway, T. S. (2015). Micro- and Nano-plastics and Human Health. In Bergmann, M., Gutow, L., & Klages, M. (Eds.), *Marine Anthropogenic Litter* (pp. 343–366). Springer International Publishing. DOI: 10.1007/978-3-319-16510-3_13

Gall, S. C., & Thompson, R. C. (2015). The impact of debris on marine life. *Marine Pollution Bulletin*, 92(1-2), 170–179. DOI: 10.1016/j.marpolbul.2014.12.041

Gao, M., Liu, Y., Dong, Y., & Song, Z. (2021). Effect of polyethylene particles on dibutyl phthalate toxicity in lettuce (Lactuca sativa L.). *Journal of Hazardous Materials*, 401, 123422. DOI: 10.1016/j.jhazmat.2020.123422

Gao, M., Wang, Z., Jia, Z., Zhang, H., & Wang, T. (2023). Brassinosteroids alleviate nanoplastic toxicity in edible plants by activating antioxidant defense systems and suppressing nanoplastic uptake. *Environment International*, 174, 107901. DOI: 10.1016/j.envint.2023.107901

Gao, Y., & Collins, C. D. (2009). Uptake Pathways of Polycyclic Aromatic Hydrocarbons in White Clover. *Environmental Science & Technology*, 43(16), 6190–6195. DOI: 10.1021/es900662d

Gao, Y., Wang, Q., Ji, G., & Li, A. (2022). Degradation of antibiotic pollutants by persulfate activated with various carbon materials. *Chemical Engineering Journal*, 429, 132387. DOI: 10.1016/j.cej.2021.132387

Garaba, S. P., & Dierssen, H. M. (2018). An airborne remote sensing case study of synthetic hydrocarbon detection using short wave infrared absorption features identified from marine-harvested macro-and microplastics. *Remote Sensing of Environment*, 205, 224–235. DOI: 10.1016/j.rse.2017.11.023

Garrido, I., Pastor-Belda, M., Campillo, N., Viñas, P., Yañez, M. J., Vela, N., Navarro, S., & Fenoll, J. (2019). Photooxidation of insecticide residues by ZnO and TiO2 coated magnetic nanoparticles under natural sunlight. *Journal of Photochemistry and Photobiology A Chemistry*, 372, 245–253. DOI: 10.1016/j.jphotochem.2018.12.027

Gasperi, J., Wright, S. L., Dris, R., Collard, F., Mandin, C., Guerrouache, M., Langlois, V., Kelly, F. J., & Tassin, B. (2018). Microplastics in air: Are we breathing it in? *Current Opinion in Environmental Science & Health*, 1, 1–5. DOI: 10.1016/j.coesh.2017.10.002

Geyer, R., Jambeck, J. R., & Law, K. L. (2017). Production, use, and fate of all plastics ever made. *Science Advances*, 3(7), e1700782. DOI: 10.1126/sciadv.1700782

Ghosh, K., & Jones, B. H. (2021). Roadmap to Biodegradable Plastics—Current State and Research Needs. *ACS Sustainable Chemistry & Engineering*, 9(18), 6170–6187. DOI: 10.1021/acssuschemeng.1c00801

Giardino, M., Balestra, V., Janner, D., & Bellopede, R. (2023). Automated method for routine microplastic detection and quantification. *The Science of the Total Environment*, 859, 160036. DOI: 10.1016/j.scitotenv.2022.160036

Giller, K. E., Delaune, T., Silva, J. V., van Wijk, M., Hammond, J., Descheemaeker, K., van de Ven, G., Schut, A. G. T., Taulya, G., Chikowo, R., & Andersson, J. A. (2021). Small farms and development in sub-Saharan Africa: Farming for food, for income or for lack of better options? *Food Security*, 13(6), 1431–1454. DOI: 10.1007/s12571-021-01209-0

Giorgetti, L., Spanò, C., Muccifora, S., Bottega, S., Barbieri, F., Bellani, L., & Ruffini Castiglione, M. (2020). Exploring the interaction between polystyrene nanoplastics and Allium cepa during germination: Internalization in root cells, induction of toxicity and oxidative stress. *Plant Physiology and Biochemistry*, 149, 170–177. DOI: 10.1016/j.plaphy.2020.02.014

Giri, S., Dimkpa, C. O., Ratnasekera, D., & Mukherjee, A. (2024). Impact of micro and nano plastics on phototrophic organisms in freshwater and terrestrial ecosystems: A review of exposure, internalization, toxicity mechanisms, and eco-corona-dependent mitigation. *Environmental and Experimental Botany*, 219, 105666. DOI: 10.1016/j.envexpbot.2024.105666

Gouin, T. (2020). Toward an improved understanding of the ingestion and trophic transfer of microplastic particles: Critical review and implications for future research. *Environmental Toxicology and Chemistry*, 39(6), 1119–1137. DOI: 10.1002/etc.4718

Gouin, T. (2021). Addressing the importance of microplastic particles as vectors for long range transport of chemical contaminants: Perspective in relation to prioritizing research and regulatory actions. *Microplastics and Nanoplastics*, 1(1), 14. DOI: 10.1186/s43591-021-00016-w

Gramentz, D. (1988). Involvement of loggerhead turtle with the plastic, metal, and hydrocarbon pollution in the central Mediterranean. *Marine Pollution Bulletin*, 19(1), 11–13. DOI: 10.1016/0025-326X(88)90746-1

Granek, E. F., Brander, S. M., & Holland, E. B. (2020). Microplastics in aquatic organisms: Improving understanding and identifying research directions for the next decade. *Limnology and Oceanography Letters*, 5(1), 1–4. DOI: 10.1002/lol2.10145

Grause, G., Kuniyasu, Y., Chien, M.-F., & Inoue, C. (2021). Separation of microplastic from soil by centrifugation and its application to agricultural soil. *Chemosphere*, 132654. Advance online publication. DOI: 10.1016/j.chemosphere.2021.132654 PMID: 34718018

Green, D. S. (2016). Effects of microplastics on European flat oysters, Ostrea edulis and their associated benthic communities. *Environmental Pollution*, 216, 95–103. DOI: 10.1016/j.envpol.2016.05.043

Gregory, M. R. (1991). The hazards of persistent marine pollution: Drift plastics and conservation islands. *Journal of the Royal Society of New Zealand*, 21(2), 83–100. DOI: 10.1080/03036758.1991.10431398

Gui, J., Sun, Y., Wang, J., Chen, X., Zhang, S., & Wu, D. (2021). Microplastics in composting of rural domestic waste: abundance, characteristics, and release from the surface of macroplastics. *Environmental pollution, 274*, 116553. DOI: 10.1016/j.envpol.2021.116553

Guilhermino, L., Martins, A., Cunha, S., & Fernandes, J. O. (2021). Science of the Total Environment Long-term adverse effects of microplastics on Daphnia magna reproduction and population growth rate at increased water temperature and light intensity : Combined effects of stressors and interactions. *The Science of the Total Environment*, 784, 147082. DOI: 10.1016/j.scitotenv.2021.147082 PMID: 33894603

Gu, J., Chen, L., Wan, Y., Teng, Y., Yan, S., & Hu, L. (2022). Experimental Investigation of Water-Retaining and Unsaturated Infiltration Characteristics of Loess Soils Imbued with Microplastics. *Sustainability (Basel)*, 15(1), 62. DOI: 10.3390/su15010062

Guo, J. J., Huang, X. P., Xiang, L., Wang, Y. Z., Li, Y. W., Li, H., Cai, Q.-Y., Mo, C.-H., & Wong, M. H. (2020). Source, migration and toxicology of microplastics in soil. *Environment International*, 137, 105263. DOI: 10.1016/j.envint.2019.105263

Gupta, N., Parsai, T., & Kulkarni, H. V. (2024). A review on the fate of micro and nano plastics (MNPs) and their implication in regulating nutrient cycling in constructed wetland systems. *Journal of Environmental Management*, 350, 119559. DOI: 10.1016/j.jenvman.2023.119559

Gupta, S., Kumar, R., Rajput, A., Gorka, R., Gupta, A., Bhasin, N., Yadav, S., Verma, A., Ram, K., & Bhagat, M. (2023). Atmospheric microplastics: Perspectives on origin, abundances, ecological and health risks. *Environmental Science and Pollution Research International*, 30(49), 107435–107464. DOI: 10.1007/s11356-023-28422-y

Guzzetti, E., Sureda, A., Tejada, S., & Faggio, C. (2018). Microplastic in marine organism: Environmental and toxicological effects. *Environmental Toxicology and Pharmacology*, 64, 164–171. DOI: 10.1016/j.etap.2018.10.009

Haapkylä, J., Unsworth, R. K., Flavell, M., Bourne, D. G., Schaffelke, B., & Willis, B. L. (2011). Seasonal rainfall and runoff promote coral disease on an inshore reef. *PLoS One*, 6(2), e16893. DOI: 10.1371/journal.pone.0016893

Habib, R. Z., Aldhanhani, J. A. K., Ali, A. H., Ghebremedhin, F., Elkashlan, M., Mesfun, M., Kittaneh, W., Al Kindi, R., & Thiemann, T. (2022). Trends of microplastic abundance in personal care products in the United Arab Emirates over the period of 3 years (2018-2020). *Environmental Science and Pollution Research International*, 29(59), 89614–89624. DOI: 10.1007/s11356-022-21773-y

Hahladakis, J. N., Velis, C. A., Weber, R., Iacovidou, E., & Purnell, P. (2018). An overview of chemical additives present in plastics: Migration, release, fate and environmental impact during their use, disposal and recycling. *Journal of Hazardous Materials*, 344, 179–199. DOI: 10.1016/j.jhazmat.2017.10.014

Hale, R. C., Seeley, M. E., La Guardia, M. J., Mai, L., & Zeng, E. Y. (2020). A global perspective on microplastics. *Journal of Geophysical Research: Oceans, 125*(1), e2018JC014719.

Hamilton, L. A., Feit, S., Muffett, C., Kelso, M., Rubright, S. M., Bernhardt, C., . . . Labbé-Bellas, R. (2019). Plastic and Climate: The hidden costs of a plastic planet. *Center for International Environmental Law (CIEL)*. Available online at www.ciel .org/plasticandclimate

Hao, T., Miao, M., Wang, T., Xiao, Y., Yu, B., Zhang, M., Ning, X., & Li, Y. (2023). Physicochemical changes in microplastics and formation of DBPs under ozonation. *Chemosphere*, 327, 138488. DOI: 10.1016/j.chemosphere.2023.138488

Haque, F., & Fan, C. (2023). Fate and impacts of microplastics in the environment: Hydrosphere, pedosphere, and atmosphere. *Environments (Basel, Switzerland)*, 10(5), 70. DOI: 10.3390/environments10050070

Hargreaves, S. K., & Hofmockel, K. S. (2014). Physiological shifts in the microbial community drive changes in enzyme activity in a perennial agroecosystem. *Biogeochemistry*, 117(1), 67–79. DOI: 10.1007/s10533-013-9893-6

Harley-Nyang, D., Memon, F. A., Jones, N., & Galloway, T. (2022). Investigation and analysis of microplastics in sewage sludge and biosolids: A case study from one wastewater treatment works in the UK. *The Science of the Total Environment*, 823, 153735. DOI: 10.1016/j.scitotenv.2022.153735

Hassan, F., Daffa, K., Nabilah, J., & Manh, H. (2023). Microplastic Contamination in Sewage Sludge: Abundance, Characteristics, and Impacts on the Environment and Human Health. *Environmental Technology & Innovation*, 31, 103176. DOI: 10.1016/j.eti.2023.103176

He, J., Han, L., Wang, F., Ma, C., Cai, Y., Ma, W., Xu, E. G., Xing, B., & Yang, Z. (2023). Photocatalytic strategy to mitigate microplastic pollution in aquatic environments: Promising catalysts, efficiencies, mechanisms, and ecological risks. *Critical Reviews in Environmental Science and Technology*, 53(4), 504–526. DOI: 10.1080/10643389.2022.2072658

Helcoski, R., Yonkos, L. T., Sanchez, A., & Baldwin, A. H. (2020). Wetland soil microplastics are negatively related to vegetation cover and stem density. *Environmental Pollution*, 256, 113391. DOI: 10.1016/j.envpol.2019.113391

Heléne Österlund. (2023) Microplastics in urban catchments: Review of sources, pathways, and entry into stormwater. Science of The Total Environment, 858(1), 159781. DOI: 10.1016/j.scitotenv.2022.159781

Henderson, L., & Green, C. (2020). Making sense of microplastics? Public understandings of plastic pollution. *Marine Pollution Bulletin*, 152, 110908. DOI: 10.1016/j.marpolbul.2020.110908 PMID: 32479284

Henoumont, C., Devreux, M., & Laurent, S. (2023). Mn-based MRI contrast agents: An overview. *Molecules (Basel, Switzerland)*, 28(21), 7275. DOI: 10.3390/molecules28217275

Hentati-Sundberg, J., Raymond, C., Sköld, M., Svensson, O., Gustafsson, B., & Bonaglia, S. (2020). Fueling of a marine-terrestrial ecosystem by a major seabird colony. *Scientific Reports*, 10(1), 15455. DOI: 10.1038/s41598-020-72238-6 PMID: 32963305

Heo, Y., Lee, E.-H., & Lee, S.-W. (2022). Adsorptive removal of micron-sized polystyrene particles using magnetic iron oxide nanoparticles. *Chemosphere*, 307, 135672. DOI: 10.1016/j.chemosphere.2022.135672

Hermabessiere, L., Dehaut, A., Paul-Pont, I., Lacroix, C., Jezequel, R., Soudant, P., & Duflos, G. (2017). Occurrence and effects of plastic additives on marine environments and organisms: A review. *Chemosphere*, 182, 781–793. DOI: 10.1016/j.chemosphere.2017.05.096 PMID: 28545000

Hermsen, E., Mintenig, S. M., Besseling, E., & Koelmans, A. A. (2018). Quality Criteria for the Analysis of Microplastic in Biota Samples. *Critical Reviews in Environmental Science and Technology*, 52(18), 10230–10240. DOI: 10.1021/acs.est.8b01611 PMID: 30137965

Hettiarachchi, H., & Meegoda, J. N. (2023). Microplastic pollution prevention: The need for robust policy interventions to close the loopholes in current waste management practices. *International Journal of Environmental Research and Public Health*, 20(14), 6434. DOI: 10.3390/ijerph20146434 PMID: 37510666

Hidalgo-ruz, V., Gutow, L., Thompson, R. C., & Thiel, M. (2012). *Microplastics in the Marine Environment: A Review of the Methods Used for Identification and Quantification*. DOI: 10.1021/es2031505

Hidayaturrahman, H., & Lee, T.-G. (2019). A study on characteristics of microplastic in wastewater of South Korea: Identification, quantification, and fate of microplastics during treatment process. *Marine Pollution Bulletin*, 146, 696–702. DOI: 10.1016/j.marpolbul.2019.06.071

Hirai, H., Takada, H., Ogata, Y., Yamashita, R., Mizukawa, K., Saha, M., Kwan, C., Moore, C., Gray, H., Laursen, D., Zettler, E. R., Farrington, J. W., Reddy, C. M., Peacock, E. E., & Ward, M. W. (2011). Organic micropollutants in marine plastics debris from the open ocean and remote and urban beaches. *Marine Pollution Bulletin*, 62(8), 1683–1692. DOI: 10.1016/j.marpolbul.2011.06.004 PMID: 21719036

Hoegh-Guldberg, O., Jacob, D., Taylor, M., Guillén Bolaños, T., Bindi, M., Brown, S., Camilloni, I. A., Diedhiou, A., Djalante, R., Ebi, K., Engelbrecht, F., Guiot, J., Hijioka, Y., Mehrotra, S., Hope, C. W., Payne, A. J., Pörtner, H.-O., Seneviratne, S. I., Thomas, A., & Zhou, G. (2019). The human imperative of stabilizing global climate change at 1.5°C. *Science*, 365(6459), eaaw6974. DOI: 10.1126/science.aaw6974

Høiberg, M. A., Woods, J. S., & Verones, F. (2022). Global distribution of potential impact hotspots for marine plastic debris entanglement. *Ecological Indicators*, 135, 108509. DOI: 10.1016/j.ecolind.2021.108509

Hong, P., Xiao, J., Liu, H., Niu, Z., Ma, Y., Wang, Q., Zhang, D., & Ma, Y. (2024). An inversion model of microplastics abundance based on satellite remote sensing: A case study in the Bohai Sea. *The Science of the Total Environment*, 909, 168537. DOI: 10.1016/j.scitotenv.2023.168537

Hopewell, J., Dvorak, R., & Kosior, E. (2009). Plastics recycling: Challenges and opportunities. *Philosophical Transactions of the Royal Society of London. Series B, Biological Sciences*, 364(1526), 2115–2126. DOI: 10.1098/rstb.2008.0311

Horton, A. A. (2022). Plastic pollution: When do we know enough? *Journal of Hazardous Materials*, 422, 126885. DOI: 10.1016/j.jhazmat.2021.126885

Horton, A. A., & Dixon, S. J. (2018). MPs: An introduction to environmental transport processes. *WIREs. Water*, 5(2), e1268. DOI: 10.1002/wat2.1268

Horton, A. A., Walton, A., Spurgeon, D. J., Lahive, E., & Svendsen, C. (2017). Microplastics in freshwater and terrestrial environments: Evaluating the current understanding to identify the knowledge gaps and future research priorities. *The Science of the Total Environment*, 586, 127–141. DOI: 10.1016/j.scitotenv.2017.01.190

Huang, G., Quershi, M., Song, L., & Di, S. H. (2023, July 2). Yao, Sun w. *The Science of the Total Environment*, 896. Advance online publication. DOI: 10.1016/j.scitotenv.2023.165308 PMID: 37414186

Huerta Lwanga, E., Mendoza Vega, J., Ku Quej, V., Chi, J. L. A., Sanchez Del Cid, L., Chi, C., Escalona Segura, G., Gertsen, H., Salánki, T., van der Ploeg, M., Koelmans, A. A., & Geissen, V. (2017). Field evidence for transfer of plastic debris along a terrestrial food chain. *Scientific Reports*, 7(1), 14071. DOI: 10.1038/s41598-017-14588-2

Hufnagl, B., Stibi, M., Martirosyan, H., Wilczek, U., Möller, J. N., Löder, M. G. J., Laforsch, C., & Lohninger, H. (2022). Computer-Assisted Analysis of Microplastics in Environmental Samples Based on μFTIR Imaging in Combination with Machine Learning. *Environmental Science & Technology Letters*, 9(1), 90–95. DOI: 10.1021/acs.estlett.1c00851

Hurley, R., & Nizzetto, L. (2017). Fate and occurrence of micro (nano) plastics in soils: Knowledge gaps and possible risks. *Current Opinion in Environmental Science & Health*, 1, 6–11. DOI: 10.1016/j.coesh.2017.10.006

ICES2018. Report of the Joint OSPAR/HELCOM/ICES Working Group on Marine Birds (JWGBIRD). 1–5 October 2018, Ostende, Belgium. icescm2017/acom:24, 75 pp. (link) 2018

Iheanacho, S., Ogbu, M., Bhuyan, M. S., & Ogunji, J. (2023). Microplastic pollution: An emerging contaminant in aquaculture. *Aquaculture and Fisheries*, 8(6), 603–616. DOI: 10.1016/j.aaf.2023.01.007

Iqbal, S., Xu, J., Allen, S. D., Khan, S., Nadir, S., Arif, M. S., & Yasmeen, T. (2020). Unraveling consequences of soil micro- and nano-plastic pollution for soil-plant system with implications for nitrogen (N) cycling and soil microbial activity. .DOI: 10.1016/j.chemosphere.2020.127578

Isari, E. A., Papaioannou, D., Kalavrouziotis, I. K., & Karapanagioti, H. K. (2021). Microplastics in Agricultural Soils: A Case Study in Cultivation of Watermelons and Canning Tomatoes. *Water (Basel)*, 13(16), 2168. DOI: 10.3390/w13162168

Issac, M. N., & Kandasubramanian, B. (2021). Effect of microplastics in water and aquatic systems. *Environmental Science and Pollution Research International*, 28(16), 19544–19562. DOI: 10.1007/s11356-021-13184-2

Jackson, J. B. C. (2008). Ecological extinction and evolution in the brave new ocean. *Proceedings of the National Academy of Sciences of the United States of America*, 105(Supplement 1), 11458–11465. DOI: 10.1073/pnas.0802812105 PMID: 18695220

Jamieson, A. J., Brooks, L. S. R., Reid, W. D. K., Piertney, S. B., Narayanaswamy, B. E., & Linley, T. D. (2019). Microplastics and synthetic particles ingested by deep-sea amphipods in six of the deepest marine ecosystems on Earth. *Royal Society Open Science*, 6(2), 180667. DOI: 10.1098/rsos.180667 PMID: 30891254

Janakiraman, V., Manjunathan, J., SampathKumar, B., Thenmozhi, M., Ramasamy, P., Kannan, K., Ahmad, I., Asdaq, S. M. B., & Sivaperumal, P. (2024). Applications of fungal based nanoparticles in cancer therapy-A review. *Process Biochemistry (Barking, London, England)*, 140, 10–18. DOI: 10.1016/j.procbio.2024.02.002

Jang, S., Kim, J. H., & Kim, J. (2021). Detection of Microplastics in Water and Ice. *Remote Sensing (Basel)*, 13(17), 3532. DOI: 10.3390/rs13173532

Jenssen, B. M., Aarnes, J. B., Murvoll, K. M., Herzke, D., & Nygård, T. (2010). Fluctuating wing asymmetry and hepatic concentrations of persistent organic pollutants are associated in European shag (Phalacrocorax aristotelis) chicks. *The Science of the Total Environment*, 408(3), 578–585. DOI: 10.1016/j.scitotenv.2009.10.036 PMID: 19896702

Jeong, Y., Gong, G., Lee, H.-J., Seong, J., Hong, S. W., & Lee, C. (2023). Transformation of microplastics by oxidative water and wastewater treatment processes: A critical review. *Journal of Hazardous Materials*, 443, 130313. DOI: 10.1016/j.jhazmat.2022.130313

Jia, L., Liu, L., Zhang, Y., Fu, W., Liu, X., Wang, Q., Tanveer, M., & Huang, L. (2023). Microplastic stress in plants: Effects on plant growth and their remediations. *Frontiers in Plant Science*, 14, 1226484. DOI: 10.3389/fpls.2023.1226484

Jiang, L., Ye, Y., Han, Y., Wang, Q., Lu, H., Li, J., Qian, W., Zeng, X., Zhang, Z., Zhao, Y., Shi, J., Luo, Y., Qiu, Y., Sun, J., Sheng, J., Huang, H., & Qian, P. (2024). Microplastics dampen the self-renewal of hematopoietic stem cells by disrupting the gut microbiota-hypoxanthine-Wnt axis. *Cell Discovery*, 10(1), 35. DOI: 10.1038/s41421-024-00665-0

Jiang, R., Lu, G., Yan, Z., Liu, J., Wu, D., & Wang, Y. (2021). Microplastic degradation by hydroxy-rich bismuth oxychloride. *Journal of Hazardous Materials*, 405, 124247. DOI: 10.1016/j.jhazmat.2020.124247

Jiang, X., Chen, H., Liao, Y., Ye, Z., Li, M., & Klobučar, G. (2019). Ecotoxicity and genotoxicity of polystyrene microplastics on higher plant Vicia faba. *Environmental Pollution*, 250, 831–838. DOI: 10.1016/j.envpol.2019.04.055

Jia, W., Karapetrova, A., Zhang, M., Xu, L., Li, K., Huang, M., Wang, J., & Huang, Y. (2022). Automated identification and quantification of invisible microplastics in agricultural soils. *The Science of the Total Environment*, 844, 156853. DOI: 10.1016/j.scitotenv.2022.156853

Jing, M., Zhang, H., Wei, M., Tang, Y., Xia, Y., Chen, Y., Shen, Z., & Chen, C. (2022). Reactive Oxygen Species Partly Mediate DNA Methylation in Responses to Different Heavy Metals in Pokeweed. *Frontiers in Plant Science*, 13, 845108. DOI: 10.3389/fpls.2022.845108

Jong, M. C., Tong, X., Li, J., Xu, Z., Chng, S. H. Q., He, Y., & Gin, K. Y. H. (2022). Microplastics in equatorial coasts: Pollution hotspots and spatiotemporal variations associated with tropical monsoons. *Journal of Hazardous Materials*, 424, 127626. DOI: 10.1016/j.jhazmat.2021.127626

Ju, S., Shin, G., Lee, M., Koo, J. M., Jeon, H., Ok, Y. S., Hwang, D. S., Hwang, S. Y., Oh, D. X., & Park, J. (2021). *Biodegradable chito-beads replacing non-biodegradable microplastics for cosmetics*. DOI: 10.1039/D1GC01588E

Jubsilp, C., Asawakosinchai, A., Mora, P., Saramas, D., & Rimdusit, S. (2021). Effects of organic based heat stabilizer on properties of polyvinyl chloride for pipe applications: A comparative study with Pb and CaZn systems. *Polymers*, 14(1), 133. DOI: 10.3390/polym14010133

Jung, M. R., Horgen, F. D., Orski, S. V., Rodriguez, V., Beers, K. L., Balazs, G. H., Jones, T. T., Work, T. M., Brignac, K. C., Royer, S. J., & Hyrenbach, K. D. (2018). Validation of ATR FT-IR to identify polymers of plastic marine debris, including those ingested by marine organisms. *Marine Pollution Bulletin*, 127, 704–716. DOI: 10.1016/j.marpolbul.2017.12.061

Jung, S., Raghavendra, A. J., & Patri, A. K. (2023). Comprehensive analysis of common polymers using hyphenated TGA-FTIR-GC/MS and Raman spectroscopy towards a database for micro- and nanoplastics identification, characterization, and quantitation. *NanoImpact*, 30, 100467. DOI: 10.1016/j.impact.2023.100467

Júnior, R. (2024). Towards to Battery Digital Passport: Reviewing Regulations and Standards for Second-Life Batteries. *Batteries*, 10(4), 115. DOI: 10.3390/batteries10040115

Juying, C., Lead, W., Kiho, K., Ofiara, D., Zhao, Y., Bera, A., Lohmann, R., & Baker, M. C. (2016). Part V. Assessment of Other Human Activities and the Marine Environment. [First World Ocean Assessment]. *First Global Integrated Marine Assessment*, 2011, 1–34.

Kalčíková, G., Gotvajn, A. Ž., Kladnik, A., & Jemec, A. (2017, November). Impact of Polyethylene Microbeads on the Floating Freshwater Plant Duckweed Lemna Minor. *Environmental Pollution*, 230, 1108–1115. DOI: 10.1016/j.envpol.2017.07.050

Kang, J., Zhou, L., Duan, X., Sun, H., Ao, Z., & Wang, S. (2019). Degradation of cosmetic microplastics via functionalized carbon nanosprings. *Matter*, 1(3), 745–758. DOI: 10.1016/j.matt.2019.06.004

Kannankai, M. P., Alex, R. K., Muralidharan, V. V., Nazeerkhan, N. P., Radhakrishnan, A., & Devipriya, S. P. (2022). Urban mangrove ecosystems are under severe threat from microplastic pollution.

Kanwar, M. K., Yu, J., & Zhou, J. (2018). Phytomelatonin: Recent advances and future prospects. *Journal of Pineal Research*, 65(4), e12526. DOI: 10.1111/jpi.12526

Karasik, R., Vegh, T., Diana, Z., Bering, J., Caldas, J., Pickle, A., Rittschof, D., & Virdin, J. (n.d.). *20 Years of Government Responses to the Global Plastic Pollution Problem.*

Karbalaei, S., Golieskardi, A., Watt, D. U., Boiret, M., Hanachi, P., Walker, T. R., & Karami, A. (2020). Analysis and inorganic composition of microplastics in commercial Malaysian fish meals. *Marine Pollution Bulletin*, 150, 110687. DOI: 10.1016/j.marpolbul.2019.110687

Katsumi, N., Kusube, T., Nagao, S., & Okochi, H. (2021). The input-output balance of microplastics derived from coated fertilizer in paddy fields and the timing of their discharge during the irrigation season. *Chemosphere*, 279, 130574. DOI: 10.1016/j.chemosphere.2021.130574

Katyal, D., Kong, E., & Villanueva, J. (2020). *Microplastics in the environment : impact on human health and future mitigation strategies*. DOI: 10.5864/d2020-005

Kaur, M., Xu, M., & Wang, L. (2022). Cyto–Genotoxic Effect Causing Potential of Polystyrene Micro-Plastics in Terrestrial Plants. *Nanomaterials (Basel, Switzerland)*, 12(12), 2024. DOI: 10.3390/nano12122024

Kedzierski, M., Falcou-Préfol, M., Kerros, M. E., Henry, M., Pedrotti, M. L., & Bruzaud, S. (2019). A machine learning algorithm for high throughput identification of FTIR spectra: Application on microplastics collected in the Mediterranean Sea. *Chemosphere*, 234, 242–251. DOI: 10.1016/j.chemosphere.2019.05.113

Kedzierski, M., Palazot, M., Soccalingame, L., Falcou-Préfol, M., Gorsky, G., Galgani, F., Bruzaud, S., & Pedrotti, M. L. (2022). Chemical composition of microplastics floating on the surface of the Mediterranean Sea. *Marine Pollution Bulletin*, 174, 113284. DOI: 10.1016/j.marpolbul.2021.113284 PMID: 34995887

Keith, A. (2002, February). Grasman, Assessing Immunological Function in Toxicological Studies of Avian Wildlife. *Integrative and Comparative Biology*, 42(1), 34–42. DOI: 10.1093/icb/42.1.34 PMID: 21708692

Keogh, E. (2010). Instance-Based Learning. In Sammut, C., & Webb, G. I. (Eds.), *Encyclopedia of Machine Learning* (pp. 549–550). Springer US. DOI: 10.1007/978-0-387-30164-8_409

Khairudin, K., Bakar, N. F. A., & Osman, M. S. (2022). Magnetically recyclable flake-like BiOI-Fe3O4 microswimmers for fast and efficient degradation of microplastics. *Journal of Environmental Chemical Engineering*, 10(5), 108275. DOI: 10.1016/j.jece.2022.108275

Khan, A., Qadeer, A., Wajid, A., Ullah, Q., Rahman, S. U., Ullah, K., Safi, S. Z., Ticha, L., Skalickova, S., Chilala, P., Bernatova, S., Samek, O., & Horky, P. (2024). Microplastics in animal nutrition: Occurrence, spread, and hazard in animals. *Journal of Agriculture and Food Research*, 17, 101258. DOI: 10.1016/j.jafr.2024.101258

Khan, N. A., Khan, A. H., López-Maldonado, E. A., Alam, S. S., López López, J. R., Méndez Herrera, P. F., Mohamed, B. A., Mahmoud, A. E. D., Abutaleb, A., & Singh, L. (2022). Microplastics: Occurrences, treatment methods, regulations and foreseen environmental impacts. *Environmental Research*, 215, 114224. DOI: 10.1016/j.envres.2022.114224 PMID: 36058276

Kibria, M., Masuk, N. I., Safayet, R., Nguyen, H. Q., & Mourshed, M. (2023). Plastic Waste: Challenges and Opportunities to Mitigate Pollution and Effective Management. *International Journal of Environmental Research*, 17(1), 20. DOI: 10.1007/s41742-023-00507-z PMID: 36711426

Kiendrebeogo, M., Estahbanati, M. K., Mostafazadeh, A. K., Drogui, P., & Tyagi, R. D. (2021). Treatment of microplastics in water by anodic oxidation: A case study for polystyrene. *Environmental Pollution*, 269, 116168. DOI: 10.1016/j.envpol.2020.116168

Kim, S., Hyeon, Y., & Park, C. (2023). Microplastics' Shape and Morphology Analysis in the Presence of Natural Organic Matter Using Flow Imaging Microscopy. *Molecules (Basel, Switzerland)*, 28(19), 6913. DOI: 10.3390/molecules28196913

Kim, S., Sin, A., Nam, H., Park, Y., Lee, H., & Han, C. (2022). Advanced oxidation processes for microplastics degradation: A recent trend. *Chemical Engineering Journal Advances*, 9, 100213. DOI: 10.1016/j.ceja.2021.100213

Kingsford, R. T., Watson, J. E. M., Lundquist, C. J., Venter, O., Hughes, L., Johnston, E. L., Atherton, J., Gawel, M., Keith, D. A., Mackey, B. G., Morley, C., Possingham, H. P., Raynor, B., Recher, H. F., & Wilson, K. A. (2009). Major conservation policy issues for biodiversity in Oceania. *Conservation Biology*, 23(4), 834–840. DOI: 10.1111/j.1523-1739.2009.01287.x PMID: 19627315

Kiran, B. R., Kopperi, H., & Venkata Mohan, S. (2022). Micro/nano-plastics occurrence, identification, risk analysis and mitigation: Challenges and perspectives. *Reviews in Environmental Science and Biotechnology*, 21(1), 169–203. DOI: 10.1007/s11157-021-09609-6

Kleit, A. N. (1992). Enforcing time-inconsistent regulation. *Economic Inquiry*, 30(4), 639–648. DOI: 10.1111/j.1465-7295.1992.tb01286.x

Koelmans, A. A., Mohamed Nor, N. H., Hermsen, E., Kooi, M., Mintenig, S. M., & De France, J. (2019). Microplastics in freshwaters and drinking water: Critical review and assessment of data quality. *Water Research*, 155, 410–422. DOI: 10.1016/j.watres.2019.02.054 PMID: 30861380

Kole P.J. (2017). Wear and Tear of Tyres: A Stealthy Source of Microplastics in the Environment. Int J Environ Res Public Health, 20(14), 1265. DOI: 10.3390/ijerph14101265

Kouchakipour, S., Hosseinzadeh, M., Qaretapeh, M. Z., & Dashtian, K. (2024). Sustainable large-scale Fe3O4/carbon for enhanced polystyrene nanoplastics removal through magnetic adsorption coagulation. *Journal of Water Process Engineering*, 58, 104919. DOI: 10.1016/j.jwpe.2024.104919

Krystosik, A., Njoroge, G., Odhiambo, L., Forsyth, J. E., Mutuku, F., & LaBeaud, A. D. (2020). Solid wastes provide breeding sites, burrows, and food for biological disease vectors, and urban zoonotic reservoirs: A call to action for solutions-based research. *Frontiers in Public Health*, 7, 405. DOI: 10.3389/fpubh.2019.00405

Kukulka, T., Proskurowski, G., Morét-Ferguson, S., Meyer, D. W., & Law, K. L. (2012). The effect of wind mixing on the vertical distribution of buoyant plastic debris. *Geophysical Research Letters*, 39(7), 2012GL051116. DOI: 10.1029/2012GL051116

Kumar, M., Chen, H., Sarsaiya, S., Qin, S., Liu, H., Awasthi, M. K., Kumar, S., Singh, L., Zhang, Z., Bolan, N. S., Pandey, A., Varjani, S., & Taherzadeh, M. J. (2021). Current research trends on micro-and nano-plastics as an emerging threat to global environment: A review. *Journal of Hazardous Materials*, 409, 124967. DOI: 10.1016/j.jhazmat.2020.124967 PMID: 33517026

Kumar, R., Verma, A., Shome, A., Sinha, R., Sinha, S., Jha, P. K., Kumar, R., Kumar, P., Shubham, , Das, S., Sharma, P., & Vara Prasad, P. V. (2021). Impacts of Plastic Pollution on Ecosystem Services, Sustainable Development Goals, and Need to Focus on Circular Economy and Policy Interventions. *Sustainability (Basel)*, 13(17), 9963. DOI: 10.3390/su13179963

Kumar, V., Singh, E., Singh, S., Pandey, A., & Bhargava, P. C. (2023). Micro-and nano-plastics (MNPs) as emerging pollutant in ground water: Environmental impact, potential risks, limitations and way forward towards sustainable management. *Chemical Engineering Journal*, 459, 141568. DOI: 10.1016/j.cej.2023.141568

Kurniawan, T. A., Haider, A., Mohyuddin, A., Fatima, R., Salman, M., Shaheen, A., Ahmad, H. M., Al-Hazmi, H. E., Othman, M. H. D., Aziz, F., Anouzla, A., & Ali, I. (2023). Tackling microplastics pollution in global environment through integration of applied technology, policy instruments, and legislation. *Journal of Environmental Management*, 346, 118971. DOI: 10.1016/j.jenvman.2023.118971

Kwon, B. G., Saido, K., Koizumi, K., Sato, H., Ogawa, N., Chung, S.-Y., Kusui, T., Kodera, Y., & Kogure, K. (2014). Regional distribution of styrene analogues generated from polystyrene degradation along the coastlines of the North-East Pacific Ocean and Hawaii. *Environmental Pollution*, 188, 45–49. DOI: 10.1016/j.envpol.2014.01.019 PMID: 24553245

Laist, D. W. (1987). Overview of the biological effects of lost and discarded plastic debris in the marine environment. *Marine Pollution Bulletin*, 18(6, Suppl. B), 319–326. DOI: 10.1016/S0025-326X(87)80019-X

Laist, D. W. (1997). Impacts of marine debris: Entanglement of marine life in marine debris including a comprehensive list of species with entanglement and ingestion records. *Marine Pollution Bulletin*, 18(6, Suppl. B), 99–139.

Landrigan P.J. (2023) The Minderoo-Monaco Commission on Plastics and Human Health. Ann Glob Health, 21(89), 23. DOI: 10.5334/aogh.4056

Lapyote Prasittisopin. (2023) Microplastics in construction and built environment. Developments in the Built Environment, 15. DOI: 10.1016/j.dibe.2023.100188

Lascelles, B. G., Langham, G. M., Ronconi, R. A., & Reid, J. B. (2012). From hotspots to site protection: Identifying Marine Protected Areas for seabirds around the globe. *Biological Conservation*, 156, 5–14. DOI: 10.1016/j.biocon.2011.12.008

Lavers, J. L., & Bond, A. L. (2023). Long-term decline in fledging body condition of Flesh-footed Shearwaters (Ardenna carneipes). *ICES Journal of Marine Science*, 0(4), 1–7. DOI: 10.1093/icesjms/fsad048

Lavers, J. L., Bond, A. L., & Huton, I. (2014). Plastic ingestion by flesh-footed shearwaters (Puffinus carneipes): Implications for fledgling body condition and the accumulation of plastic-derived chemicals. *Environmental Pollution*, 187, 124–129. DOI: 10.1016/j.envpol.2013.12.020 PMID: 24480381

Lavers, J. L., Hutton, I., & Bond, A. L. (2019). Clinical pathology of plastic ingestion in marine birds and relationships with blood chemistry. *Environmental Science & Technology*, 53(15), 9224–9231. DOI: 10.1021/acs.est.9b02098 PMID: 31304735

Lehmann, A., Leifheit, E. F., Feng, L., Bergmann, J., Wulf, A., & Rillig, M. C. (2020). Microplastic fiber and drought effects on plants and soil are only slightly modified by arbuscular mycorrhizal fungi. *Soil Ecology Letters*, 1–13.

Lei, B., Bissonnette, J. R., Hogan, Ú. E., Bec, A. E., Feng, X., & Smith, R. D. L. (2022). Customizable Machine-Learning Models for Rapid Microplastic Identification Using Raman Microscopy. *Analytical Chemistry*, 94(49), 17011–17019. DOI: 10.1021/acs.analchem.2c02451

Lenz, R., Enders, K., Stedmon, C. A., MacKenzie, D. M. A., & Nielsen, T. G. (2015). A critical assessment of visual identification of marine microplastic using Raman spectroscopy for analysis improvement. *Marine Pollution Bulletin*, 100(1), 82–91. DOI: 10.1016/j.marpolbul.2015.09.026 PMID: 26455785

Letcher, R. J., Bustnes, J. O., Dietz, R., Jenssen, B. M., Jørgensen, E. H., Sonne, C., Verreault, J., Vijayan, M. M., & Gabrielsen, G. W. (2010). Exposure and effects assessment of persistent organohalogen contaminants in arctic wildlife and fish. *The Science of the Total Environment*, 408(15), 2995–3043. DOI: 10.1016/j.scitotenv.2009.10.038 PMID: 19910021

Le, V. G., Nguyen, M. K., Nguyen, H. L., Lin, C., Hadi, M., Hung, N. T. Q., Hoang, H.-G., Nguyen, K. N., Tran, H.-T., Hou, D., Zhang, T., & Bolan, N. S. (2023). A comprehensive review of micro-and nano-plastics in the atmosphere: Occurrence, fate, toxicity, and strategies for risk reduction. *The Science of the Total Environment*, 904, 166649. DOI: 10.1016/j.scitotenv.2023.166649 PMID: 37660815

Li, H., Luo, Q. P., Zhao, S., Zhou, Y. Y., Huang, F. Y., Yang, X. R., & Su, J. Q. (2023). Effect of phenol formaldehyde-associated microplastics on soil microbial community, assembly, and functioning. *Journal of hazardous materials, 443*(Pt B), 130288. DOI: 10.1016/j.jhazmat.2022.130288

Li, S., Ding, F., Flury, M., Wang, Z., Xu, L., Li, S., Jones, D. L., & Wang, J. (2022). Macro- and microplastic accumulation in soil after 32 years of plastic film mulching. *Environmental pollution, 300*, 118945. DOI: 10.1016/j.envpol.2022.118945

Liang, Y., Lehmann, A., Yang, G., Leifheit, E. F., & Rillig, M. C. (2021). Effects of Microplastic Fibers on Soil Aggregation and Enzyme Activities Are Organic Matter Dependent. *Frontiers in Environmental Science*, 9, 650155. DOI: 10.3389/fenvs.2021.650155

Lian, J., Liu, W., Meng, L., Wu, J., Zeb, A., Cheng, L., & Sun, H. (2021). Effects of microplastics derived from polymer-coated fertilizer on maize growth, rhizosphere, and soil properties. *Journal of Cleaner Production*, 318, 128571. DOI: 10.1016/j.jclepro.2021.128571

Liao, H., Gao, D., Kong, C., Junaid, M., Li, Y., Chen, X., Zheng, Q., Chen, G., & Wang, J. (2023). Trophic transfer of nanoplastics and di (2-ethylhexyl) phthalate in a freshwater food chain (Chlorella pyrenoidosa-Daphnia magna-Micropterus salmoides) induced disturbance of lipid metabolism in fish. *Journal of Hazardous Materials*, 459, 132294. DOI: 10.1016/j.jhazmat.2023.132294

Li, J., Liu, H., & Paul Chen, J. (2018). Microplastics in freshwater systems: A review on occurrence, environmental effects, and methods for microplastics detection. *Water Research*, 137(May), 362–374. DOI: 10.1016/j.watres.2017.12.056 PMID: 29580559

Li, J., Li, Y., Xiong, Z., Yao, G., & Lai, B. (2019). The electrochemical advanced oxidation processes coupling of oxidants for organic pollutants degradation: A mini-review. *Chinese Chemical Letters*, 30(12), 2139–2146. DOI: 10.1016/j.cclet.2019.04.057

Li, J., Song, Y., & Cai, Y. (2020). Focus topics on microplastics in soil: Analytical methods, occurrence, transport, and ecological risks. *Environmental Pollution*, 257, 113570. DOI: 10.1016/j.envpol.2019.113570

Li, K., Jia, W., Xu, L., Zhang, M., & Huang, Y. (2023). The plastisphere of biodegradable and conventional microplastics from residues exhibit distinct microbial structure, network and function in plastic-mulching farmland. *Journal of Hazardous Materials*, 442, 130011. DOI: 10.1016/j.jhazmat.2022.130011

Lin, D., Yang, G., Dou, P., Qian, S., Zhao, L., Yang, Y., & Fanin, N. (2020). Microplastics negatively affect soil fauna but stimulate microbial activity: insights from a field-based microplastic addition experiment. *Proceedings. Biological sciences, 287*(1934), 20201268. https://doi.org/DOI: 10.1098/rspb.2020.1268

Lin, D., Yang, G., Dou, P., Qian, S., Zhao, L., Yang, Y., Fanin, N. (2020). MPs negatively affect soil fauna but stimulate microbial activity: Insights from a field-based microplastic addition experiment. *Proc. r. Soc*, 287(1934), 20201268.

Lin, J., Liu, H., & Zhang, J. (2022). Recent advances in the application of machine learning methods to improve identification of the microplastics in environment. *Chemosphere*, 307, 136092. DOI: 10.1016/j.chemosphere.2022.136092

Lin, J., Yan, D., Fu, J., Chen, Y., & Ou, H. (2020). Ultraviolet-C and vacuum ultraviolet inducing surface degradation of microplastics. *Water Research*, 186, 116360. DOI: 10.1016/j.watres.2020.116360

Lin, Z., Jin, T., Zou, T., Xu, L., Xi, B., Xu, D., & Peng, J. (2022). Current progress on plastic/microplastic degradation: Fact influences and mechanism. *Environmental Pollution*, 304, 119159. DOI: 10.1016/j.envpol.2022.119159

Li, P. C., Li, X. N., Du, Z. H., Wang, H., Yu, Z.-R., & Li, J.-L. (2018). Di(2-ethylhexyl) phthalate (DEHP) induced kidney injury in quail (Coturnix Japonica) via inhibiting HSF1/ HSF3-dependent heat shock response. *Chemosphere*, 209, 981–988. DOI: 10.1016/j.chemosphere.2018.06.158 PMID: 30114749

Li, S., Guo, J., Wang, T., Gong, L., Liu, F., Brestic, M., Liu, S., Song, F., & Li, X. (2021). Melatonin reduces nanoplastic uptake, translocation, and toxicity in wheat. *Journal of Pineal Research*, 71(3), e12761. DOI: 10.1111/jpi.12761

Li, S., Wang, T., Guo, J., Dong, Y., Wang, Z., Gong, L., & Li, X. (2021). Polystyrene microplastics disturb the redox homeostasis, carbohydrate metabolism and phytohormone regulatory network in barley. *Journal of Hazardous Materials*, 415, 125614. DOI: 10.1016/j.jhazmat.2021.125614

Li, T., Cao, X., Zhao, R., & Cui, Z. (2023). Stress response to nano-plastics with different charges in Brassica napus L. during seed germination and seedling growth stages. *Frontiers of Environmental Science & Engineering*, 17(4), 43. DOI: 10.1007/s11783-023-1643-y

Li, T., Cui, L., Xu, Z., Liu, H., Cui, X., & Fantke, P. (2023). Micro-and nanoplastics in soil: Linking sources to damage on soil ecosystem services in life cycle assessment. *The Science of the Total Environment*, 904, 166925. DOI: 10.1016/j.scitotenv.2023.166925

Liu, Z., Wang, G., Sheng, C., Zheng, Y., Tang, D., Zhang, Y., Hou, X., Yao, M., Zong, Q. & Zhou, Z. (2024). Intracellular Protein Adsorption Behavior and Biological Effects of Polystyrene Nanoplastics in THP-1 Cells. *Environmental Science & Technology*.

Liu, F., Olesen, K. B., Borregaard, A. R., & Vollertsen, J. (2019). Microplastics in urban and highway stormwater retention ponds. *The Science of the Total Environment*, 671, 992–1000. DOI: 10.1016/j.scitotenv.2019.03.416

Liu, H., Yang, X., Liu, G., Liang, C., Xue, S., Chen, H., Ritsema, C. J., & Geissen, V. (2017). Response of soil dissolved organic matter to microplastic addition in Chinese loess soil. *Chemosphere*, 185, 907–917. DOI: 10.1016/j.chemosphere.2017.07.064

Liu, J., Ma, Y., Zhu, D., Xia, T., Qi, Y., Yao, Y., Guo, X., Ji, R., & Chen, W. (2018). Polystyrene nanoplastics-enhanced contaminant transport: Role of irreversible adsorption in glassy polymeric domain. *Environmental Science & Technology*, 52(5), 2677–2685. DOI: 10.1021/acs.est.7b05211

Liu, K., Pang, X., Chen, H., & Jiang, L. (2024). Visual detection of microplastics using Raman spectroscopic imaging. *Analyst*, 149(1), 161–168. DOI: 10.1039/D3AN01270K

Liu, P., Qian, L., Wang, H., Zhan, X., Lu, K., Gu, C., & Gao, S. (2019). New insights into the aging behavior of microplastics accelerated by advanced oxidation processes. *Environmental Science & Technology*, 53(7), 3579–3588. DOI: 10.1021/acs.est.9b00493

Liu, P., Zhan, X., Wu, X., Li, J., Wang, H., & Gao, S. (2020). Effect of weathering on environmental behavior of microplastics: Properties, sorption and potential risks. *Chemosphere*, 242, 125193. DOI: 10.1016/j.chemosphere.2019.125193

Liu, W., Pan, T., Liu, H., Jiang, M., & Zhang, T. (2023). Adsorption behavior of imidacloprid pesticide on polar microplastics under environmental conditions: Critical role of photo-aging. *Frontiers of Environmental Science & Engineering*, 17(4), 41. DOI: 10.1007/s11783-023-1641-0

Liu, W., Ye, T., Jägermeyr, J., Müller, C., Chen, S., Liu, X., & Shi, P. (2021). Future climate change significantly alters interannual wheat yield variability over half of harvested areas. *Environmental Research Letters*, 16(9), 094045. DOI: 10.1088/1748-9326/ac1fbb

Liu, X., Sun, P., Qu, G., Jing, J., Zhang, T., Shi, H., & Zhao, Y. (2021). Insight into the characteristics and sorption behaviors of aged polystyrene microplastics through three type of accelerated oxidation processes. *Journal of Hazardous Materials*, 407, 124836. DOI: 10.1016/j.jhazmat.2020.124836

Liu, Z., Demeestere, K., & Van Hulle, S. (2021). Comparison and performance assessment of ozone-based AOPs in view of trace organic contaminants abatement in water and wastewater: A review. *Journal of Environmental Chemical Engineering*, 9(4), 105599. DOI: 10.1016/j.jece.2021.105599

Li, W., Luo, Y., & Pan, X. (2020). Identification and Characterization Methods for Microplastics Basing on Spatial Imaging in Micro-/Nanoscales. In He, D., & Luo, Y. (Eds.), *Microplastics in Terrestrial Environments: Emerging Contaminants and Major Challenges* (pp. 25–37). Springer International Publishing. DOI: 10.1007/698_2020_446

Li, X., Wang, R., Dai, W., Luan, Y., & Li, J. (2023). Impacts of Micro(nano)plastics on Terrestrial Plants: Germination, Growth, and Litter. *Plants*, 12(20), 3554. DOI: 10.3390/plants12203554

Li, Y., Li, J., Ding, J., Song, Z., Yang, B., Zhang, C., & Guan, B. (2022). Degradation of nano-sized polystyrene plastics by ozonation or chlorination in drinking water disinfection processes. *Chemical Engineering Journal*, 427, 131690. DOI: 10.1016/j.cej.2021.131690

Li, Y., Liu, Y., Liu, S., Zhang, L., Shao, H., Wang, X., & Zhang, W. (2022). Photo-aging of baby bottle-derived polyethersulfone and polyphenylsulfone microplastics and the resulting bisphenol S release. *Environmental Science & Technology*, 56(5), 3033–3044. DOI: 10.1021/acs.est.1c05812

Li, Y., Zhang, C., Tian, Z., Cai, X., & Guan, B. (2024). Identification and quantification of nanoplastics (20–1000 nm) in a drinking water treatment plant using AFM-IR and Pyr-GC/MS. *Journal of Hazardous Materials*, 463, 132933. DOI: 10.1016/j.jhazmat.2023.132933

Li, Y., Zhu, Y., Huang, J., Ho, Y.-W., Fang, J. K.-H., & Lam, E. Y. (2024). High-throughput microplastic assessment using polarization holographic imaging. *Scientific Reports*, 14(1), 2355. DOI: 10.1038/s41598-024-52762-5

Li, Z., Li, Q., Li, R., Zhou, J., & Wang, G. (2021). The distribution and impact of polystyrene nano-plastics on cucumber plants. *Environmental Science and Pollution Research International*, 28(13), 16042–16053. DOI: 10.1007/s11356-020-11702-2

Li, Z., Li, R., Li, Q., Zhou, J., & Wang, G. (2020). Physiological response of cucumber (Cucumis sativus L.) leaves to polystyrene nano-plastics pollution. *Chemosphere*, 255, 127041. DOI: 10.1016/j.chemosphere.2020.127041

Llorente-García, B. E., Hernández-López, J. M., Zaldívar-Cadena, A. A., Siligardi, C., & Cedillo-González, E. I. (2020). First insights into photocatalytic degradation of HDPE and LDPE microplastics by a mesoporous N–TiO2 coating: Effect of size and shape of microplastics. *Coatings*, 10(7), 658. DOI: 10.3390/coatings10070658

Long B. (2023) Impact of plastic film mulching on microplastic in farmland soils in Guangdong province. China. Heliyon, 27(9), e16587. DOI: 10.1016/j.heliyon.2023.e16587

Long, M., Paul-Pont, I., Hegaret, H., Moriceau, B., Lambert, C., Huvet, A., & Soudant, P. (2017). Interactions between polystyrene microplastics and marine phytoplankton lead to species-specific hetero-aggregation. *Environmental Pollution*, 228, 454–463. DOI: 10.1016/j.envpol.2017.05.047

Long, Y., Zhou, Z., Wen, X., Wang, J., Xiao, R., Wang, W., & Deng, C. (2023). Microplastics removal and characteristics of a typical multi-combination and multi-stage constructed wetlands wastewater treatment plant in Changsha, China. *Chemosphere*, 312, 137199. DOI: 10.1016/j.chemosphere.2022.137199

Lorenzo-Navarro, J., Castrillón-Santana, M., Santesarti, E., De Marsico, M., Martínez, I., Raymond, E., Gómez, M., & Herrera, A. (2020). SMACC: A System for Microplastics Automatic Counting and Classification. *IEEE Access, 8*, 25249–25261. DOI: 10.1109/ACCESS.2020.2970498

Lorenzo-Navarro, J., Castrillón-Santana, M., Sánchez-Nielsen, E., Zarco, B., Herrera, A., Martínez, I., & Gómez, M. (2021). Deep learning approach for automatic microplastics counting and classification. *The Science of the Total Environment*, 765, 142728. DOI: 10.1016/j.scitotenv.2020.142728

Louzao, M., Becares, J., Rodriguez, B., Hyrenbach, K. D., Ruiz, A., & Arcos, J. M. (2009). Combining vessel-based surveys and tracking data to identify key marine areas for seabirds. *Marine Ecology Progress Series*, 391, 183–197. DOI: 10.3354/meps08124

Lozano, Y. M., Aguilar-Trigueros, C. A., Onandia, G., Maaß, S., Zhao, T., & Rillig, M. C. (2021). Effects of microplastics and drought on soil ecosystem functions and multifunctionality. *Journal of Applied Ecology*, 58(5), 988–996. DOI: 10.1111/1365-2664.13839

Lozano, Y. M., Lehnert, T., Linck, L. T., Lehmann, A., & Rillig, M. C. (2021). Microplastic Shape, Polymer Type, and Concentration Affect Soil Properties and Plant Biomass. *Frontiers in Plant Science*, 12, 616645. DOI: 10.3389/fpls.2021.616645

Lozano, Y. M., & Rillig, M. C. (2020). Effects of microplastic fibers and drought on plant communities. *Environmental Science & Technology*, 54(10), 6166–6173. DOI: 10.1021/acs.est.0c01051

Lu, K., Zhan, D., Fang, Y., Li, L., Chen, G., Chen, S., & Wang, L. (2022, September 28). Microplastics, potential threat to patients with lung diseases. *Frontiers in Toxicology*, 4, 958414. DOI: 10.3389/ftox.2022.958414

Luo, H., Liu, C., He, D., Sun, J., Zhang, A., Li, J., & Pan, X. (2022). Interactions between polypropylene microplastics (PP-MPs) and humic acid influenced by aging of MPs. *Water Research*, 222, 118921. DOI: 10.1016/j.watres.2022.118921

Luo, H., Zeng, Y., Zhao, Y., Xiang, Y., Li, Y., & Pan, X. (2021). Effects of advanced oxidation processes on leachates and properties of microplastics. *Journal of Hazardous Materials*, 413, 125342. DOI: 10.1016/j.jhazmat.2021.125342

Luo, H., Zhao, Y., Li, Y., Xiang, Y., He, D., & Pan, X. (2020). Aging of microplastics affects their surface properties, thermal decomposition, additives leaching and interactions in simulated fluids. *The Science of the Total Environment*, 714, 136862. DOI: 10.1016/j.scitotenv.2020.136862

Luoto, T. P., Rantala, M. V., Kivilä, E. H., Nevalainen, L., & Ojala, A. E. (2019). Biogeochemical cycling and ecological thresholds in a High Arctic lake (Svalbard). *Aquatic Sciences*, 81(2), 1–16. DOI: 10.1007/s00027-019-0630-7

Luo, Y., Su, W., Xu, X., Xu, D., Wang, Z., Wu, H., Chen, B., & Wu, J. (2023). Raman Spectroscopy and Machine Learning for Microplastics Identification and Classification in Water Environments. *IEEE Journal of Selected Topics in Quantum Electronics*, 29(4), 1–8. Advance online publication. DOI: 10.1109/JSTQE.2022.3222065

Luo, Y., & Zhou, X. (2006). *Soil Respiration and the Environment*. Academic Press., DOI: 10.1016/B978-0-12-088782-8.X5000-1

Lu, S., Chen, J., Wang, J., Wu, D., Bian, H., Jiang, H., Sheng, L., & He, C. (2023). Toxicological effects and transcriptome mechanisms of rice (Oryza sativa L.) under stress of quinclorac and polystyrene nanoplastics. *Ecotoxicology and Environmental Safety*, 249, 114380. DOI: 10.1016/j.ecoenv.2022.114380

Lusher, A. L., Hurley, R., Arp, H. P. H., Booth, A. M., Bråte, I. L. N., Gabrielsen, G. W., Gomiero, A., Gomes, T., Grøsvik, B. E., Green, N., Haave, M., Hallanger, I. G., Halsband, C., Herzke, D., Joner, E. J., Kögel, T., Rakkestad, K., Ranneklev, S. B., Wagner, M., & Olsen, M. (2021). Moving forward in microplastic research: A Norwegian perspective. *Environment International*, 157, 106794. DOI: 10.1016/j.envint.2021.106794 PMID: 34358913

Lusher, A. L., McHugh, M., & Thompson, R. C. (2013). Occurrence of microplastics in the gastrointestinal tract of pelagic and demersal fish from the English Channel. *Marine Pollution Bulletin*, 67(1-2), 94–99. DOI: 10.1016/j.marpolbul.2012.11.028

Lu, Y., Zhang, Y., Deng, Y., Jiang, W., Zhao, Y., Geng, J., Ding, L., & Ren, H. (2016). Uptake and accumulation of polystyrene microplastics in zebrafish (*Danio rerio*) and toxic effects in liver. *Environmental Science & Technology*, 50(7), 4054–4060. DOI: 10.1021/acs.est.6b00183

Lv, X., Dong, Q., Zuo, Z., Liu, Y., Huang, X., & Wu, W.-M. (2019). Microplastics in a municipal wastewater treatment plant: Fate, dynamic distribution, removal efficiencies, and control strategies. *Journal of Cleaner Production*, 225, 579–586. DOI: 10.1016/j.jclepro.2019.03.321

Lwanga, E. H., Beriot, N., Corradini, F., Silva, V., Yang, X., Baartman, J., Rezaei, M., van Schaik, L., Riksen, M., & Geissen, V. (2022). Review of microplastic sources, transport pathways and correlations with other soil stressors: A journey from agricultural sites into the environment. *Chemical and Biological Technologies in Agriculture*, 9(1), 20. DOI: 10.1186/s40538-021-00278-9

Lwanga, E. H., Gertsen, H., Gooren, H., Peters, P., Salánki, T., Van Der Ploeg, M., & Geissen, V. (2016). Microplastics in the terrestrial ecosystem: Implications for *Lumbricus terrestris* (Oligochaeta, Lumbricidae*).Environmental Science & Technology*, 50(5), 2685–2691. DOI: 10.1021/acs.est.5b05478

Lwanga, E. H., Gertsen, H., Gooren, H., Peters, P., Salánki, T., van der Ploeg, M., & Geissen, V. (2017). Incorporation of microplastics from litter into burrows of *Lumbricus terrestris.Environmental Pollution*, 220, 523–531. DOI: 10.1016/j.envpol.2016.09.096

Maceda-veiga, A., & Dom, O. (2016). *The aquarium hobby : can sinners become saints in freshwater fish conservation?* DOI: 10.1111/faf.12097

MacLeod, M., Arp, H. P. H., Tekman, M. B., & Jahnke, A. (2021). The global threat from plastic pollution. *Science*, 373(6550), 61–65. DOI: 10.1126/science.abg5433 PMID: 34210878

Maddela, N. R., Ramakrishnan, B., Kadiyala, T., Venkateswarlu, K., & Megharaj, M. (2023). Do Microplastic and Nanoplastics Pose Risks to Biota in Agricultural Ecosystems? *Soil Systems*, 7(1), 19. DOI: 10.3390/soilsystems7010019

Maes, T., Jessop, R., Wellner, N., Haupt, K., & Mayes, A. G. (2017). A rapid-screening approach to detect and quantify microplastics based on fluorescent tagging with Nile Red. *Scientific Reports*, 7(March), 1–10. DOI: 10.1038/srep44501 PMID: 28300146

Mah, A. (2022). *Plastic unlimited: How corporations are fuelling the ecological crisis and what we can do about it*. John Wiley & Sons.

Mahesh, S., Gowda, N. K., & Mahesh, S. (2023). Identification of microplastics from urban informal solid waste landfill soil; MP associations with COD and chloride. *Water science and technology: a journal of the International Association on Water Pollution Research, 87*(1), 115–129. https://doi.org/DOI: 10.2166/wst.2022.412

Maity, S., Chatterjee, A., Guchhait, R., De, S., & Pramanick, K. (2020). Cytogenotoxic potential of a hazardous material, polystyrene microparticles on Allium cepa L. *Journal of Hazardous Materials*, 385, 121560. DOI: 10.1016/j.jhazmat.2019.121560

Malik, S., Bora, J., Nag, S., Sinha, S., Mondal, S., Rustagi, S., & Minkina, T. (2023). Fungal-based remediation in the treatment of anthropogenic activities and pharmaceutical-pollutant-contaminated wastewater. *Water (Basel)*, 15(12), 2262. DOI: 10.3390/w15122262

Malinowska, K., Bukowska, B., Piwoński, I., Foksiński, M., Kisielewska, A., Zarakowska, E., Gackowski, D., & Sicińska, P. (2022). Polystyrene nanoparticles: The mechanism of their genotoxicity in human peripheral blood mononuclear cells. *Nanotoxicology*, 16(6-8), 791–811. DOI: 10.1080/17435390.2022.2149360

Malinowski, C. R., Searle, C. L., Schaber, J., & Höök, T. O. (2023). Microplastics impact simple aquatic food web dynamics through reduced zooplankton feeding and potentially releasing algae from consumer control. *The Science of the Total Environment*, 904(August), 166691. Advance online publication. DOI: 10.1016/j.scitotenv.2023.166691 PMID: 37659532

Marchio, E. A. (2018). The Art of Aquarium Keeping Communicates Science and Conservation. *Frontiers in Communication*, 3(April), 1–9. DOI: 10.3389/fcomm.2018.00017

Mariano, S., Tacconi, S., Fidaleo, M., Rossi, M., & Dini, L. (2021). Micro and nanoplastics identification: Classic methods and innovative detection techniques. *Frontiers in Toxicology*, 3, 636640. DOI: 10.3389/ftox.2021.636640

Marine Debris Program. (2015). *Laboratory Methods for the Analysis of Microplastics in the Marine Environment: Recommendations for quantifying synthetic particles in waters and sediments.*

Martin, R. G. (1975). Sexual and Aggressive Behavior, Density and Social Structure in A Natural Population of Mosquitofish, Gambusia affinis holbrooki. *Copeia*, 1975(3), 445. DOI: 10.2307/1443641

Masnadi, M. S., El-Houjeiri, H. M., Schunack, D., Li, Y., Englander, J. G., Badahdah, A., Monfort, J.-C., Anderson, J. E., Wallington, T. J., Bergerson, J. A., Gordon, D., Koomey, J., Przesmitzki, S., Azevedo, I. L., Bi, X. T., Duffy, J. E., Heath, G. A., Keoleian, G. A., McGlade, C., & Brandt, A. R. (2018). Global carbon intensity of crude oil production. *Science*, 361(6405), 851–853. DOI: 10.1126/science.aar6859

Mason, S. A., Welch, V. G., & Neratko, J. (2018). Synthetic polymer contamination in bottled water. *Frontiers in Chemistry*, 6, 389699. DOI: 10.3389/fchem.2018.00407

Massarelli, C., Campanale, C., & Uricchio, V. F. (2021). A Handy Open-Source Application Based on Computer Vision and Machine Learning Algorithms to Count and Classify Microplastics. *Water (Basel)*, 13(15), 15. Advance online publication. DOI: 10.3390/w13152104

Masura, J., Baker, J. E., Foster, G. D., Arthur, C., & Herring, C. (2015). Laboratory methods for the analysis of microplastics in the marine environment: recommendations for quantifying synthetic particles in waters and sediments. *NOAA Technical Memorandum NOS-OR&R-48.*

Matos, D. M., Ramos, J. A., Bessa, F., Silva, V., Rodrigues, I., Antunes, S., dos Santos, I., Coentro, J., Brandão, A. L. C., Batista de Carvalho, L. A. E., Marques, M. P. M., Santos, S., & Paiva, V. H. (2023). Anthropogenic debris ingestion in a tropical seabird community: Insights from taxonomy and foraging distribution. *The Science of the Total Environment*, 898, 165437. DOI: 10.1016/j.scitotenv.2023.165437 PMID: 37437636

Matthews, S., Mai, L., Jeong, C. B., Lee, J. S., Zeng, E. Y., & Xu, E. G. (2021). Key mechanisms of micro-and nanoplastic (MNP) toxicity across taxonomic groups. *Comparative Biochemistry and Physiology. Toxicology & Pharmacology : CBP*, 247, 109056. DOI: 10.1016/j.cbpc.2021.109056

Mattonai, M., Watanabe, A., & Ribechini, E. (2020). Characterization of volatile and non-volatile fractions of spices using evolved gas analysis and multi-shot analytical pyrolysis. *Microchemical Journal*, 159, 105321. DOI: 10.1016/j.microc.2020.105321

Maulana, D. A., & Ibadurrohman, M. (2021). *Synthesis of nano-composite Ag/TiO2 for polyethylene microplastic degradation applications.* Paper presented at the IOP Conference Series: Materials Science and Engineering. DOI: 10.1088/1757-899X/1011/1/012054

McCarty, J. P., & Second, A. L. (2000). Possible effects of PCB contamination on female plum age color and reproductive success in Hudson River Tree Swallows. *The Auk*, 117(4), 987–995. DOI: 10.1093/auk/117.4.987

McCauley, D. J., Pinsky, M. L., Palumbi, S. R., Estes, J. A., Joyce, F. H., & Warner, R. R. (2015). Marine defaunation: Animal loss in the global ocean. *Science*, 347(6219), 1255641–1255647. DOI: 10.1126/science.1255641 PMID: 25593191

Mccormick, A. A. (2022). *Microplastic contamination and possible sources in a small public aquarium.*

Mcdevitt, J. P., Criddle, C. S., Morse, M., Hale, R. C., Bott, C. B., & Rochman, C. M. (2017). *Addressing the Issue of Microplastics in the Wake of the Microbead-Free Waters Act: A New Standard Can Facilitate Improved Policy.* DOI: 10.1021/acs.est.6b05812

Mecozzi, M., Pietroletti, M., & Monakhova, Y. B. (2016). FTIR spectroscopy supported by statistical techniques for the structural characterization of plastic debris in the marine environment: Application to monitoring studies. *Marine Pollution Bulletin*, 106(1-2), 155–161. DOI: 10.1016/j.marpolbul.2016.03.012

Meegoda, J. N., & Hettiarachchi, M. C. (2023). A path to a reduction in micro and nano-plastics pollution. *International Journal of Environmental Research and Public Health*, 20(8), 5555. DOI: 10.3390/ijerph20085555 PMID: 37107837

Mehdinia, A., Dehbandi, R., Hamzehpour, A., & Rahnama, R. (2020). Identification of microplastics in the sediments of southern coasts of the Caspian Sea, north of Iran. *Environmental Pollution*, 258, 113738. DOI: 10.1016/j.envpol.2019.113738

Meides, N., Mauel, A., Menzel, T., Altstädt, V., Ruckdäschel, H., Senker, J., & Strohriegl, P. (2022). Quantifying the Fragmentation of Polypropylene upon Exposure to Accelerated Weathering. *Macroplastics and Nanoplastics*, 2, 1–13.

Meides, N., Menzel, T., Poetzschner, B., Löder, M. G. J., Mansfeld, U., Strohriegl, P., Altstaedt, V., & Senker, J. (2021). Reconstructing the Environmental Degradation of Polystyrene by Accelerated Weathering. *Environmental Science & Technology*, 55(12), 7930–7938. DOI: 10.1021/acs.est.0c07718

Melillo, J. M., Richmond, T. T., & Yohe, G. W. (Eds.). (2014). *Climate change impacts in the United States: Third National Climate Assessment*. U.S. Global Change Research Program. DOI: 10.7930/J0Z31WJ2

Melo-Merino, S. M., Reyes-Bonilla, H., & Lira-Noriega, A. (2020). Ecological niche models and species distribution models in marine environments: A literature review and spatial analysis of evidence. *Ecological Modelling, 415*(September), 108837. DOI: 10.1016/j.ecolmodel.2019.108837

Meng, F., Yang, X., Riksen, M., Xu, M., & Geissen, V. (2021). Response of common bean (Phaseolus vulgaris L.) growth to soil contaminated with microplastics. *The Science of the Total Environment*, 755, 142516. DOI: 10.1016/j.scitotenv.2020.142516

Meng, X., Zhang, J., Wang, W., Gonzalez-Gil, G., Vrouwenvelder, J. S., & Li, Z. (2022). Effects of nano- and microplastics on kidney: Physicochemical properties, bioaccumulation, oxidative stress and immunoreaction. *Chemosphere*, 288(Pt 3), 132631. DOI: 10.1016/j.chemosphere.2021.132631

Miao, F., Liu, Y., Gao, M., Yu, X., Xiao, P., Wang, M., Wang, S., & Wang, X. (2020). Degradation of polyvinyl chloride microplastics via an electro-Fenton-like system with a TiO2/graphite cathode. *Journal of Hazardous Materials*, 399, 123023. DOI: 10.1016/j.jhazmat.2020.123023

Microplastics in environment: Global concern, challenges, and controlling measures—PMC. (n.d.). Retrieved July 1, 2024, from https://www.ncbi.nlm.nih.gov/pmc/articles/PMC9135010/

Mihai, , Gündoğdu, S., Markley, L. A., Olivelli, A., Khan, F. R., Gwinnett, C., Gutberlet, J., Reyna-Bensusan, N., Llanquileo-Melgarejo, P., Meidiana, C., Elagroudy, S., Ishchenko, V., Penney, S., Lenkiewicz, Z., & Molinos-Senante, M. (2022). Plastic Pollution, Waste Management Issues, and Circular Economy Opportunities in Rural Communities. *Sustainability (Basel)*, 14(1), 20. DOI: 10.3390/su14010020

Mintenig, S. M., Int-Veen, I., Löder, M. G. J., Primpke, S., & Gerdts, G. (2017). Identification of microplastic in effluents of waste water treatment plants using focal plane array-based micro-Fourier-transform infrared imaging. *Water Research*, 108, 365–372. DOI: 10.1016/j.watres.2016.11.015 PMID: 27838027

Mitrano, D. M., Wick, P., & Nowack, B. (2021). Placing nano-plastics in the context of global plastic pollution. *Nature Nanotechnology*, 16(5), 491–500. DOI: 10.1038/s41565-021-00888-2 PMID: 33927363

Mittler, R. (2017). ROS Are Good. *Trends in Plant Science*, 22(1), 11–19. DOI: 10.1016/j.tplants.2016.08.002

Moeck, C., Davies, G., Krause, S., Schneidewind, U. (2023). Microplastics and nanoplastics in agriculture—A potential source of soil and groundwater contamination? Grundwasser - Zeitschrift der Fachsektion Hydrogeologie, 23-35. DOI: 10.1007/s00767-022-00533-2

Mohamed Noor, M., Wong, S., Ngadi, N., Mohammed Inuwa, I., & Opotu, L. (2022). Assessing the effectiveness of magnetic nanoparticles coagulation/flocculation in water treatment: A systematic literature review. *International Journal of Environmental Science and Technology*, 19(7), 6935–6956. DOI: 10.1007/s13762-021-03369-0

Mohanan, N., Montazer, Z., Sharma, P. K., & Levin, D. B. (2020). Microbial and Enzymatic Degradation of Synthetic Plastics. *Frontiers in Microbiology*, 11, 580709. DOI: 10.3389/fmicb.2020.580709

Mokhtarzadeh, Z., Keshavarzi, B., Moore, F., Busquets, R., Rezaei, M., Padoan, E., & Ajmone-Marsan, F. (2022). Microplastics in industrial and urban areas in South-West Iran. *International Journal of Environmental Science and Technology*, 19(10), 10199–10210. DOI: 10.1007/s13762-022-04223-7

Monclús L, McCann Smith E, Ciesielski TM, Wagner M, Jaspers VLB. (2022). Microplastic Ingestion Induces Size-Specific Effects in Japanese Quail. Environ Sci Technol., 56(22), 15902-15911.

Monira, S., Roychand, R., Hai, F. I., Bhuiyan, M., Dhar, B. R., & Pramanik, B. K. (2023). Nano and microplastics occurrence in wastewater treatment plants: A comprehensive understanding of microplastics fragmentation and their removal. *Chemosphere*, 334, 139011. DOI: 10.1016/j.chemosphere.2023.139011

Montevecchi, W. A. (1993). Birds as indicators of change in marine prey stocks. In Furness, W. R., & Greenwood, J. J. D. (Eds.), *Birds as Monitors of Environmental Change* (pp. 217–266). Chapman and Hall. DOI: 10.1007/978-94-015-1322-7_6

Moore, C. J. (2008). Synthetic polymers in the marine environment: A rapidly increasing, long-term threat. *Environmental Research*, 108(2), 131–139. DOI: 10.1016/j.envres.2008.07.025

Moss, R. H., Edmonds, J. A., Hibbard, K. A., Manning, M. R., Rose, S. K., Van Vuuren, D. P., Carter, T. R., Emori, S., Kainuma, M., Kram, T., Meehl, G. A., Mitchell, J. F. B., Nakicenovic, N., Riahi, K., Smith, S. J., Stouffer, R. J., Thomson, A. M., Weyant, J. P., & Wilbanks, T. J. (2010). The next generation of scenarios for climate change research and assessment. *Nature*, 463(7282), 747–756. DOI: 10.1038/nature08823

Mullen, G. R., & Durden, L. A. (2019). Medical and Veterinary Entomology (Third Edition). Academin Press.

Müller, Y. K., Wernicke, T., Pittroff, M., Witzig, C. S., Storck, F. R., Klinger, J., & Zumbülte, N. (2020). *Microplastic analysis — are we measuring the same ? Results on the first global comparative study for microplastic analysis in a water sample.*

Munari, C., Infantini, V., Scoponi, M., Rastelli, E., Corinaldesi, C., & Mistri, M. (2017). Microplastics in the sediments of Terra Nova bay (Ross sea, Antarctica). *Marine Pollution Bulletin*, 122(1-2), 161–165. DOI: 10.1016/j.marpolbul.2017.06.039

Murphy, F., Ewins, C., Carbonnier, F., & Quinn, B. (2016). Wastewater treatment works (WwTW) as a source of microplastics in the aquatic environment. *Environmental Science & Technology*, 50(11), 5800–5808. DOI: 10.1021/acs.est.5b05416

Muthulakshmi, L., Mohan, S., & Tatarchuk, T. (2023). Microplastics in water: Types, detection, and removal strategies. *Environmental Science and Pollution Research International*, 30(36), 84933–84948. DOI: 10.1007/s11356-023-28460-6

Muyle, A. M., Seymour, D. K., Lv, Y., Huettel, B., & Gaut, B. S. (2022). Gene Body Methylation in Plants: Mechanisms, Functions, and Important Implications for Understanding Evolutionary Processes. *Genome Biology and Evolution*, 14(4), evac038. Advance online publication. DOI: 10.1093/gbe/evac038

Nabi, I., Ahmad, F., & Zhang, L. (2021). Application of titanium dioxide for the photocatalytic degradation of macro-and micro-plastics: A review. *Journal of Environmental Chemical Engineering*, 9(5), 105964. DOI: 10.1016/j.jece.2021.105964

Naji, A., Nuri, M., Amiri, P., & Niyogi, S. (2019). Small microplastic particles (S-MPPs) in sediments of mangrove ecosystem on the northern coast of the Persian Gulf. *Marine Pollution Bulletin*, 146, 305–311. DOI: 10.1016/j.marpolbul.2019.06.033

Nakanishi, Y., Yamaguchi, H., Hirata, Y., Nakashima, Y., & Fujiwara, Y. (2021). Micro-abrasive glass surface for producing microplastics for biological tests. *Wear*, 477, 203816. DOI: 10.1016/j.wear.2021.203816

Napper, I. E., Baroth, A., Barrett, A. C., Bhola, S., Chowdhury, G. W., Davies, B. F., Duncan, E. M., Kumar, S., Nelms, S. E., Hasan Niloy, M. N., Nishat, B., Maddalene, T., Thompson, R. C., & Koldewey, H. (2021). The abundance and characteristics of microplastics in surface water in the transboundary Ganges River. *Environmental Pollution*, 274, 116348. DOI: 10.1016/j.envpol.2020.116348

Napper, I. E., Davies, B. F. R., Clifford, H., Elvin, S., Koldewey, H. J., Mayewski, P. A., Miner, K. R., Potocki, M., Elmore, A. C., Gajurel, A. P., & Thompson, R. C. (2020). Reaching new heights in plastic pollution-preliminary findings of microplasties on mount everest. *One Earth*, 3(5), 621–630. DOI: 10.1016/j.oneear.2020.10.020

National Audubon Society. (2012b). *Important Bird Areas Program: A Global Currency for Bird Conservation*. National Audubon Society.

National Oceanic and Atmospheric Administration (NOAA). (2014). NOAA Marine Debris Program. 2014 Report on the Entanglement of Marine Species in Marine Debris With an Emphasis on Species in the United States. (Silver Spring, MD. 28).

Nelson, J. B. (1978). *The Sulidae: Gannets and boobies*. Oxford University Press.

Netthong, R., Khumsikiew, J., Donsamak, S., Navabhatra, A., Yingngam, K., & Yingngam, B. (2024). Bibliometric analysis of antibacterial drug resistance: An overview. In *Frontiers in Combating Antibacterial Resistance* (pp. 196–245). Current Perspectives and Future Horizons. DOI: 10.4018/979-8-3693-4139-1.ch009

Neumanna, S., Harju, M., & Herzke, D. (2021). Ingested plastics in northern fulmars (Fulmarus glacialis): A pathway for polybrominated diphenyl ether (PBDE) exposure? *The Science of the Total Environment*, 778, 146313. DOI: 10.1016/j.scitotenv.2021.146313 PMID: 33721646

Newman, J., Fletcher, D., Moller, H., Bragg, C., Scott, D., & McKechnie, S. (2009). Estimates of productivity and detection probabilities of breeding attempts in the sooty shearwater (Puffinus griseus). *Wildlife Research*, 36(2), 159–168. DOI: 10.1071/WR06074

Ng, E. L., Lwanga, E. H., Eldridge, S. M., Johnston, P., Hu, H. W., Geissen, V., & Chen, D. (2018). An overview of microplastic and nanoplastic pollution in agroecosystems. *The Science of the Total Environment*, 627, 1377–1388. DOI: 10.1016/j.scitotenv.2018.01.341

Niccolai E. (2023) Adverse Effects of Micro- and Nanoplastics on Humans and the Environment. Int J Mol Sci., 31(24), 15822. .DOI: 10.3390/ijms242115822

Niedzielska. (2015) Oxidative Stress in Neurodegenerative Diseases. Mol Neurobiol., 53(6), 4094-4125. DOI: 10.1007/s12035-015-9337-5

Nielsen, M. B., Clausen, L. P. W., Cronin, R., Hansen, S. F., Oturai, N. G., & Syberg, K. (2023). Unfolding the science behind policy initiatives targeting plastic pollution. *Microplastics and Nanoplastics*, 3(1), 3. DOI: 10.1186/s43591-022-00046-y PMID: 36748026

Niu, J., Xu, D., Wu, W., & Gao, B. (2024). Tracing microplastic sources in urban water bodies combining their diversity, fragmentation and stability. *npj Clean Water, 7*(1), 37. DOI: 10.1038/s41545-024-00329-2

NOAA. (2024). National Centers for Environmental Information, Monthly Global Climate Report for March 2024. Retrieved 03 April 2024, from https://www.ncei.noaa.gov/access/monitoring/monthlyreport/global/202403/supplemental/page-1

Nobre, C. R., Santana, M. F. M., Maluf, A., Cortez, F. S., Cesar, A., Pereira, C. D. S., & Turra, A. (2015). Assessment of microplastic toxicity to embryonic development of the sea urchin *Lytechinus variegatus* (Echinodermata: Echinoidea). *Marine Pollution Bulletin*, 92(1-2), 99–104. DOI: 10.1016/j.marpolbul.2014.12.050

Nolte, T. M., Hartmann, N. B., Kleijn, J. M., Garnæs, J., Van De Meent, D., Hendriks, A. J., & Baun, A. (2017). The toxicity of plastic nanoparticles to green algae as influenced by surface modification, medium hardness and cellular adsorption. *Aquatic Toxicology (Amsterdam, Netherlands)*, 183, 11–20. DOI: 10.1016/j.aquatox.2016.12.005

Nosike, E. I., Zhang, Y., & Wu, A. (2021). Magnetic hybrid nanoparticles for environmental remediation *Magnetic Nanoparticle-BasedHybrid Materials*, 591–615.

Nur, N., Jahncke, J., Herzog, M. P., Howar, J., Hyrenbach, K. D., Zamon, J. E., Ainley, D. G., Wiens, J. A., Morgan, K., Ballance, L. T., & Stralberg, D. (2011). Where the wild things are: Predicting hotspots of seabird aggregations in the California Current System. *Ecological Applications*, 21(6), 2241–2257. DOI: 10.1890/10-1460.1 PMID: 21939058

Nyadjro, E. S., Webster, J. A. B., Boyer, T. P., Cebrian, J., Collazo, L., Kaltenberger, G., Larsen, K., Lau, Y. H., Mickle, P., Toft, T., & Wang, Z. (2023). The NOAA NCEI marine microplastics database. *Scientific Data*, 10(1), 726. DOI: 10.1038/s41597-023-02632-y

Nzeyimana, B. S., & Mary, A. D. C. (2024). Sustainable sewage water treatment based on natural plant coagulant: Moringa oleifera. *Discover Water*, 4(1), 15. Advance online publication. DOI: 10.1007/s43832-024-00069-x

O'Connor, D., Pan, S., Shen, Z., Song, Y., Jin, Y., Wu, W. M., & Hou, D. (2019). Microplastics undergo accelerated vertical migration in sand soil due to small size and wet-dry cycles. *Environmental Pollution*, 249, 527–534. DOI: 10.1016/j.envpol.2019.03.092

O'Donovan, S., Mestre, N. C., Abel, S., Fonseca, T. G., Carteny, C. C., Cormier, B., Keiter, S. H., & Bebianno, M. J. (2018). Ecotoxicological effects of chemical contaminants adsorbed to microplastics in the clam Scrobicularia plana. *Frontiers in Marine Science*, 5, 143. DOI: 10.3389/fmars.2018.00143

O'Shea, T. J., & Stafford, C. J. (1980). Phthalate plasticizers: Accumulation and effects on weight and food consumption in captive starlings. *Bulletin of Environmental Contamination and Toxicology*, 25(1), 345–352. DOI: 10.1007/BF01985536 PMID: 7426782

Obbard, R. W., Sadri, S., Wong, Y. Q., Khitun, A. A., Baker, I., & Thompson, R. C. (2014). Global warming releases microplastic legacy frozen in Arctic Sea ice. *Earth's Future*, 2(6), 315–320. DOI: 10.1002/2014EF000240

OECD. (2022). Plastic Pollution Is Growing Relentlessly as Waste Management and Recycling Fall Short. OECD Rep. Available online: https://www.oecd.org/environment/plastics/

Oehlmann, J., Schulte-Oehlmann, U., Kloas, W., Jagnytsch, O., Lutz, I., Kusk, K. O., Wollenberger, L., Santos, E. M., Paull, G. C., Van Look, K. J. W., & Tyler, C. R. (2009). A critical analysis of the biological impacts of plasticizers on wildlife. *Philosophical Transactions of the Royal Society of London. Series B, Biological Sciences*, 364(1526), 2047–2062. DOI: 10.1098/rstb.2008.0242 PMID: 19528055

Ogunola, O. S., Onada, O. A., & Falaye, A. E. (2018). *Mitigation measures to avert the impacts of plastics and microplastics in the marine environment (a review).* DOI: 10.1007/s11356-018-1499-z

Oliveira, Y. M., Vernin, N. S., Maia Bila, D., Marques, M., & Tavares, F. W. (2022). Pollution caused by nanoplastics: Adverse effects and mechanisms of interaction via molecular simulation. *PeerJ*, 10, e13618. DOI: 10.7717/peerj.13618

Omidi, A., Naeemipoor, H., & Hosseini, M. (2012). Plastic debris in the digestive tract of sheep and goats: An increasing environmental contamination in Birjand, Iran. *Bulletin of Environmental Contamination and Toxicology*, 88(5), 691–694. DOI: 10.1007/s00128-012-0587-x

Onyena, A. P., Aniche, D. C., Ogbolu, B. O., Rakib, M. R. J., Uddin, J., & Walker, T. R. (2022). Governance Strategies for Mitigating Microplastic Pollution in the Marine Environment: A Review. *Microplastics*, 1(1), 1. Advance online publication. DOI: 10.3390/microplastics1010003

Onyena, A. P., Aniche, D. C., Ogbolu, B. O., Rakib, R. J., Uddin, J., & Walker, T. R. (2022). Governance Strategies for Mitigating Microplastic Pollution in the Marine Environment. *RE:view*, 15–46.

Ortiz, D., Munoz, M., Nieto-Sandoval, J., Romera-Castillo, C., de Pedro, Z. M., & Casas, J. A. (2022). Insights into the degradation of microplastics by Fenton oxidation: From surface modification to mineralization. *Chemosphere*, 309, 136809. DOI: 10.1016/j.chemosphere.2022.136809

OSPAR Commission. (2008). Background Document for the EcoQO on Plastic Particles in Stomachs of Seabirds.

Ouyang, Z., Li, S., Xue, J., Liao, J., Xiao, C., Zhang, H., & Guo, X. (2023). Dissolved organic matter derived from biodegradable microplastic promotes photo-aging of coexisting microplastics and alters microbial metabolism. *Journal of Hazardous Materials*, 445, 130564. DOI: 10.1016/j.jhazmat.2022.130564

Ouyang, Z., Li, S., Zhao, M., Wangmu, Q., Ding, R., Xiao, C., & Guo, X. (2022). The aging behavior of polyvinyl chloride microplastics promoted by UV-activated persulfate process. *Journal of Hazardous Materials*, 424, 127461. DOI: 10.1016/j.jhazmat.2021.127461

Ouyang, Z., Zhang, Z., Jing, Y., Bai, L., Zhao, M., Hao, X., Li, X., & Guo, X. (2022). The photo-aging of polyvinyl chloride microplastics under different UV irradiations. *Gondwana Research*, 108, 72–80. DOI: 10.1016/j.gr.2021.07.010

Padervand, M., Lichtfouse, E., Robert, D., Wang, C., Padervand, M., Lichtfouse, E., Robert, D., & Wang, C. (2020). Removal of microplastics from the environment. A review To cite this version : HAL Id : hal-02562545 Removal of microplastics from the environment. A review. *Environmental Chemistry Letters*, 18(3), 807–828. DOI: 10.1007/s10311-020-00983-1

Pandey, K. A. (2024). A Comprehensive Exploration of Soil, Water & Air Pollution in Agriculture, BFC Publications.

Pan, Z., Guo, H., Chen, H., Wang, S., Sun, X., Zou, Q., Zhang, Y., Lin, H., Cai, S., & Huang, J. (2019). Microplastics in the Northwestern Pacific: Abundance, distribution, and characteristics. *The Science of the Total Environment*, 650, 1913–1922. DOI: 10.1016/j.scitotenv.2018.09.244

Pasaribu, B., Acosta, K., Aylward, A., Liang, Y., Abramson, B. W., Colt, K., Hartwick, N. T., Shanklin, J., Michael, T. P., & Lam, E. (2023). Genomics of turions from the Greater Duckweed reveal its pathways for dormancy and re-emergence strategy. *The New Phytologist*, 239(1), 116–131. DOI: 10.1111/nph.18941

Pathan, S., Arfaioli, P., Bardelli, T., Ceccherini, M., Nannipieri, P., & Pietramellara, G. (2020). Soil pollution from micro-and nanoplastic debris: A hidden and unknown biohazard. *Sustainability (Basel)*, 12(18), 7255. DOI: 10.3390/su12187255

Patil, S. S., Bhagwat, R. V., Kumar, V., & Durugkar, T. (2019). Megaplastics to nanoplastics: emerging environmental pollutants and their environmental impacts. In *Environmental Contaminants: Ecological Implications and Management,* (pp. 205-235). Springer. DOI: 10.1007/978-981-13-7904-8_10

Paździor, K., Bilińska, L., & Ledakowicz, S. (2019). A review of the existing and emerging technologies in the combination of AOPs and biological processes in industrial textile wastewater treatment. *Chemical Engineering Journal*, 376, 120597. DOI: 10.1016/j.cej.2018.12.057

Peng, J., Wang, J., & Cai, L. (2017). Current understanding of microplastics in the environment: Occurrence, fate, risks, and what we should do. *Integrated Environmental Assessment and Management*, 13(3), 476–482. DOI: 10.1002/ieam.1912

Pereira, J. M., Rodríguez, Y., Blasco-Monleon, S., Porter, A., Lewis, C., & Pham, C. K. (2020). Microplastic in the stomachs of open-ocean and deep-sea fishes of the North-East Atlantic. *Environmental Pollution*, 265, 115060. DOI: 10.1016/j.envpol.2020.115060

Pérez-Reverón, R., Álvarez-Méndez, S. J., González-Sálamo, J., Socas-Hernández, C., Díaz-Peña, F. J., Hernández-Sánchez, C., & Hernández-Borges, J. (2023). Nanoplastics in the soil environment: Analytical methods, occurrence, fate and ecological implications. *Environmental Pollution*, 317, 120788. DOI: 10.1016/j.envpol.2022.120788

Periyasamy, A. P., & Tehrani-Bagha, A. (2022). A review on microplastic emission from textile materials and its reduction techniques. *Polymer Degradation & Stability*, 199, 109901. DOI: 10.1016/j.polymdegradstab.2022.109901

Peters, K. A., & Otis, D. L. (2007). Shorebird roost-site selection at two temporal scales: Is human disturbance a factor? *Journal of Applied Ecology*, 44(1), 196–209. DOI: 10.1111/j.1365-2664.2006.01248.x

Peters, R. J. B., Relou, E., Sijtsma, E. L. E., & Undas, A. K. (2022). *Evaluation of Nanoparticle Tracking Analysis (NTA) for the Measurement of Nanoplastics in Drinking Water*. DOI: 10.21203/rs.3.rs-1809144/v1

Pflugmacher, S., Tallinen, S., Mitrovic, S. M., Penttinen, O.-P., Kim, Y.-J., Kim, S., & Esterhuizen, M. (2021). Case Study Comparing Effects of Microplastic Derived from Bottle Caps Collected in Two Cities on Triticum aestivum (Wheat). *Environments (Basel, Switzerland)*, 8(7), 64. DOI: 10.3390/environments8070064

Pfohl, P., Wagner, M., Meyer, L., Domercq, P., Praetorius, A., Hu, T., Hofmann, T., & Wohlleben, W. (2022). *Environmental Degradation of Microplastics: How to Measure Fragmentation Rates to Secondary Micro- and Nanoplastic Fragments and Dissociation into Dissolved Organics*. DOI: 10.1021/acs.est.2c01228

Phan, S., & Luscombe, C. K. (2023). Recent trends in marine microplastic modeling and machine learning tools: Potential for long-term microplastic monitoring. *Journal of Applied Physics*, 133(2), 020701. DOI: 10.1063/5.0126358

Phan, S., Padilla-Gamiño, J. L., & Luscombe, C. K. (2022). The effect of weathering environments on microplastic chemical identification with Raman and IR spectroscopy: Part I. polyethylene and polypropylene. *Polymer Testing*, 116, 107752. DOI: 10.1016/j.polymertesting.2022.107752

Phillips, R., Fort, J., & Dias, M. (2022). Conservation status and overview of threats to seabirds. In *Conservation of Marine Birds* (pp. 33–56). Elsevier.

Piazza, V., Uheida, A., Gambardella, C., Garaventa, F., Faimali, M., & Dutta, J. (2022). Ecosafety screening of photo-fenton process for the degradation of microplastics in water. *Frontiers in Marine Science*, 8, 791431. DOI: 10.3389/fmars.2021.791431

Picó, Y., & Barceló, D. (2022). Micro (Nano) plastic analysis: A green and sustainable perspective. *Journal of Hazardous Materials Advances*, 6, 100058. DOI: 10.1016/j.hazadv.2022.100058

Pieniazek, L. S., McKinney, M., & Carr, J. 2023, October. Quantification of Microplastics in Soil Using UV-VIS-NIR Spectroscopy with Ultra High Resolution. In *ASA, CSSA, SSSA International Annual Meeting*. ASA-CSSA-SSSA.

Pierce, K. E., Harris, R. J., Larned, L. S., & Pokras, M. A. (2004). Obstruction and starvation associ ated with plastic ingestion in a northern gannet morus bassanus and a greater shearwater Puffinus Gravis. *Marine Ornithology*, 32, 187–189.

Pivato, A. F., Miranda, G. M., Prichula, J., Lima, J. E., Ligabue, R. A., Seixas, A., & Trentin, D. S. (2022). Hydrocarbon-based plastics: Progress and perspectives on consumption and biodegradation by insect larvae. *Chemosphere*, 293, 133600. DOI: 10.1016/j.chemosphere.2022.133600

PlasticsEurope. (2016). *Plastics – the facts 2016. An analysis of European plastics production, demand and waste data*. PlasticsEurope.

Poma, A. M. G., Morciano, P., & Aloisi, M. (2023). Beyond genetics: Can micro and nano-plastics induce epigenetic and gene-expression modifications? *Frontiers in Epigenetics and Epigenomics*, 1, 1241583. DOI: 10.3389/freae.2023.1241583

Prata, J. C., Patr, A. L., Mouneyrac, C., Walker, T. R., Duarte, A. C., & Rocha-santos, T. (2019). *Solutions and Integrated Strategies for the Control and Mitigation of Plastic and Microplastic Pollution*.

Primpke, S., Cross, R. K., Mintenig, S. M., Simon, M., Vianello, A., Gerdts, G., & Vollertsen, J. (2020). Toward the Systematic Identification of Microplastics in the Environment: Evaluation of a New Independent Software Tool (siMPle) for Spectroscopic Analysis. *Applied Spectroscopy*, 74(9), 1127–1138. DOI: 10.1177/0003702820917760

Primpke, S., Godejohann, M., & Gerdts, G. (2020). Rapid identification and quantification of microplastics in the environment by quantum cascade laser-based hyperspectral infrared chemical imaging. *Environmental Science & Technology*, 54(24), 15893–15903. DOI: 10.1021/acs.est.0c05722

Primpke, S., Lorenz, C., Rascher-Friesenhausen, R., & Gerdts, G. (2017). An automated approach for microplastics analysis using focal plane array (FPA) FTIR microscopy and image analysis. *Analytical Methods*, 9(9), 1499–1511. DOI: 10.1039/C6AY02476A

Provencher, J. F., Bond, A. L., Avery-Gomm, S., Borrelle, S. B., Bravo Rebolledo, E. L., Hammer, S., Kühn, S., Lavers, J. L., Mallory, M. L., Trevail, A., & van Franeker, J. A. (2017). Quantifying ingested debris in marine megafauna: A review and recommendations for standardization. *Analytical Methods*, 9(9), 1454–1469. DOI: 10.1039/C6AY02419J

Provencher, J. F., Gaston, A. J., & Mallory, M. L. (2009). Evidence for increased ingestion of plastics by northern fulmars (Fulmarus glacialis) in the Canadian Arctic. *Marine Pollution Bulletin*, 58(7), 1092–1095. DOI: 10.1016/j.marpolbul.2009.04.002 PMID: 19403145

Puskic, P. S., Lavers, J. L., & Bond, A. L. (2020). A critical review of harm associated with plastic ingestion on vertebrates. *The Science of the Total Environment*, 743, 140666. DOI: 10.1016/j.scitotenv.2020.140666

Qi, G., Zhao, L., Liu, J., Tian, C., & Zhang, S. (2024). Single particle detection of micro/nano plastics based on recyclable SERS sensor with two-dimensional AuNPs thin films. *Materials Today. Communications*, 38, 108293. DOI: 10.1016/j.mtcomm.2024.108293

Qiu, G., Han, Z., Wang, Q., Wang, T., Sun, Z., Yu, Y., Han, X., & Yu, H. (2023). Toxicity effects of nanoplastics on soybean (Glycine max L.): Mechanisms and transcriptomic analysis. *Chemosphere*, 313, 137571. DOI: 10.1016/j.chemosphere.2022.137571

Raamsdonk, L. W. D. Van, Zande, M. Van Der, & Koelmans, A. A. (2020). *and Potential Health E ff ects of Microplastics Present in the Food Chain.*

Rai, M., Pant, G., Pant, K., Aloo, B. N., Kumar, G., Singh, H. B., & Tripathi, V. (2023). Microplastic Pollution in Terrestrial Ecosystems and Its Interaction with Other Soil Pollutants: A Potential Threat to Soil Ecosystem Sustainability. *Resources*, 12(6), 67. DOI: 10.3390/resources12060067

Rai, P. K., Lee, J., Brown, R. J., & Kim, K. H. (2021). Micro-and nano-plastic pollution: Behavior, microbial ecology, and remediation technologies. *Journal of Cleaner Production*, 291, 125240. DOI: 10.1016/j.jclepro.2020.125240

Rajabi, M., Ebrahimi, P., & Aryankhesal, A. (2021). Collaboration between the government and nongovernmental organizations in providing health-care services: A systematic review of challenges. *Journal of Education and Health Promotion*, 10(1), 242. DOI: 10.4103/jehp.jehp_1312_20 PMID: 34395679

Raji, M., Mirbagheri, S. A., Ye, F., & Dutta, J. (2021). Nano zero-valent iron on activated carbon cloth support as Fenton-like catalyst for efficient color and COD removal from melanoidin wastewater. *Chemosphere*, 263, 127945. DOI: 10.1016/j.chemosphere.2020.127945

Rajpar, M. N., Ozdemir, I., Zakaria, M., Sheryar, S., & Rab, A. (2018). *Seabirds as bioindicators of marine ecosystems*. InTech. DOI: 10.5772/intechopen.75458

Raju, S., Carbery, M., Kuttykattil, A., Senathirajah, K., Subashchandrabose, S. R., Evans, G., & Thavamani, P. (2018). Transport and fate of microplastics in wastewater treatment plants: Implications to environmental health. *Reviews in Environmental Science and Biotechnology*, 17(4), 637–653. DOI: 10.1007/s11157-018-9480-3

Ramasamy, R., Aragaw, T. A., & Balasaraswathi Subramanian, R. (2022). Wastewater treat- ment plant effluent and microfiber pollution: Focus on industry-specific wastewater. *Environmental Science and Pollution Research International*, 29(34), 51211–51233. DOI: 10.1007/s11356-022-20930-7 PMID: 35606585

Randhawa, J. S. (2023). Advanced analytical techniques for microplastics in the environment: A review. *Bulletin of the National Research Center*, 47(1), 174. DOI: 10.1186/s42269-023-01148-0

Rathod, N. B., Xavier, K. M., Özogul, F., & Phadke, G. G. (2023). Impacts of nano/micro-plastics on safety and quality of aquatic food products. *Advances in Food and Nutrition Research*, 103, 1–40. DOI: 10.1016/bs.afnr.2022.07.001

Raymond, B., Shaffer, S., Sokolov, S., Woehler, E., Costa, D., Einoder, L., Hindell, M., Hosie, G., Pinkerton, M., Sagar, P. M., Scott, D., Smith, A., Thompson, D. R., Vertigan, C., & Weimerskirch, H. (2010). Shearwater foraging in the Southern Ocean: The roles of prey availability and winds. *PLoS One*, 5(6, e10960), e10960. DOI: 10.1371/journal.pone.0010960 PMID: 20532034

Reddy, A. V. B., Moniruzzaman, M., & Aminabhavi, T. M. (2019). Polychlorinated biphenyls (PCBs) in the environment: Recent updates on sampling, pretreatment, cleanup technologies and their analysis. *Chemical Engineering Journal*, 358, 1186–1207. DOI: 10.1016/j.cej.2018.09.205

Reisser, J., Shaw, J., Hallegraeff, G., Proietti, M., Barnes, D. K., Thums, M., Wilcox, C., Hardesty, B. D., & Pattiaratchi, C. (2014). Millimeter-sized marine plastics: A new pelagic habitat for microorganisms and invertebrates. *PLoS One*, 9(6), e100289. DOI: 10.1371/journal.pone.0100289

Ren, F., Huang, J., & Yang, Y. (2024). Unveiling the impact of microplastics and nanoplastics on vascular plants: A cellular metabolomic and transcriptomic review. *Ecotoxicology and Environmental Safety*, 279, 116490. DOI: 10.1016/j.ecoenv.2024.116490

Renner, G., Sauerbier, P., Schmidt, T. C., & Schram, J. (2019). Robust Automatic Identification of Microplastics in Environmental Samples Using FTIR Microscopy. *Analytical Chemistry*, 91(15), 9656–9664. DOI: 10.1021/acs.analchem.9b01095

Renner, G., Schmidt, T. C., & Schram, J. (2018). Analytical methodologies for monitoring micro (nano) plastics: Which are fit for purpose? *Current Opinion in Environmental Science & Health*, 1, 55–61. DOI: 10.1016/j.coesh.2017.11.001

Renner, G., Schmidt, T. C., & Schram, J. (2020). Automated rapid & intelligent microplastics mapping by FTIR microscopy: A Python–based workflow. *MethodsX*, 7, 100742. DOI: 10.1016/j.mex.2019.11.015

Ren, X., Tang, J., Wang, L., & Liu, Q. (2021). Microplastics in soil-plant system: Effects of nano/microplastics on plant photosynthesis, rhizosphere microbes and soil properties in soil with different residues. *Plant and Soil*, 462(1–2), 561–576. DOI: 10.1007/s11104-021-04869-1

Reproducibility, M., Al-azzawi, M. S. M., Kefer, S., Weißer, J., Reichel, J., Schwaller, C., Glas, K., Knoop, O., & Drewes, J. E. (2020). *Validation of Sample Preparation Methods for Microplastic Analysis in Wastewater.*

Revell, L. E., Kuma, P., Le Ru, E. C., Somerville, W. R., & Gaw, S. (2021). Direct radiative effects of airborne microplastics. *Nature*, 598(7881), 462–467. DOI: 10.1038/s41586-021-03864-x

Rillig, M. C., & Bonkowski, M. (2018). Microplastic and soil protists: A call for research. *Environmental Pollution*, 241, 1128–1131. DOI: 10.1016/j.envpol.2018.04.147

Rillig, M. C., Lehmann, A., De Souza Machado, A. A., & Yang, G. (2019). Microplastic effects on plants. *The New Phytologist*, 223(3), 1066–1070. DOI: 10.1111/nph.15794

Rillig, M. C., Ziersch, L., & Hempel, S. (2017). Microplastic transport in soil by earthworms. *Scientific Reports*, 7(1), 1362. DOI: 10.1038/s41598-017-01594-7

Ritchie, H. (2023). How much of global greenhouse gas emissions come from plastics? Retrieved 03 April 2024, from: https://ourworldindata.org/ghg-emissions-plastics

Robards, M. D., Piatt, J. F., & Wohl, K. D. (1995). Increasing frequency of plastic particles ingested by seabirds in the subarctic north Pacific. *Marine Pollution Bulletin*, 30(2), 151–157. DOI: 10.1016/0025-326X(94)00121-O

Rodríguez, A., Rodríguez, B., & Nazaret Carrasco, M. (2012, October). High prevalence of parental delivery of plastic debris in Cory's shearwaters (Calonectris diomedea). *Marine Pollution Bulletin*, 64(10), 2219–2223. DOI: 10.1016/j.marpolbul.2012.06.011 PMID: 22784377

Rogers, K. (2024). Microplastics. Encyclopedia Britannica. https://www.britannica.com/technology/microplastic

Roman, L., Hardesty, B. D., Hindell, M. A., & Wilcox, C. (2019). A quantitative analysis linking seabird mortality and marine debris ingestion. *Scientific Reports*, 9(1), 3202. DOI: 10.1038/s41598-018-36585-9 PMID: 30824751

Roman, L., Lowenstine, L., Parsley, L. M., Wilcox, C., Hardesty, B. D., Gilardi, K., & Hindell, M. (2019). Is plastic ingestion in birds as toxic as we think? Insights from a plastic feeding experiment. *The Science of the Total Environment*, 665, 660–667. DOI: 10.1016/j.scitotenv.2019.02.184 PMID: 30776638

Ronconi, R. A., Lascelles, B. G., Langham, G. M., Reid, J. B., & Oro, D. (2012). The role of seabirds in marine protected area identification, delineation, and monitoring: Introduction and synthesis. *Biological Conservation*, 156, 1–4. DOI: 10.1016/j.biocon.2012.02.016

Rose, P. K., Jain, M., Kataria, N., Sahoo, P. K., Garg, V. K., & Yadav, A. (2023). Microplastics in multimedia environment: A systematic review on its fate, transport, quantification, health risk, and remedial measures. *Groundwater for Sustainable Development*, 20, 100889. DOI: 10.1016/j.gsd.2022.100889

Rostami, S., Talaie, M. R., Talaiekhozani, A., & Sillanpää, M. (2021). *Evaluation of the available strategies to control the emission of microplastics into the aquatic environment.*

Rowe, C. L. (2008). "The calamity of so long life": Life histories, contaminants, and potential emerging threats to long-lived vertebrates. *A.I.B.S. Bulletin*, 58(7), 623–631.

Rowlands, E., Galloway, T., Cole, M., Lewis, C., Peck, V., Thorpe, S., & Manno, C. (2021). The effects of combined ocean acidification and nanoplastic exposures on the embryonic development of Antarctic krill. *Frontiers in Marine Science*, 8, 709763. DOI: 10.3389/fmars.2021.709763

Royer, S. J., Ferrón, S., Wilson, S. T., & Karl, D. M. (2018). Production of methane and ethylene from plastic in the environment. *PLoS One*, 13(8), e0200574. DOI: 10.1371/journal.pone.0200574

Ruan, X., Xie, L., Liu, J., Ge, Q., Liu, Y., Li, K., You, W., Huang, T., & Zhang, L. (2024). Rapid detection of nanoplastics down to 20 nm in water by surface-enhanced raman spectroscopy. *Journal of Hazardous Materials*, 462, 132702. DOI: 10.1016/j.jhazmat.2023.132702

Rubin, A. E., & Zucker, I. (2022). Interactions of microplastics and organic compounds in aquatic environments: A case study of augmented joint toxicity. *Chemosphere*, 289, 133212. DOI: 10.1016/j.chemosphere.2021.133212

Rummel, C. (2022). *Ecotoxicological and Microbial Studies on Weathering Plastic*.

Russo, M., Oliva, M., Hussain, M. I., & Muscolo, A. (2023). The hidden impacts of micro/nanoplastics on soil, crop and human health. *Journal of Agriculture and Food Research*, 14, 100870. DOI: 10.1016/j.jafr.2023.100870

Ryan, P. G. (2016). Ingestion of plastics by marine organisms. Hazardous Chemicals Associated With Plastics in the Marine Environment 235-238.

Sander, M., Weber, M., Lott, C., Zumstein, M., Künkel, A., & Battagliarin, G. (2023). Polymer Biodegradability 2.0: A Holistic View on Polymer Biodegradation in Natural and Engineered Environments. *Advances in Polymer Science*, 293, 65–110. DOI: 10.1007/12_2023_163

Sandhya. (2024). A Comprehensive Exploration of Soil, Water & Air Pollution in Agriculture, BFC Publications.

Sangkham, S., Faikhaw, O., Munkong, N., Sakunkoo, P., Arunlertaree, C., Chavali, M., Mousazadeh, M., & Tiwari, A. (2022). A review on microplastics and nanoplastics in the environment: Their occurrence, exposure routes, toxic studies, and potential effects on human health. *Marine Pollution Bulletin*, 181, 113832. DOI: 10.1016/j.marpolbul.2022.113832

Sathish, M. N., Jeyasanta, I., & Patterson, J. (2020). Occurrence of microplastics in epipelagic and mesopelagic fishes from Tuticorin, Southeast coast of India. *The Science of the Total Environment*, 720, 137614. DOI: 10.1016/j.scitotenv.2020.137614

Sathyanarayana, S., Karr, C. J., Lozano, P., Brown, E., Calafat, A. M., Liu, F., & Swan, S. H. (2008). Baby care products: Possible sources of infant phthalate exposure. *Pediatrics*, 121(2), e260–e268. DOI: 10.1542/peds.2006-3766

Satti, S. M., & Shah, A. A. (2020). Polyester-based biodegradable plastics: An approach towards sustainable development. *Letters in Applied Microbiology*, 70(6), 413–430. DOI: 10.1111/lam.13287

Sayed, A. E. D. H., Emeish, W. F. A., Bakry, K. A., Al-Amgad, Z., Lee, J. S., & Mansour, S. (2024). Polystyrene nanoplastic and engine oil synergistically intensify toxicity in *Nile tilapia, Oreochromis niloticus*: Polystyrene nanoplastic and engine oil toxicity in *Nile tilapia.BMC Veterinary Research*, 20(1), 143. DOI: 10.1186/s12917-024-03987-z

Schmidt, N., Fauvelle, V., Ody, A., Castro-Jiménez, J., Jouanno, J., Changeux, T., Thibaut, T., & Sempéré, R. (2019). The Amazon River: A Major Source of Organic Plastic Additives to the Tropical North Atlantic? *Environmental Science & Technology*, 53(13), 7513–7521. DOI: 10.1021/acs.est.9b01585 PMID: 31244083

Schröder B. (2024). From the Automated Calculation of Potential Energy Surfaces to Accurate Infrared Spectra. J Phys Chem Lett., 21(15), 3159-3169. .DOI: 10.1021/acs.jpclett.4c00186

Schymanski, D., Goldbeck, C., Humpf, H.-U., & Fürst, P. (2018). Analysis of microplastics in water by micro-Raman spectroscopy: Release of plastic particles from different packaging into mineral water. *Water Research*, 129, 154–162. DOI: 10.1016/j.watres.2017.11.011 PMID: 29145085

Serratosa, J., Hyrenbach, K. D., Miranda-Urbina, D., Portflitt-Toro, M., Luna, N., & Luna-Jorquera, G. (2020). Environmental drivers of seabird at-sea distribution in the Eastern South Pacific Ocean: Assemblage composition across a longitudinal productivity gradient. *Frontiers in Marine Science*, 6, 838. DOI: 10.3389/fmars.2019.00838

Setälä, O., Lehtiniemi, M., Coppock, R., & Cole, M. (2018). Microplastics in marine food webs. In Zeng, E. Y. (Ed.), *Microplastic Contamination in Aquatic Environments: An Emerging Matter of Environmental Urgency* (pp. 339–363). Elsevier. DOI: 10.1016/B978-0-12-813747-5.00011-4

Setälä, O., Norkko, J., & Lehtiniemi, M. (2016). Feeding type affects microplastic ingestion in a coastal invertebrate community. *Marine Pollution Bulletin*, 102(1), 95–101. DOI: 10.1016/j.marpolbul.2015.11.053

Shafiuddin Ahmed. (2023). *Microplastics in aquatic environments: A comprehensive review of toxicity, removal, and remediation strategie.*

Shah, T., Ali, A., Haider, G., Asad, M., & Munsif, F. (2023). Microplastics alter soil enzyme activities and microbial community structure without negatively affecting plant growth in an agroecosystem. *Chemosphere*, 322, 138188. DOI: 10.1016/j.chemosphere.2023.138188

Sharma, S., Sharma, V., & Chatterjee, S. (2023). Contribution of plastic and microplastic to global climate change and their conjoining impacts on the environment: A review. *The Science of the Total Environment*, 875, 162627. DOI: 10.1016/j.scitotenv.2023.162627

Shen, M., Huang, W., Chen, M., Song, B., Zeng, G., & Zhang, Y. (2020). (Micro) plastic crisis: Un-ignorable contribution to global greenhouse gas emissions and climate change. *Journal of Cleaner Production*, 254, 120138. DOI: 10.1016/j.jclepro.2020.120138

Shepherd, P. C. F., & Boates, J. S. (1999). Effects of a commercial baitworm harvest on Semipalmated Sandpipers and their prey in the Bay of Fundy hemispheric shorebird reserve. *Conservation Biology*, 13(2), 347–356. DOI: 10.1046/j.1523-1739.1999.013002347.x

Shi, B., Patel, M., Yu, D., Yan, J., Li, Z., Petriw, D., Pruyn, T., Smyth, K., Passeport, E., Miller, R. J. D., & Howe, J. Y. (2022). Automatic quantification and classification of microplastics in scanning electron micrographs via deep learning. *The Science of the Total Environment*, 825, 153903. DOI: 10.1016/j.scitotenv.2022.153903

Shim, W. J., Song, Y. K., Hong, S. H., & Jang, M. (2016). Identification and quantification of microplastics using Nile Red staining. *Marine Pollution Bulletin*, 113(1-2), 469–476. DOI: 10.1016/j.marpolbul.2016.10.049

Shirazimoghaddam, S., Amin, I., Faria Albanese, J. A., & Shiju, N. R. (2023). Chemical Recycling of Used PET by Glycolysis Using Niobia-Based Catalysts. *ACS Engineering Au*, 3(1), 37–44. DOI: 10.1021/acsengineeringau.2c00029 PMID: 36820227

Shi, Y., Liu, P., Wu, X., Shi, H., Huang, H., Wang, H., & Gao, S. (2021). Insight into chain scission and release profiles from photodegradation of polycarbonate microplastics. *Water Research*, 195, 116980. DOI: 10.1016/j.watres.2021.116980

Silva, A. B., Costa, M. F., & Duarte, A. C. (2018). Biotechnology advances for dealing with environmental pollution by micro (nano) plastics: Lessons on theory and practices. *Current Opinion in Environmental Science & Health*, 1, 30–35. DOI: 10.1016/j.coesh.2017.10.005

Silva, A. L. P., Silva, S. A., Duarte, A., Barceló, D., & Rocha-Santos, T. (2022). Analytical methodologies used for screening micro (nano) plastics in (eco) toxicity tests. *Green Analytical Chemistry*, 3, 100037. DOI: 10.1016/j.greeac.2022.100037

Singh, A., Singh, J., Vasishth, A., Kumar, A., & Pattnaik, S. S. (2024). *Emerging Materials in Advanced Oxidation Processes for Micropollutant Treatment Process. Advanced Oxidation Processes for Micropollutant Remediation*. CRC Press.

Singh, B., & Kumar, A. (2024). Advances in microplastics detection: A comprehensive review of methodologies and their effectiveness. *Trends in Analytical Chemistry*, 170, 117440. DOI: 10.1016/j.trac.2023.117440

Sjollema, S. B., Redondo-Hasselerharm, P., Leslie, H. A., Kraak, M. H., & Vethaak, A. D. (2016). Do plastic particles affect microalgal photosynthesis and growth? *Aquatic Toxicology (Amsterdam, Netherlands)*, 170, 259–261. DOI: 10.1016/j.aquatox.2015.12.002

Skocaj, M., Filipic, M., Petkovic, J., & Novak, S. (2011). Titanium dioxide in our everyday life; is it safe? *Radiology and Oncology*, 45(4), 227–247. DOI: 10.2478/v10019-011-0037-0

Smith, M. A., Walker, N. J., Free, C. M., Kirchhoff, M. J., Drew, G. S., Warnock, N., & Stenhouse, I. J. (2014). Identifying marine Important Bird Areas using at-sea survey data. *Biological Conservation*, 172, 180–189. DOI: 10.1016/j.biocon.2014.02.039

Snelgrove, P. V. R., Austen, M. C., Boucher, G., Heip, C., Hutchings, P. A., King, G. M., Koike, I., Lambshead, P. J. D., & Smith, C. R. (2000). Linking biodiversity above and below the marine sediment-water interface. *Bioscience*, 50(12), 1076–1088. DOI: 10.1641/0006-3568(2000)050[1076:LBAABT]2.0.CO;2

Sohail, M., Urooj, Z., Noreen, S., Baig, M. M. F. A., Zhang, X., & Li, B. (2023). Micro-and nanoplastics: Contamination routes of food products and critical interpretation of detection strategies. *The Science of the Total Environment*, 891, 164596. DOI: 10.1016/j.scitotenv.2023.164596

Solís-Balbín, C., Sol, D., Laca, A., Laca, A., & Díaz, M. (2023). Destruction and entrainment of microplastics in ozonation and wet oxidation processes. *Journal of Water Process Engineering*, 51, 103456. DOI: 10.1016/j.jwpe.2022.103456

Song, Y. K., Hong, S. H., Jang, M., Han, G. M., Rani, M., Lee, J., & Shim, W. J. (2015). A comparison of microscopic and spectroscopic identification methods for analysis of microplastics in environmental samples. *Marine Pollution Bulletin*, 93(1–2), 202–209. DOI: 10.1016/j.marpolbul.2015.01.015 PMID: 25682567

Staehelin, J., & Hoigne, J. (1982). Decomposition of ozone in water: Rate of initiation by hydroxide ions and hydrogen peroxide. *Environmental Science & Technology*, 16(10), 676–681. DOI: 10.1021/es00104a009

Steinmetz, Z., Kintzi, A., Muñoz, K., & Schaumann, G. E. (2020). A simple method for the selective quantification of polyethylene, polypropylene, and polystyrene plastic debris in soil by pyrolysis-gas chromatography/mass spectrometry. *Journal of Analytical and Applied Pyrolysis*, 147, 104803. DOI: 10.1016/j.jaap.2020.104803

Stoks, R., Geerts, A. N., & De Meester, L. (2014). Evolutionary and plastic responses of freshwater invertebrates to climate change: Realized patterns and future potential. *Evolutionary Applications*, 7(1), 42–55. DOI: 10.1111/eva.12108

Su, J., Zhang, F., Yu, C., Zhang, Y., Wang, J., Wang, C., Wang, H., & Jiang, H. (2023). Machine learning: Next promising trend for microplastics study. *Journal of Environmental Management*, 344, 118756. DOI: 10.1016/j.jenvman.2023.118756

Su, L., Xiong, X., Zhang, Y., Wu, C., Xu, X., Sun, C., & Shi, H. (2022). Global transportation of plastics and microplastics: A critical review of pathways and influences. *The Science of the Total Environment*, 831, 154884. DOI: 10.1016/j.scitotenv.2022.154884

Sulpizio, J. (2019). Microplastics in our waters, an unquestionable concern. *York Daily Record*, 24–26. https://www.proquest.com/newspapers/microplastics-our-waters-unquestionable-concern/docview/2322667200/se-2?accountid=206735

Sun, C., Yang, X., Gu, Q., Jiang, G., Shen, L., Zhou, J., Li, L., Chen, H., Zhang, G., & Zhang, Y. (2023). Comprehensive analysis of nanoplastic effects on growth phenotype, nanoplastic accumulation, oxidative stress response, gene expression, and metabolite accumulation in multiple strawberry cultivars. *The Science of the Total Environment*, 897, 165432. DOI: 10.1016/j.scitotenv.2023.165432

Sun, X.-D., Yuan, X.-Z., Jia, Y., Feng, L.-J., Zhu, F.-P., Dong, S.-S., Liu, J., Kong, X., Tian, H., Duan, J.-L., Ding, Z., Wang, S.-G., & Xing, B. (2020). Differentially charged nano-plastics demonstrate distinct accumulation in Arabidopsis thaliana. *Nature Nanotechnology*, 15(9), 755–760. DOI: 10.1038/s41565-020-0707-4

Surgun-Acar, Y. (2022). Response of soybean (Glycine max L.) seedlings to polystyrene nanoplastics: Physiological, biochemical, and molecular perspectives. *Environmental Pollution*, 314, 120262. DOI: 10.1016/j.envpol.2022.120262

Susanti, N. K., Mardiastuti, A., & Wardiatno, Y. (2020). Microplastics and the impact of plastic on wildlife: A literature review. *Environmental Earth Sciences*, 528, 012013.

Swan, S. H. (2008). Environmental phthalate exposure in relation to reproductive outcomes and other health endpoints in humans. *Environmental Research*, 108(2), 177–184. DOI: 10.1016/j.envres.2008.08.007

Tadsuwan, K., & Babel, S. (2022). Unraveling microplastics removal in wastewater treatment plant: A comparative study of two wastewater treatment plants in Thailand. *Chemosphere*, 307, 135733. DOI: 10.1016/j.chemosphere.2022.135733

Tagg, A. S., Brandes, E., Fischer, F., Fischer, D., Brandt, J., & Labrenz, M. (2022). Agricultural application of microplastic-rich sewage sludge leads to further uncontrolled contamination. *The Science of the Total Environment*, 806(Pt 4), 150611. DOI: 10.1016/j.scitotenv.2021.150611

Talsness, C. E., Andrade, A. J., Kuriyama, S. N., Taylor, J. A., & Vom Saal, F. S. (2009). Components of plastic: Experimental studies in animals and relevance for human health. *Philosophical Transactions of the Royal Society of London. Series B, Biological Sciences*, 364(1526), 2079–2096. DOI: 10.1098/rstb.2008.0281

Tamayo-Belda, M., Pulido-Reyes, G., González-Pleiter, M., Martín-Betancor, K., Leganés, F., Rosal, R., & Fernández-Piñas, F. (2022). Identification and toxicity towards aquatic primary producers of the smallest fractions released from hydrolytic degradation of polycaprolactone microplastics. *Chemosphere*, 303(Pt 1), 134966. DOI: 10.1016/j.chemosphere.2022.134966

Tanaka, K., van Franeker, J. A., Deguchi, T., & Takada, H. (2019). Piece-by-piece analysis of additives and manufacturing byproducts in plastics ingested by seabirds: Implication for risk of exposure to seabirds. *Marine Pollution Bulletin*, 145, 36–41. DOI: 10.1016/j.marpolbul.2019.05.028 PMID: 31590798

Tanaka, K., Watanuki, Y., Takada, H., Ishizuka, M., Yamashita, R., Kazama, M., Hiki, N., Kashiwada, F., Mizukawa, K., Mizukawa, H., Hyrenbach, D., Hester, M., Ikenaka, Y., & Nakayama, S. M. M. (2020). In vivo accumulation of plastic-derived chemicals into seabird tissues. *Current Biology*, 30(4), 723–728. DOI: 10.1016/j.cub.2019.12.037 PMID: 32008901

Tang D. (2020). Health, Safety and Environment Program, Curtin University. DOI: 10.9734/ajee/2020/v13i130170

Tang, P., Forster, R., McCumskay, R., Rogerson, M. & Waller, C. (2019). Handheld FT-IR spectroscopy for the triage of micro-and meso-sized plastics in the marine environment incorporating an accelerated weathering study and an aging estimation.

Tanner, D. J. (1974). *Plastics waste: A technological and economic study*. University of Surrey.

Tavares, D. C., de Moura, J. F., Merico, A., & Siciliano, S. (2017). Incidence of marine debris in seabirds feeding at different water depths. *Marine Pollution Bulletin*, 119(2), 68–73. DOI: 10.1016/j.marpolbul.2017.04.012 PMID: 28431744

Teuten, E. L., Saquing, J. M., Knappe, D. R. U., Barlaz, M. A., Jonsson, S., Björn, A., Rowland, S. J., Thompson, R. C., Galloway, T. S., Yamashita, R., Ochi, D., Watanuki, Y., Moore, C., Viet, P. H., Tana, T. S., Prudente, M., Boonyatumanond, R., Zakaria, M. P., Akkhavong, K., & Takada, H. (2009). Transport and release of chemicals from plastics to the environment and to wildlife. *Philosophical Transactions of the Royal Society of London. Series B, Biological Sciences*, 364(1526), 2027–2045. DOI: 10.1098/rstb.2008.0284

Thammasanya, T., Patiam, S., Rodcharoen, E., & Chotikarn, P. (2024). A new approach to classifying polymer type of microplastics based on Faster-RCNN-FPN and spectroscopic imagery under ultraviolet light. *Scientific Reports*, 14(1), 3529. DOI: 10.1038/s41598-024-53251-5

Thiounn, T., & Smith, R. C. (2020). Advances and approaches for chemical recycling of plastic waste. *Journal of Polymer Science*, 58(10), 1347–1364. DOI: 10.1002/pol.20190261

Thompson, J. R., Wilder, L. M., & Crooks, R. M. (2021). Filtering and continuously separating microplastics from water using electric field gradients formed electrochemically in the absence of buffer. *Chemical Science (Cambridge)*, 12(41), 13744–13755. DOI: 10.1039/D1SC03192A

Thushari, G. G. N., & Senevirathna, J. D. M. (2020). Plastic pollution in the marine environment. *Heliyon*, 6(8), e04709. DOI: 10.1016/j.heliyon.2020.e04709 PMID: 32923712

Tian, H., Zhang, L., Dong, J., Wu, L., Fang, F., Wang, Y., Li, H., Xie, C., Li, W., Wei, Z., Liu, Z., & Zhang, M. (2022). A One-Step Surface Modification Technique Improved the Nutrient Release Characteristics of Controlled-Release Fertilizers and Reduced the Use of Coating Materials. *Journal of Cleaner Production*, 369, 133331. DOI: 10.1016/j.jclepro.2022.133331

Tian, L., Chen, Q., Jiang, W., Wang, L., Xie, H., Kalogerakis, N., Ma, Y., & Ji, R. (2019). A carbon-14 radiotracer-based study on the phototransformation of polystyrene nanoplastics in water versus in air. *Environmental Science. Nano*, 6(9), 2907–2917. DOI: 10.1039/C9EN00662A

Tian, X., Beén, F., & Bäuerlein, P. S. (2022). Quantum cascade laser imaging (LDIR) and machine learning for the identification of environmentally exposed microplastics and polymers. *Environmental Research*, 212, 113569. DOI: 10.1016/j.envres.2022.113569

Tirkey, A., & Upadhyay, L. S. B. (2021). Microplastics: An overview on separation, identification and characterization of microplastics. *Marine Pollution Bulletin*, 170, 112604. DOI: 10.1016/j.marpolbul.2021.112604

Tiwari, M., Rathod, T. D., Ajmal, P. Y., Bhangare, R. C., & Sahu, S. K. (2019). Distribution and characterization of microplastics in beach sand from three different Indian coastal environments. *Marine Pollution Bulletin*, 140, 262–273. DOI: 10.1016/j.marpolbul.2019.01.055

Todd, E. C. D. (2014). Foodborne Diseases: Overview of Biological Hazards and Foodborne Diseases. Encyclopedia of Food Safety, 221–42. DOI: 10.1016/B978-0-12-378612-8.00071-8

Toropova, C., Meliane, I., Laffoley, D., Matthews, E., & Spalding, M. (Eds.). (2010). *Global Ocean Protection: Present Status and Future Possibilities. Brest, France: Agence des aires marines protégées; Gland, Switzerland: IUCN WCPA; Cambridge, UK: UNEP-WCMC; Arlington, USA: TNC; Tokyo, Japan: UNU*. WCS.

Tripathy, B. C., & Oelmüller, R. (2012). Reactive oxygen species generation and signaling in plants. *Plant Signaling & Behavior*, 7(12), 1621–1633. DOI: 10.4161/psb.22455

Tsakona, M., Baker, E., Rucevska, I., Maes, T., Appelquist, L. R., Macmillan-Lawler, M., Harris, P., Raubenheimer, K., Langeard, R., Savelli-Soderberg, H., Woodall, K. O., Dittkrist, J., Zwimpfer, T. A., Aidis, R., Mafuta, C., & Schoolmeester, T. (2021). Marine Litter and Plastic Waste Vital Graphics. *UN environment programme*.

Turner, A., & Holmes, L. (2011). Occurrence, distribution and characteristics of beached plastic production pellets on the island of Malta (central Mediterranean). *Marine Pollution Bulletin*, 62(2), 377–381. DOI: 10.1016/j.marpolbul.2010.09.027 PMID: 21030052

Tuyen, P. T. T. (n.d.). *International legal framework for ocean plastic pollution prevention and enforcement in Vietnam*.

Ullah, R., Tsui, M. T.-K., Chow, A., Chen, H., Williams, C., & Ligaba-Osena, A. (2023). Micro(nano)plastic pollution in terrestrial ecosystem: Emphasis on impacts of polystyrene on soil biota, plants, animals, and humans. *Environmental Monitoring and Assessment*, 195(1), 252. DOI: 10.1007/s10661-022-10769-3

Unuofin, J. O. (2020). Garbage in garbage out: The contribution of our industrial advancement to wastewater degeneration. *Environmental Science and Pollution Research International*, 27(18), 22319–22335. DOI: 10.1007/s11356-020-08944-5

Uogintė, I., Pleskytė, S., Skapas, M., Stanionytė, S., & Lujanienė, G. (2023). Degradation and optimization of microplastic in aqueous solutions with graphene oxide-based nanomaterials. *International Journal of Environmental Science and Technology*, 20(9), 9693–9706. DOI: 10.1007/s13762-022-04657-z

Upadhyay, R. K. (2020). Markers for global climate change and its impact on social, biological and ecological systems: A review. *American Journal of Climate Change*, 9(03), 159–203. DOI: 10.4236/ajcc.2020.93012

Urli, S., Corte Pause, F., Crociati, M., Baufeld, A., Monaci, M., & Stradaioli, G. (2023). Impact of Microplastics and Nanoplastics on Livestock Health: An Emerging Risk for Reproductive Efficiency. *Animals (Basel)*, 13(7), 1132. DOI: 10.3390/ani13071132 PMID: 37048387

Usman, S., Abdull Razis, A. F., Shaari, K., Azmai, M. N. A., Saad, M. Z., Mat Isa, N., & Nazarudin, M. F. (2022). The Burden of Microplastics Pollution and Contending Policies and Regulations. *International Journal of Environmental Research and Public Health*, 19(11), 6773. DOI: 10.3390/ijerph19116773 PMID: 35682361

Usman, S., Muhammad, Y., & Chiroman, A. (2016). Roles of soil biota and biodiversity in soil environment–A concise communication. *Eurasian Journal of Soil Science*, 5(4), 255–265. DOI: 10.18393/ejss.2016.4.255-265

Valavanidis, A., Vlahogianni, T., Dassenakis, M., & Scoullos, M. (2006). Molecular biomarkers of oxidative stress in aquatic organisms in relation to toxic environmental pollutants. *Ecotoxicology and Environmental Safety*, 64(2), 178–189. DOI: 10.1016/j.ecoenv.2005.03.013 PMID: 16406578

Valiyaveettil Salimkumar, A., Kurisingal Cleetus, M. C., Ehigie, J. O., Onogbosele, C. O., Nisha, P., Kumar, B. S., Prabhakaran, M. P., & Rejish Kumar, V. J. (2024). *Adsorption Behavior and Interaction of Micro-Nanoplastics in Soils and Aquatic Environment. Management of Micro and Nano-plastics in Soil and Biosolids: Fate.* Occurrence, Monitoring, and Remedies.

Van Cauwenberghe, L., & Janssen, C. R. (2014). Microplastics in bivalves cultured for human consumption. *Environmental Pollution*, 193, 65–70. DOI: 10.1016/j.envpol.2014.06.010

van Emmerik, T., & Schwarz, A. (2020). Plastic debris in rivers. *WIREs. Water*, 7(1), e1398. DOI: 10.1002/wat2.1398

Van Franeker, J. A. (2004). *Save the North Sea e Fulmar Study Manual 1: Collection and Dissection Procedures. Alterra Rapport 672*. Alterra.

van Franeker, J. A., Blaize, C., Danielsen, J., Fairclough, K., Gollan, J., Guse, N., Hansen, P.-L., Heubeck, M., Jensen, J.-K., Gilles, L. G., Olsen, B., Olsen, K.-O., Pedersen, J., Stienen, E. W. M., & Turner, D. M. (2011). Monitoring plastic ingestion by the northern fulmar (Fulmarus glacialis) in the North Sea. *Environmental Pollution*, 159(10), 2609–2615. DOI: 10.1016/j.envpol.2011.06.008 PMID: 21737191

van Schothorst, B., Beriot, N., Huerta Lwanga, E., & Geissen, V. (2021). Sources of Light Density Microplastic Related to Two Agricultural Practices: The Use of Compost and Plastic Mulch. *Environments (Basel, Switzerland)*, 8(4), 36. DOI: 10.3390/environments8040036

Van Sebille, E., Wilcox, C., Lebreton, L., Maximenko, N., Hardesty, B. D., van Franeker, J. A., Eriksen, M., Siegel, D., Galgani, F., & Law, K. L. (2015). A global inventory of small floating plastic debris. *Environmental Research Letters*, 10(12), 124006. DOI: 10.1088/1748-9326/10/12/124006

Vandermeersch, G., Van Cauwenberghe, L., Janssen, C. R., Marques, A., Granby, K., Fait, G., Kotterman, M. J. J., Diogène, J., Bekaert, K., Robbens, J., & Devriese, L. (2015). A critical view on microplastic quantification in aquatic organisms. *Environmental Research*, 143, 46–55. Advance online publication. DOI: 10.1016/j.envres.2015.07.016 PMID: 26249746

VanderWerf, E. A., & Young, L. C. (2017). *A summary and gap analysis of seabird monitoring in the US Tropical Pacific. Report prepared for the US Fish and Wildlife Service, Region 1*. Pacific Rim Conservation.

Verma, K. K., Song, X.-P., Xu, L., Huang, H.-R., Liang, Q., Seth, C. S., & Li, Y.-R. (2023). Nano-microplastic and agro-ecosystems: A mini-review. *Frontiers in Plant Science*, 14, 1283852. DOI: 10.3389/fpls.2023.1283852 PMID: 38053770

Vethaak, A. D., & Leslie, H. A. (2016). Plastic debris is a human health issue. *Environmental Science & Technology*, 50(13), 6825–6826. DOI: 10.1021/acs.est.6b02569

Vidal, C., & Pasquini, C. (2021). A comprehensive and fast microplastics identification based on near-infrared hyperspectral imaging (HSI-NIR) and chemometrics. *Environmental Pollution*, 285, 117251. DOI: 10.1016/j.envpol.2021.117251

Vidayanti, V., & Retnaningdyah, C. (2024). Microplastic pollution in the surface waters, sediments, and wild crabs of mangrove ecosystems of East Java, Indonesia. *Emerging Contaminants*, 10(4), 100343. DOI: 10.1016/j.emcon.2024.100343

Vighi, M., Bayo, J., Fernández-Piñas, F., Gago, J., Gómez, M., Hernández-Borges, J., Herrera, A., Landaburu, J., Muniategui-Lorenzo, S., Muñoz, A.-R., Rico, A., Romera-Castillo, C., Viñas, L., & Rosal, R. (2021). Micro and Nano-Plastics in the Environment: Research Priorities for the Near Future. In de Voogt, P. (Ed.), *Reviews of Environmental Contamination and Toxicology* (Vol. 257, pp. 163–218). Springer International Publishing. DOI: 10.1007/398_2021_69

Villa, K., Děkanovský, L., Plutnar, J., Kosina, J., & Pumera, M. (2020). Swarming of perovskite-like Bi2WO6 microrobots destroy textile fibers under visible light. *Advanced Functional Materials*, 30(51), 2007073. DOI: 10.1002/adfm.202007073

Vincze, É.-B., Becze, A., Laslo, É., & Mara, G. (2024). Beneficial Soil Microbiomes and Their Potential Role in Plant Growth and Soil Fertility. *Agriculture*, 14(1), 152. DOI: 10.3390/agriculture14010152

Vital-Grappin, A. D., Ariza-Tarazona, M. C., Luna-Hernández, V. M., Villarreal-Chiu, J. F., Hernández-López, J. M., Siligardi, C., & Cedillo-González, E. I. (2021). The role of the reactive species involved in the photocatalytic degradation of hdpe microplastics using c, n-tio2 powders. *Polymers*, 13(7), 999. DOI: 10.3390/polym13070999

Vos, P., Hogers, R., Bleeker, M., Reijans, M., Lee, T. V. D., Hornes, M., Friters, A., Pot, J., Paleman, J., Kuiper, M., & Zabeau, M. (1995). AFLP: A new technique for DNA fingerprinting. *Nucleic Acids Research*, 23(21), 4407–4414. DOI: 10.1093/nar/23.21.4407

Wagner, M., Scherer, C., Alvarez-Muñoz, D., Brennholt, N., Bourrain, X., Buchinger, S., Fries, E., Grosbois, C., Klasmeier, J., Marti, T., Rodriguez-Mozaz, S., Urbatzka, R., Vethaak, A. D., Winther-Nielsen, M., & Reifferscheid, G. (2014). Microplastics in freshwater ecosystems: What we know and what we need to know. *Environmental Sciences Europe*, 26(1), 12. DOI: 10.1186/s12302-014-0012-7

Walker, T. R., & Fequet, L. (2023). Current trends of unsustainable plastic production and micro (nano) plastic pollution. *Trends in Analytical Chemistry*, 160, 116984. DOI: 10.1016/j.trac.2023.116984

Walters, D. R., & Kingham, G. (1990). Uptake and translocation of α-difluoromethylornithine, a polyamine biosynthesis inhibitor, by barley seedlings: Effects on mildew infection. *The New Phytologist*, 114(4), 659–665. DOI: 10.1111/j.1469-8137.1990.tb00437.x

Wang, F., Gao, J., Zhai, W., Liu, D., Zhou, Z., & Wang, P. (2020). The influence of polyethylene microplastics on pesticide residue and degradation in the aquatic environment. *Journal of Hazardous Materials*, 394, 122517. DOI: 10.1016/j.jhazmat.2020.122517

Wang, J., Li, J., Liu, S., Li, H., Chen, X., Peng, C., Zhang, P., & Liu, X. (2021). Distinct microplastic distributions in soils of different land-use types: A case study of Chinese farmlands. *Environmental Pollution*, 269, 116199. DOI: 10.1016/j.envpol.2020.116199

Wang, J., Lu, L., Wang, M., Jiang, T., Liu, X., & Ru, S. (2019). Typhoons increase the abundance of microplastics in the marine environment and cultured organisms: A case study in Sanggou Bay, China. *The Science of the Total Environment*, 667, 1–8. DOI: 10.1016/j.scitotenv.2019.02.367

Wang, J., Lu, S., Bian, H., Xu, M., Zhu, W., Wang, H., He, C., & Sheng, L. (2022, October). Effects of individual and combined polystyrene nanoplastics and phenanthrene on the enzymology, physiology, and transcriptome parameters of rice (Oryza sativa L.). *Chemosphere*, 304, 135341. DOI: 10.1016/j.chemosphere.2022.135341

Wang, J., Wang, H., Huang, L., & Wang, C. (2017). Surface treatment with Fenton for separation of acrylonitrile-butadiene-styrene and polyvinylchloride waste plastics by flotation. *Waste Management (New York, N.Y.)*, 67, 20–26. DOI: 10.1016/j.wasman.2017.05.009

Wang, J., Zheng, L., & Li, J. (2018). A critical review on the sources and instruments of marine microplastics and prospects on the relevant management in China. *Waste management & research: The journal of the International Solid Wastes and Public Cleansing Association. Waste Management & Research*, 36(10), 898–911. DOI: 10.1177/0734242X18793504

Wang, K., Du, Y., Li, P., Guan, C., Zhou, M., Wu, L., Liu, Z., & Huang, Z. (2024). Nanoplastics causes heart aging/myocardial cell senescence through the Ca2+/mtDNA/cGAS-STING signaling cascade. *Journal of Nanobiotechnology*, 22(1), 96. DOI: 10.1186/s12951-024-02375-x

Wang, L., Kaeppler, A., Fischer, D., & Simmchen, J. (2019). Photocatalytic TiO2 micromotors for removal of microplastics and suspended matter. *ACS Applied Materials & Interfaces*, 11(36), 32937–32944. DOI: 10.1021/acsami.9b06128

Wang, L., Wu, W. M., Bolan, N. S., Tsang, D. C., Li, Y., Qin, M., & Hou, D. (2021). Environmental fate, toxicity and risk management strategies of nano-plastics in the environment: Current status and future perspectives. *Journal of Hazardous Materials*, 401, 123415. DOI: 10.1016/j.jhazmat.2020.123415

Wang, W. (2021, February 15). A. Adams, Sun, Zhang, Qingdao University of Science and Technology. *Journal of Hazardous Materials*, 424(Part C). Advance online publication. DOI: 10.1016/j.jhazmat.2021.127531 PMID: 34740160

Wang, W., Yuan, W., Xu, E. G., Li, L., Zhang, H., & Yang, Y. (2022). Uptake, translocation, and biological impacts of micro (nano) plastics in terrestrial plants: Progress and prospects. *Environmental Research*, 203, 111867. DOI: 10.1016/j.envres.2021.111867

Wang, W., Zhao, Y., Bai, H., Zhang, T., Ibarra-Galvan, V., & Song, S. (2018). Methylene blue removal from water using the hydrogel beads of poly(vinyl alcohol)-sodium alginate-chitosan-montmorillonite. *Carbohydrate Polymers*, 198, 518–528. DOI: 10.1016/j.carbpol.2018.06.124

Wang, X., Bolan, N., Tsang, D. C., Sarkar, B., Bradney, L., & Li, Y. (2021). A review of microplastics aggregation in aquatic environment: Influence factors, analytical methods, and environmental implications. *Journal of Hazardous Materials*, 402, 123496. DOI: 10.1016/j.jhazmat.2020.123496

Wang, X., Dai, Y., Li, Y., & Yin, L. (2023). Application of advanced oxidation processes for the removal of micro/nanoplastics from water: A review. *Chemosphere*, 140636.

Wang, X., Diao, Y., Dan, Y., Liu, F., Wang, H., Sang, W., & Zhang, Y. (2022). Effects of solution chemistry and humic acid on transport and deposition of aged microplastics in unsaturated porous media. *Chemosphere*, 309, 136658. DOI: 10.1016/j.chemosphere.2022.136658

Wang, X., Liu, L., Zheng, H., Wang, M., Fu, Y., Luo, X., Li, F., & Wang, Z. (2020). Polystyrene microplastics impaired the feeding and swimming behavior of mysid shrimp Neomysis japonica. *Marine Pollution Bulletin*, 150(September), 110660. DOI: 10.1016/j.marpolbul.2019.110660 PMID: 31727317

Wang, X., Zheng, H., Zhao, J., Luo, X., Wang, Z., & Xing, B. (2020). Photodegradation elevated the toxicity of polystyrene microplastics to grouper (Epinephelus moara) through disrupting hepatic lipid homeostasis. *Environmental Science & Technology*, 54(10), 6202–6212. DOI: 10.1021/acs.est.9b07016

Wang, Y., Huang, J., Zhu, F., & Zhou, S. (2021). Airborne microplastics: A review on the occurrence, migration and risks to humans. *Bulletin of Environmental Contamination and Toxicology*, 107(4), 657–664. DOI: 10.1007/s00128-021-03180-0 PMID: 33742221

Wang, Y., & Qian, H. (2021). Phthalates and their impacts on human health. *Healthcare (Basel)*, 9(5), 603. DOI: 10.3390/healthcare9050603

Wang, Z., Lin, T., & Chen, W. (2020). Occurrence and removal of microplastics in an advanced drinking water treatment plant (ADWTP). *The Science of the Total Environment*, 700, 134520. DOI: 10.1016/j.scitotenv.2019.134520

Wang, Z., Li, W., Li, W., Yang, W., & Jing, S. (2023). Effects of microplastics on the water characteristic curve of soils with different textures. *Chemosphere*, 317, 137762. DOI: 10.1016/j.chemosphere.2023.137762

Wang, Z., Pal, D., Pilechi, A., & Ariya, P. A. (2024). Nanoplastics in Water: Artificial Intelligence-Assisted 4D Physicochemical Characterization and Rapid In Situ Detection. *Environmental Science & Technology*, 58(20), 8919–8931. DOI: 10.1021/acs.est.3c10408

Wan, S., Wang, X., Chen, W., Wang, M., Zhao, J., Xu, Z., Wang, R., Mi, C., Zheng, Z., & Zhang, H. (2024). Exposure to high dose of polystyrene nanoplastics causes trophoblast cell apoptosis and induces miscarriage. *Particle and Fibre Toxicology*, 21(1), 13. DOI: 10.1186/s12989-024-00574-w

Waqas, M., Wong, M. S., Stocchino, A., Abbas, S., Hafeez, S., & Zhu, R. (2023). Marine plastic pollution detection and identification by using remote sensing-meta analysis. *Marine Pollution Bulletin*, 197, 115746. DOI: 10.1016/j.marpolbul.2023.115746

Watteau, F., Dignac, M. F., Bouchard, A., Revallier, A., & Houot, S. (2018). Microplastic detection in soil amended with municipal solid waste composts as revealed by transmission electronic microscopy and pyrolysis/GC/MS. *Frontiers in Sustainable Food Systems*, 2, 81. DOI: 10.3389/fsufs.2018.00081

Watts, A. J. R., Lewis, C., Goodhead, R. M., Beckett, S. J., Moger, J., Tyler, C. R., & Galloway, T. S. (2014). Uptake and retention of microplastics by the shore crab Carcinus maenas. *Environmental Science & Technology*, 48(15), 8823–8830. DOI: 10.1021/es501090e PMID: 24972075

Weber, C. J., Santowski, A., & Chifflard, P. (2022). Investigating the dispersal of macro- and microplastics on agricultural fields 30 years after sewage sludge application. *Scientific Reports*, 12(1), 6401. DOI: 10.1038/s41598-022-10294-w

Weber, F., Zinnen, A., & Kerpen, J. (2023). Development of a machine learning-based method for the analysis of microplastics in environmental samples using µ-Raman spectroscopy. *Microplastics and Nanoplastics*, 3(1), 9. DOI: 10.1186/s43591-023-00057-3

Wegner, A., Besseling, E., Foekema, E. M., Kamermans, P., & Koelmans, A. A. (2012). Effects of nanopolystyrene on the feeding behavior of the blue mussel (Mytilus edulis L.). *Environmental Toxicology and Chemistry*, 31(11), 2490–2497. DOI: 10.1002/etc.1984

Weimerskirch, H., Pinaud, D., Pawlowski, F., & Bost, C. A. (2007). Does prey capture induce area-restricted search? A fine-scale study using GPS in a marine predator, the wandering albatross. *American Naturalist*, 170(5), 734–743. DOI: 10.1086/522059 PMID: 17926295

Weis, J. S., & De Falco, F. (2022). Microfibers: Environmental Problems and Textile Solutions. *Microplastics*, 1(4), 626–639. DOI: 10.3390/microplastics1040043

Wei, X. F., Yang, W., & Hedenqvist, M. S. (2024). Plastic pollution amplified by a warming climate. *Nature Communications*, 15(1), 2052. DOI: 10.1038/s41467-024-46127-9

Wiedner, K., & Polifka, S. (2020). Effects of microplastic and microglass particles on soil microbial community structure in an arable soil (Chernozem). *Soil (Göttingen)*, 6(2), 315–324. DOI: 10.5194/soil-6-315-2020

Wieland, S., Ramsperger, A. F. R. M., Gross, W., Lehmann, M., Witzmann, T., Caspari, A., Obst, M., Gekle, S., Auernhammer, G. K., Fery, A., Laforsch, C., & Kress, H. (2024). Nominally identical microplastic models differ greatly in their particle-cell interactions. *Nature Communications*, 15(1), 922. DOI: 10.1038/s41467-024-45281-4

Wilcox, C., van Sebille, E., & Hardesty, B. D. (2015). Threat of plastic pollution to seabirds is global, pervasive, and increasing. *Proceedings of the National Academy of Sciences of the United States of America*, 112(38), 11899–11904. DOI: 10.1073/pnas.1502108112 PMID: 26324886

Williams, S., Stoskopf, M., Drive, C., City, M., Carolina, N., Francis-floyd, R., Koutsos, L., Dierenfeld, E., Dierenfeld, E. S., Drive, G., Louis, S., Cicotello, E., German, D., Semmen, K., Keaffaber, J., Olea-popelka, F., Livingston, S., Sullivan, K., & Valdes, E. (2022). *Recommendations and Action Plans to Improve Ex Situ Nutrition and Health of Marine Teleosts*. DOI: 10.1002/aah.10150

Wilson, R. P., Grémillet, D., Syder, J., Kierspel, M. A. M., Garthe, S., Weimerskirch, H., Schäfer-Neth, C., Scolaro, J. A., Bost, C. A., Plötz, J., & Nel, D. (2002). Remote-sensing systems and seabirds: Their use, abuse and potential for measuring marine environmental variables. *Marine Ecology Progress Series*, 228, 241–261. DOI: 10.3354/meps228241

Windsor, F. M., Durance, I., Horton, A. A., Thompson, R. C., Tyler, C. R., & Ormerod, S. J. (2019). A catchment-scale perspective of plastic pollution. *Global Change Biology*, 25(4), 1207–1221. DOI: 10.1111/gcb.14572 PMID: 30663840

Wolff, S., Weber, F., Kerpen, J., Winklhofer, M., Engelhart, M., & Barkmann, L. (2021). *Elimination of Microplastics by Downstream Sand Filters in Wastewater Treatment.*

Wong, E. L., Vuong, K. Q., & Chow, E. (2021). Nanozymes for environmental pollutant monitoring and remediation. *Sensors (Basel)*, 21(2), 408. DOI: 10.3390/s21020408

Wong, S. L., Ngadi, N., Abdullah, T. A. T., & Inuwa, I. M. (2015). Current state and future prospects of plastic waste as source of fuel: A review. *Renewable & Sustainable Energy Reviews*, 50, 1167–1180. DOI: 10.1016/j.rser.2015.04.063

Woodall, L. C., Sanchez-Vidal, A., Canals, M., Paterson, G. L. J., Coppock, R., Sleight, V., Calafat, A., Rogers, A. D., Narayanaswamy, B. E., & Thompson, R. C. (2014). The deep sea is a major sink for microplastic debris. *Royal Society Open Science*, 1(4), 140317. DOI: 10.1098/rsos.140317

Wooller, R. D., Bradley, J. S., & Croxall, J. P. (1992). Long-term population studies of seabirds. *Trends in Ecology & Evolution*, 7(4), 111–114. DOI: 10.1016/0169-5347(92)90143-Y PMID: 21235974

Wright, S. L., & Kelly, F. J. (2017). Plastic and human health: A micro issue? *Environmental Science & Technology*, 51(12), 6634–6647. DOI: 10.1021/acs.est.7b00423

Wright, S. L., Thompson, R. C., & Galloway, T. S. (2013). The physical impacts of microplastics on marine organisms: A review. *Environmental Pollution*, 178, 483–492. DOI: 10.1016/j.envpol.2013.02.031

Wright, S. L., Ulke, J., Font, A., Chan, K. L. A., & Kelly, F. J. (2019). Atmospheric microplastic deposition in an urban environment and an evaluation of transport. *Environment International*, 136, 105411. DOI: 10.1016/j.envint.2019.105411

Wright, S., Cassee, F. R., Erdely, A., & Campen, M. J. (2024). Micro- and nanoplastics concepts for particle and fiber toxicologists. *Particle and Fibre Toxicology*, 21(1), 18. DOI: 10.1186/s12989-024-00581-x

Wu, J., Zhang, Y., & Tang, Y. (2022). Fragmentation of microplastics in the drinking water treatment process-A case study in Yangtze River region, China. *The Science of the Total Environment*, 806, 150545. DOI: 10.1016/j.scitotenv.2021.150545

Wu, Y., Guo, P., Zhang, X., Zhang, Y., Xie, S., & Deng, J. (2019). Effect of microplastics exposure on the photosynthesis system of freshwater algae. *Journal of Hazardous Materials*, 374, 219–227. DOI: 10.1016/j.jhazmat.2019.04.039

Xia, W., Rao, Q., Deng, X., Chen, J., & Xie, P. (2020). Rainfall is a significant environmental factor of microplastic pollution in inland waters. *The Science of the Total Environment*, 732, 139065. DOI: 10.1016/j.scitotenv.2020.139065

Xie, L., Gong, K., Liu, Y., & Zhang, L. (2022). Strategies and challenges of identifying nanoplastics in environment by surface-enhanced Raman spectroscopy. *Environmental Science & Technology*, 57(1), 25–43. DOI: 10.1021/acs.est.2c07416

Xu, C., Wang, H., Zhou, L., & Yan, B. (2023). Phenotypic and transcriptomic shifts in roots and leaves of rice under the joint stress from microplastic and arsenic. *Journal of Hazardous Materials*, 447, 130770. DOI: 10.1016/j.jhazmat.2023.130770

Xu, C., Zhang, B., Gu, C., Shen, C., Yin, S., Aamir, M., & Li, F. (2020). Are we underestimating the sources of microplastic pollution in the terrestrial environment? *Journal of Hazardous Materials*, 400, 123228. Advance online publication. DOI: 10.1016/j.jhazmat.2020.123228 PMID: 32593024

Xu, Z., Zhang, Y., Lin, L., Wang, L., Sun, W., Liu, C., Yu, G., Yu, J., Lv, Y., Chen, J., Chen, X., Fu, L., & Wang, Y. (2022). Toxic effects of microplastics in plants depend more by their surface functional groups than just accumulation contents. *The Science of the Total Environment*, 833, 155097. DOI: 10.1016/j.scitotenv.2022.155097

Yadav V, Dhanger S, & Sharma J. (2022). Microplastics accumulation in agricultural soil: Evidence for the presence, potential effects, extraction, and current bioremediation approaches. DOI: 10.7324/JABB.2022.10s204

Ya, H., Xing, Y., Zhang, T., Lv, M., & Jiang, B. (2022). LDPE microplastics affect soil microbial community and form a unique plastisphere on microplastics. *Applied Soil Ecology*, 180, 104623. DOI: 10.1016/j.apsoil.2022.104623

Yamashita, R., Takada, H., Fukuwaka, M. A., & Watanuki, Y. (2011). Physical and chemical effects of ingested plastic debris on short-tailed shearwaters, Puffinus tenuirostris, in the North Pacific Ocean. *Marine Pollution Bulletin*, 62(12), 2845–2849. DOI: 10.1016/j.marpolbul.2011.10.008 PMID: 22047741

Yamauchi, J., Yamaoka, A., Ikemoto, K., & Matsui, T. (1991). Reaction mechanism for ozone oxidation of polyethylene as studied by ESR and IR spectroscopies. *Bulletin of the Chemical Society of Japan*, 64(4), 1173–1177. DOI: 10.1246/bcsj.64.1173

Yang, C., & Gao, X. (2022). Impact of microplastics from polyethylene and biodegradable mulch films on rice (Oryza sativa L.). *The Science of the Total Environment*, 828, 154579. DOI: 10.1016/j.scitotenv.2022.154579

Yang, C., Xie, J., Gowen, A., & Xu, J.-L. (2024). Machine learning driven methodology for enhanced nylon microplastic detection and characterization. *Scientific Reports*, 14(1), 3464. DOI: 10.1038/s41598-024-54003-1

Yang, H., Dong, H., Huang, Y., Chen, G., & Wang, J. (2022). Interactions of MPs and main pollutants and environmental behavior in soils. *The Science of the Total Environment*, 821, 153511. DOI: 10.1016/j.scitotenv.2022.153511

Yang, L., Luo, W., Zhao, P., Zhang, Y., Kang, S., Giesy, J. P., & Zhang, F. (2021). Microplastics in the Koshi River, a remote alpine river crossing the Himalayas from China to Nepal. *Environmental Pollution*, 290, 118121. DOI: 10.1016/j.envpol.2021.118121

Yang, L., Zhang, Y., Kang, S., Wang, Z., & Wu, C. (2021). Microplastics in soil: A review on methods, occurrence, sources, and potential risk. *The Science of the Total Environment*, 780, 146546. DOI: 10.1016/j.scitotenv.2021.146546

Yang, Y., Zhang, W., Wang, S., Zhang, H., & Zhang, Y. (2020). Response of male reproductive function to environmental heavy metal pollution in a free-living passerine bird, *Passer montanus.The Science of the Total Environment*, 747, 141402. DOI: 10.1016/j.scitotenv.2020.141402 PMID: 32771794

Yang, Z., Lü, F., Zhang, H., Wang, W., Shao, L., Ye, J., & He, P. (2021). Is incineration the terminator of plastics and microplastics? *Journal of Hazardous Materials*, 401, 123429. DOI: 10.1016/j.jhazmat.2020.123429

Yazdian, H., Jaafarzadeh, N., & Zahraie, B. (2014). Relationship between benthic macroinvertebrate bio-indices and physicochemical parameters of water: A tool for water resources managers. *Journal of Environmental Health Science & Engineering*, 12(1), 1–9. DOI: 10.1186/2052-336X-12-30 PMID: 24410768

Yee M.S. (2021) Impact of Microplastics and Nanoplastics on Human Health. Nanomaterials (Basel), 16(11), 496. .DOI: 10.3390/nano11020496

Yi, M., Zhou, S., Zhang, L., & Ding, S. (2021). The effects of three different microplastics on enzyme activities and microbial communities in soil. *Water environment research: a research publication of the Water Environment Federation, 93*(1), 24–32. DOI: 10.1002/wer.1327

Yingngam, B. (2023a). Chemistry of Essential Oils. In *ACS Symposium Series* (Vol. 1433, pp. 189-223). Washington DC: USA, American Chemical Society. DOI: 10.1021/bk-2022-1433.ch003

Yingngam, B. (2023b). Modern solvent-free microwave extraction with essential oil optimization and structure-activity relationships. In *Studies in Natural Products Chemistry* (Vol. 77, pp. 365–420). Elsevier. DOI: 10.1016/B978-0-323-91294-5.00011-7

Yoon, J. H., Kim, B. H., & Kim, K. H. (2024). Distribution of microplastics in soil by types of land use in metropolitan area of Seoul. *Applied Biological Chemistry*, 67(1), 15. DOI: 10.1186/s13765-024-00869-8

Yoro, K. O., & Daramola, M. O. (2020). CO_2 emission sources, greenhouse gases, and the global warming effect. In *Advances in Carbon Capture* (pp. 3-28). Woodhead Publishing. DOI: 10.1016/B978-0-12-819657-1.00001-3

Younes, N. A., Dawood, M. F. A., & Wardany, A. A. (2019). Biosafety assessment of graphene nanosheets on leaf ultrastructure, physiological and yield traits of Capsicum annuum L. and Solanum melongena L. *Chemosphere*, 228, 318–327. DOI: 10.1016/j.chemosphere.2019.04.097

Young, L. C., Vanderlip, C., Duffy, D. C., Afanasyev, V., & Shaffer, S. A. (2009). Bringing home the trash: Do colony-based differences in foraging distribution lead to increased plastic ingestion in Laysan Albatrosses? *PLoS One*, 4(10), e7623. DOI: 10.1371/journal.pone.0007623 PMID: 19862322

Yu J.R., Adingo S., Liu X.L., Li X.D., Sun J., & Zhang X.N. (2022). Micro plastics in soil ecosystem – A review of sources. DOI: 10.17221/242/2021-PSE

Yu, H., Fan, P., Hou, J., Dang, Q., Cui, D., Xi, B., & Tan, W. (2020). Inhibitory effect of microplastics on soil extracellular enzymatic activities by changing soil properties and direct adsorption: An investigation at the aggregate-fraction level. *Environmental pollution, 267*, 115544. DOI: 10.1016/j.envpol.2020.115544

Yuan, D. (2023) Microplastics in the tropical Northwestern Pacific Ocean and the Indonesian seas. Journal of Sea Research, 194. DOI: 10.1016/j.seares.2023.102406

Yuan, W., Zhou, Y., Liu, X., & Wang, J. (2019). New Perspective on the Nanoplastics Distrupting the Reproduction of an Endangered Fern in Australian Freshwater. *Environmental Science & Technology*, 53(21), 12715–12724. DOI: 10.1021/acs.est.9b02882

Yue, Y., Li, X., Wei, Z., Zhang, T., Wang, H., Huang, X., & Tang, S. (2023). Recent Advances on Multilevel Effects of Micro(Nano)Plastics and Coexisting Pollutants on Terrestrial Soil-Plants System. *Sustainability (Basel)*, 15(5), 4504. DOI: 10.3390/su15054504

Yu, H., Zhang, Z., Zhang, Y., Song, Q., Fan, P., Xi, B., & Tan, W. (2021). Effects of microplastics on soil organic carbon and greenhouse gas emissions in the context of straw incorporation: A comparison with different types of soil. *Environmental Pollution*, 288, 117733. DOI: 10.1016/j.envpol.2021.117733

Yu, S.-Y., Xie, Z.-H., Wu, X., Zheng, Y.-Z., Shi, Y., Xiong, Z.-K., & Pan, Z.-C. (2024). Review of advanced oxidation processes for treating hospital sewage to achieve decontamination and disinfection. *Chinese Chemical Letters*, 35(1), 108714. DOI: 10.1016/j.cclet.2023.108714

Yuwendi, Y., Ibadurrohman, M., Setiadi, S., & Slamet, S. (2022). Photocatalytic degradation of polyethylene microplastics and disinfection of E. coli in water over Fe-and Ag-modified TiO2 nanotubes. *Bulletin of Chemical Reaction Engineering & Catalysis*, 17(2), 263–277. DOI: 10.9767/bcrec.17.2.13400.263-277

Yu, Z. F., Song, S., Xu, X. L., Ma, Q., & Lu, Y. (2021). Sources, migration, accumulation and influence of MPs in terrestrial plant communities. *Environmental and Experimental Botany*, 192, 104635. DOI: 10.1016/j.envexpbot.2021.104635

Yu, Z., Xu, X., Guo, L., Jin, R., & Lu, Y. (2024). Uptake and transport of micro/nano-plastics in terrestrial plants: Detection, mechanisms, and influencing factors. *The Science of the Total Environment*, 907, 168155. DOI: 10.1016/j.scitotenv.2023.168155

Zafar, R., Park, S. Y., & Kim, C. G. (2021). Surface modification of polyethylene microplastic particles during the aqueous-phase ozonation process. *Environmental Engineering Research*, 26(5), 200412. DOI: 10.4491/eer.2020.412

Zaghden, H., Kallel, M., Elleuch, B., Oudot, J., Saliot, A., & Sayadi, S. (2014). Evaluation of hydrocarbon pollution in marine sediments of Sfax coastal areas from the Gabes Gulf of Tunisia, Mediterranean Sea. *Environmental Earth Sciences*, 72(4), 1073–1082. DOI: 10.1007/s12665-013-3023-6

Zang, H., Zhou, J., Marshall, M. R., Chadwick, D. R., Wen, Y., & Jones, D. L. (2020). Microplastics in the agroecosystem: Are they an emerging threat to the plant-soil system? *Soil Biology & Biochemistry*, 148, 107926. DOI: 10.1016/j.soilbio.2020.107926

Zettler, E. R., Mincer, T. J., & Amaral-Zettler, L. A. (2013). Life in the "plastisphere": Microbial communities on plastic marine debris. *Environmental Science & Technology*, 47(13), 7137–7146. DOI: 10.1021/es401288x

Zhang, G. S., & Liu, Y. F. (2018). The distribution of microplastics in soil aggregate fractions in southwestern China. *The Science of the Total Environment*, 642, 12–20. DOI: 10.1016/j.scitotenv.2018.06.004

Zhang, H., Lang, Z., & Zhu, J.-K. (2018). Dynamics and function of DNA methylation in plants. *Nature Reviews. Molecular Cell Biology*, 19(8), 489–506. DOI: 10.1038/s41580-018-0016-z

Zhang, X., Chen, J., & Li, J. (2020). The removal of microplastics in the wastewater treatment process and their potential impact on anaerobic digestion due to pollutants association. *Chemosphere*, 251, 126360. DOI: 10.1016/j.chemosphere.2020.126360

Zhang, X., Wu, D., Jiang, X., Xu, J., & Liu, J. (2024). Source, environmental behavior and ecological impact of biodegradable microplastics in soil ecosystems: A review. *Reviews of Environmental Contamination and Toxicology*, 262(1), 6. DOI: 10.1007/s44169-023-00057-7

Zhang, Y., Gao, T., Kang, S., & Sillanpää, M. (2019). Importance of atmospheric transport for microplastics deposited in remote areas. *Environmental Pollution*, 254, 112953. DOI: 10.1016/j.envpol.2019.07.121

Zhang, Y., Wang, S., Olga, V., Xue, Y., Lv, S., Diao, X., Zhang, Y., Han, Q., & Zhou, H. (2021). The potential effects of microplastic pollution on human digestive tract cells. *Chemosphere*, 291(Pt 1), 132714. DOI: 10.1016/j.chemosphere.2021.132714

Zhao, K., Li, C., & Li, F. (2024). Research progress on the origin, fate, impacts and harm of microplastics and antibiotic resistance genes in wastewater treatment plants. *Scientific Reports*, 14(1), 9719. DOI: 10.1038/s41598-024-60458-z

Zhao, S., Zhu, L., & Li, D. (2016). Microscopic anthropogenic litter in terrestrial birds from Shanghai, China: Not only plastics but also natural fibers. *The Science of the Total Environment*, 550, 1110–1115. DOI: 10.1016/j.scitotenv.2016.01.112

Zhao, X., Li, Z., Chen, Y., Shi, L., & Zhu, Y. (2007). Solid-phase photocatalytic degradation of polyethylene plastic under UV and solar light irradiation. *Journal of Molecular Catalysis A Chemical*, 268(1-2), 101–106. DOI: 10.1016/j.molcata.2006.12.012

Zheng, J., & Suh, S. (2019). Strategies to reduce the global carbon footprint of plastics. *Nature Climate Change*, 9(5), 374–378. DOI: 10.1038/s41558-019-0459-z

Zhou, A., Zhang, Y., Xie, S., Chen, Y., Li, X., Wang, J., & Zou, J. (2020). Microplastics and their potential effects on the aquaculture systems: A critical review. *Reviews in Aquaculture*, 1–15. DOI: 10.1111/raq.12496

Zhou, C.-Q., Lu, C.-H., Mai, L., Bao, L.-J., Liu, L.-Y., & Zeng, E. Y. (2021). Response of rice (Oryza sativa L.) roots to nanoplastic treatment at seedling stage. *Journal of Hazardous Materials*, 401, 123412. DOI: 10.1016/j.jhazmat.2020.123412

Zhou, F., Wang, X., Wang, G., & Zuo, Y. (2022). A Rapid Method for Detecting Microplastics Based on Fluorescence Lifetime Imaging Technology (FLIM). *Toxics*, 10(3), 118. DOI: 10.3390/toxics10030118

Zhou, Q., Lian, J., Liu, W., Men, S., Wu, J., Sun, Y., Zeb, A., Yang, T., & Ma, Q. (2020). Transcriptome mechanisms underlying interaction of polystyrene nanoplastics and wheat Triticum aestivum L. DOI: 10.21203/rs.3.rs-98748/v1

Zhu, K., Jia, H., Sun, Y., Dai, Y., Zhang, C., Guo, X., Wang, T., & Zhu, L. (2020). Long-term phototransformation of microplastics under simulated sunlight irradiation in aquatic environments: Roles of reactive oxygen species. *Water Research*, 173, 115564. DOI: 10.1016/j.watres.2020.115564

Ziani, K., Ioni ă-Mîndrican, C.-B., Mititelu, M., Neac u, S. M., Negrei, C., Moro an, E., Drăgănescu, D., & Preda, O.-T. (2023). Microplastics: A Real Global Threat for Environment and Food Safety: A State of the Art Review. *Nutrients*, 15(3), 617. DOI: 10.3390/nu15030617 PMID: 36771324

Zuo, Z., Sun, L., Wang, T., Miao, P., Zhu, X., Liu, S., Song, F., Mao, H., & Li, X. (2017). Melatonin Improves the Photosynthetic Carbon Assimilation and Antioxidant Capacity in Wheat Exposed to Nano-ZnO Stress. *Molecules (Basel, Switzerland)*, 22(10), 1727. DOI: 10.3390/molecules22101727

About the Contributors

Nisha Gaur currently works at the Strategic Product Development Division at Defence Research Laboratory, DRDO, Tezpur Assam. Nisha does research in Environmental Science and Microbiology. Their current project is Drinking Water Treatment using Nanotechnology.

Eti Sharma is currently working as Assistant Professor, School of Biotechnology, Gautam Buddha University, Greater Noida, India. She holds a Ph.D. degree in Molecular Biology & Biotechnology from G.B. Pant University of Agriculture & Technology, Pantnagar. She established herself in the field of bioprospecting and metabolomics of plants. She has expertise in plants research, nanobiotechnology , biofabrication and her current research interest lies in environmental applications of plants. She is also working on more edited book proposals. She has authored many research papers, book chapters and popular articles in nationally and internationally reputed journals.

Tuan Anh Nguyen was born in Hanoi, Vietnam. He received his BSc in Physics from Hanoi Univ. in 1992, and his Ph.D in Chemistry from Paris Diderot Univ. (France)in 2003. He was Visiting Scientist at Seoul National Univ.(South Korea, 2004) and Univ. of Wollongong (Australia, 2005). He then worked as Postdoctoral Research Associate & Research Scientist in Montana State Univ. (USA), 2006-2009. In 2012, he was appointed as Head of Microanalysis Department at Institute for Tropical Technology (Vietnam Academy of Science and Technology). He is Editor-In-Chief of "Kenkyu Journal of Nanotechnology & Nanoscience" and Founding Co-Editor-In-Chief of "Current Nanotoxicity & Prevention". He is author of 3 Vietnamese books and Editor of 14 Elsevier books in Micro & Nano Technologies Series (with 930 contributors from 58countries).

Muhammad Bilal is an associate professor (Faculty of Civil and Environmental Engineering).

* * *

Nandini Arya is currently enrolled in B.Sc. (Hons.) Agriculture in School of Agriculture, Uttaranchal University, Dehradun, Uttarakhand, India.

Manjunatha Badiger has been working as an Assistant Professor in the Department of Electronics and Communication Engineering at Sahyadri College of Engineering & Management. He has 12 years of experience in Academics, Research, and Administration. He is currently pursuing his Ph.D. in the field of Deep Learning at Visvesvaraya Technological University. He has obtained an M.Tech in VLSI & Embedded System Design from Reva Institute of Technology and Management. He has obtained a B.E. in Electronics & Communication Engineering from the NMAM Institute of Technology, Nitte. His areas of interest are Machine Learning, Deep Learning & Image processing, and VLSI. He has contributed to National and International journals, and book chapters to various reputed journals and Conferences.

Anil Barla is a Post Doctoral Fellow (PDF) at the Institute of Environment and Sustainable Development (IESD), Banaras Hindu University, Varansi, Uttar Pradesh, India.

Nazuk Bhasin is Senior Research Fellow (SRF) at the Institute of Environment and Sustainable Development (IESD), Banaras Hindu University, Varansi, Uttar Pradesh, India.

Mehnaz Fathima C. is working as an Assistant Professor in the Department of Electronics & Communication Engineering, Sahyadri College of Engineering & Management. The areas of interest are VLSI & Embedded System, Signal Processing.

Swagata Chakraborty is a Research Fellow (Ph.D.) Department of Environmental Biotechnology School of Environmental Science Bharathidasan University Tiruchirappalli.

Jyoti Chawla is currently working as Professor in the Department of Applied Sciences (Chemistry), School of Engineering and Technology, Manav Rachna International Institute of Research and Studies, Faridabad. She received her Ph.D. in Chemical Sciences from Guru Nanak Dev University (GNDU) where she conducted research as junior research fellow under collaborative IUC-DAE project of GNDU, Amritsar and Bhabha Atomic Research Centre (BARC) Mumbai. She is NET qualified. Her research interest includes Water Treatment, Surface Chemistry, Risk assessment of Nanomaterials. She has contributed more than 50 papers in journals and conferences proceedings. She has also contributed 6 chapters in Books. She had guided two Ph.D. scholars. Recently she is guiding three research scholars.

Himadri Sekhar Das is presently working as Assistant Professor in Haldia Institute of Technology, Department of Electronics and Communication Engineering. He did his Post Doc from Haldia Institute of Technology. He got his Ph. D degree from Indian Institute of Engineering Science and Technology (IIEST), Shibpur in 2021, M. Tech from Maulana Abul Kalam Azad University of Technology (Haldia Institute of Technology), West Bengal, in 2010. M.Sc. from Vidyasagar University in 2006. He has more than 8 years' research and 4 years teaching experience. His research area was doped ZnO based thin films for different optoelectronic applications. Synthesis and characterization of organic-inorganic Perovskite material for Solar cell. ZnO based transparent conducting oxide thin film for mesoporous TiO2 based dye sensitized solar cell. Solid state electrolyte based dye sensitized solar cell: stability and device performance. Solvothermal synthesis of ZnO nanoparticle. Polymer based Organic Solar Cell, Photonic Materials and devices (LED & amp; Thin Film Display). He has published more than 30 research articles and 8 book chapters. He is also DST Fellow and life member and a Fellow of the Indian Chemical Society (FICS) and the Indian Photobiology Society (IPS) also a Member of the Royal Society of Chemistry-UK.

Atin Kumar is an Assistant Professor at School of Agriculture, Uttaranchal University Dehradun, Uttarakhand. He holds a B.Sc. in Agriculture and M.Sc. in Agriculture with a specialization in Agricultural Chemistry & Soil Science from CCS University, Meerut. He obtained his Ph.D. in Soil Science from Sardar Vallabhbhai Patel University of Agriculture & Technology, Meerut (U.P.) in 2019. He has actively participated in numerous national and international workshops, conferences, and training. His expertise lies in soil fertility, nutrient management, and water quality. With almost 4 years of experience in teaching and research, he has published various research papers, articles, abstracts, and book chapters. Dr. Atin honored with the Young Researcher Award, and the Global Scientist Award for his contributions to agriculture and sustainable development.

Rajeev Kumar is currently working as Associate Professor in the Department of Applied Sciences (Chemistry), School of Engineering and Technology, Manav Rachna International Institute of Research and Studies, Faridabad. He obtained his Ph.D. degree from Indian Institute of Technology, Roorkee (IIT Roorkee). He had done M.Sc. from CCS University, Meerut. He had qualified NET-JRF from CSIR-UGC. He has more than 14 years of teaching and research experience. He has published 34 research papers in national and international reputed journals. He has also contributed 7 chapters in Books. He had co-guided two Ph.D. scholars. Recently he is guiding two research scholars. His current research interests include Environmental Sciences, Organo-metallic synthesis, Analytical Chemistry, Water Treatment, and Optimization techniques.

Sudhanshu Kumar is Senior Research Fellow (SRF) at the Institute of Environment and Sustainable Development (IESD), Banaras Hindu University, Varansi, Uttar Pradesh, India.

Varuna Kumara is an author and researcher in the field of Artificial Intelligence and Machine Learning. He earned his B.E. degree in Electronics and Communication in 2009 and completed his M. Tech degree in 2012 from Srinivas Institute of Technology in Mangalore. Currently a Research Scholar at JAIN (Deemed to be University) in Bengaluru, India. His interests lie in the areas of soft computing, Image processing, and AI & ML.

Jose Alex Mathew, B.E(ECE), M-Tech (BM), Ph.D., F.I. E, M.I.S.T. E, and Senior Member IEEE, received B. E (KVGCE, Sullia) and MTech(MIT Manipal) degrees in Electronics and Communication Engineering from Mangalore University, India in 1992 and 1997 respectively. He received his PhD in Medical Image Processing and System Development from Mangalore University, India in 2012. He Completed many NPTEL courses in the Computer Science field. He worked at MESCE Kuttippuram, SJEC Mangalore, VJEC Kannur, PACE Mangalore, SCME Mangaluru and SIT Mangaluru and held different positions like PG Coordinator, HOD, Director etc. Currently, he is working as a Professor of Data Science at Alva's Institute of Engineering & Technology. His research interests include Artificial Intelligence, Machine Learning, Signal and Image Processing, VLSI, Communication and Control and Renewable Energy. He has contributed many National and International journals, and book chapters to various reputed journals and Conferences.

Pracheta N. A. is working as an Assistant Professor (since 2015) in the Department of Bioscience and Biotechnology at Banasthali Vidyapith, Rajasthan, India. She received her B.Sc. and M.Sc. degrees from C.C.S. University, Meerut, India, and her Ph.D. degree in Biotechnology from Banasthali Vidyapith, Rajasthan, India in 2014. She has published more than 85 articles in reputed national and international journals. Her current research focuses on; i) investigating the pharmacognostical and scavenging properties of medicinal plants, ii) isolation and characterization of natural bioactive ingredients from medicinal plants, iii) anticancer activity of isolated bioactive constituents against environmental carcinogens, iv) evaluating the molecular mechanisms of ingredients and crude extract to fight against various stages of cancer or slowing the progression of ageing and agerelated oxidative stress disorders. She has presented her research work orally in 11 reputed conferences and has attended about 40 National and international conferences, workshops, symposiums and seminars. She has been invited 7 times to deliver lectures on her research findings at various national and international prestigious conferences. Her work, which she presented at the "National Symposium on Biomarkers in Health and Disease: Bench to Bed Side", SGRR - IMHS, Dehradun, has been covered by one of the most insightful newsletter in the field of anticancer drugs. She has won 9 awards in National and International conferences, to name a few: Young Scientist award (2011, 2015), Women Scientist Award (2020), and Best paper presentation Awards. Her work was recognised by World Congress for Man and Nature (WCMANU-2011) where she was decorated with Young Scientist award. She has also won the Best oral presentation award at the same conference. She is a member of 3 professional societies viz. Life time APOCP MEMBER in Asian Pacific Journal of Cancer Prevention (2011- onwards), Life time member of Indian Science Congress Association (Membership No.: L 20483) and Life time member of Human Welfare Society (2002-onwards Reg. No. 1047). She has guided 27 M.Sc. dissertation students, 3 students awarded PhD degree, One submitted their PhD thesis and currently guiding 4 Ph.D. scholars in the field of toxicology and extraction and characterization of natural bioactive ingredients from medicinal plants and to check their anti-carcinogenic potential against environmental carcinogens by using specified cell lines and Swiss albino mice.

Xuan-Hoa Nghiem received his PhD in Economics from RMIT university, Melbourne, Australia. His research interests include macroeconomics, monetary economics, energy economics, environmental economics, energy finance, corporate finance, sustainable development goals (SDGs). His work has been published in high-quality journals, including Technological Forecasting and Social Change, Resources Policy, Renewable Energy, Finance Research Letters, Journal of Environmental Management, etc.

Bahati Shabani Nzeyimana is a PhD Research scholar, Department Of Environmental Science, Bishop Heber College Affiliated to Bharathidasan University/India Interested wastewater treatment, Environmental Sustainability, Waste Management, Industry safety occupational health, Natural plant Coagulation, Air Quality monitoring (Traditional Vs AI), Climate change and Meteorological parameters, stormwater management, Environmental Education.

Ashutosh Pandey earned his doctorate in marketing from ABV-Indian Institute of Information Technology and Management, Gwalior. He holds a UGC-NET Management qualification and has more than a decade of experience in industry and academia.. He has taught postgraduate and executive students about business research methods, advanced marketing research, services marketing, marketing management, consumer behaviour, and structural equation modeling. His research has appeared in many prestigious journals, including Environment, Development, and Sustainability, Journal of Quality Assurance in Hospitality & Tourism, and International Journal of Tourism Cities. He has worked as a visiting professor at IMT Ghaziabad, Indian Institute of Foreign Trade, New Delhi and the Indian Institute of Travel and Tourism Management in Gwalior. Dr. Pandey has been a resource person for several qualitative and quantitative research workshops hosted at reputed organisations.

Gourisankar Roymahapatra is currently working as an Associate Professor of Chemistry at the School of Applied Science and Humanities, Haldia Institute of Technology (HIT) Haldia, India. He completed his Ph.D. from Jadavpur University, Kolkata (2014) with N-Heterocyclic Carbene Chemistry as a major field under the mentoring of Dr. Joydev Dinda of HIT Haldia, and Prof. Ambikesh Mahapatra of Jadavpur University. Before joining to Global Institute of Science and Technology (GIST Haldia), as a Lecturer in Chemistry in the year 2011, he worked at MCC PTA Chem. Corp. Pvt. Ltd India (MCPI), Haldia, (2003 ~ 2011) as a Senior Chemist. He joined HIT Haldia in 2015 as an Assistant Professor of Chemistry. His area of research interest includes NHC complexes, organometallics, catalysis, antibiotics, anti-carcinogenic, gas adsorption, super alkali, computational chemistry and DFT. For his contribution in research, he got the 'Distinguished Young Scientist Award in Chemistry -2014'; 'Bharat Gaurabh Award – 2018'; and 'Indian Chemical Society Research Excellent Award 2021'. Dr. Roymahapatra is a Life Fellow of the Indian Chemical Society (FICS), and a Life Fellow of the Institutes of Chemistry.

Ganesh Srinivasa Shetty has received his Bachelor of Engineering from Malnad College Of Engineering, Hassan and completed his M.Tech degree from Sri Jayachamarajendra College Of Engineering Mysore. He is a part time research scholar at JAIN (Deemed to be University) in Bengaluru, India. Currently, he is working as Senior Assistant Professor in Shri Madhwa Vadiraja Institute of Technology and Management (SMVITM) Vishwothama Nagar, Bantakal-5764115, Udupi, Kanataka, India. His interests lie in the areas of Cryptograhy and Artificial Intelligence and Machine Learning. He has authored 5+ research articles/conference.

Savidhan Shetty received the master's degree from Visvesvaraya Technological University, Karnataka, India. He is currently pursuing a Ph.D. degree in the Department of Electronics and Communication Engineering from the National Institute of Technology Karnataka, Surathkal, India. His research interests include RF wireless communication, free-space optical communication, underwater optical wireless communication, Robotics and Environmental Engineering.

Gopal Singh is Professor at the Institute of Environment and Sustainable Development (IESD), Banaras Hindu University, Varansi, Uttar Pradesh, India.

Jyoti Syal is Assistant Professor of English in Department of Mathematics and Humanities, Maharishi Markandeshwar Engineering College, Maharishi Markandeshwar Deemed to be University, Mullana, Ambala, Haryana, India. She has 16 years of enriched experience of teaching UG classes. She did her Ph.D from Gurukul Kangri Vishwavidalaya, Haridwar in the year 2012. 3 Ph. Ds have been awarded under her supervision and 4 Ph. Ds are going on. She has 37 publications to her credit in various national and international journals of repute.

Amit Tiwari is Senior Research Fellow (SRF) at the Institute of Environment and Sustainable Development (IESD), Banaras Hindu University, Varansi, Uttar Pradesh, India.

Mohit Yadav is an Associate Professor in the area of Human Resource Management at Jindal Global Business School (JGBS). He has a rich blend of work experience from both Academics as well as Industry. Prof. Mohit holds a Ph.D. from Department of Management Studies, Indian Institute of Technology Roorkee (IIT Roorkee) and has completed Master of Human Resource and Organizational Development (MHROD) from prestigious Delhi School of Economics, University of Delhi. He also holds a B.Com (Hons.) degree from University of Delhi and UGC-JRF scholarship. He has published various research papers and book chapters with reputed publishers like Springer, Sage, Emerald, Elsevier, Inderscience etc. and presented research papers in national and International conferences both in India and abroad. He has many best paper awards on his credit too. He is reviewer of various international journals like Computers in Human Behavior, Policing etc. His areas of interest are Organizational Behavior, HRM, Recruitment and Selection, Organizational Citizenship Behavior, Quality of work life and role.

Bancha Yingngam, an Associate Professor at Ubon Ratchathani University in Thailand, specializes in Pharmaceutical Sciences and Technology. His academic journey began in 2005 when he earned a bachelor's degree in pharmacy with second-class honors. Motivated by his interest in phytomedicines and community pharmacy, Dr. Yingngam pursued and completed his Ph.D. at the same institution in 2011. Following his doctoral studies, Dr. Yingngam's research career advanced significantly. He spent several years at Karl-Franzens-Universität Graz, Austria, initially as a postdoctoral fellow and later in other research capacities. During this period, which extended from 2013 to 2021, he broadened his research scope to include optimization and machine learning algorithms, applying these to the formulation of pharmaceuticals, cosmetics, perfumes, and nutraceuticals. Throughout his career, Dr. Yingngam has made substantial contributions to his field, authoring over thirty scholarly articles, books, book chapters, and conference papers. His research primarily focuses on optimization problems and the application of machine learning in healthcare product formulations, reflecting his commitment to advancing pharmaceutical science and technology.

Index

A

B

C

D

E

N

P

R

S

T

W

Milton Keynes UK
Ingram Content Group UK Ltd.
UKHW050804081024
449245UK00008BA/79

9 798369 334478